Precalculus 2

An Investigation of Functions

Chapters 5-9

Edition 2.1

David Lippman
Melonie Rasmussen

This book is also available to read free online at
http://www.opentextbookstore.com/precalc/
If you want a printed copy, buying from the bookstore is cheaper than printing yourself.

Front cover photo by Alexander Baxevanis. Flickr. Licensed CC-BY 2.0.

Back and inside cover photos by David Lippman, of artwork by John Rogers.
Lituus, 2010. Dichromatic glass and aluminum.
Washington State Arts Commission in partnership with Pierce College

About the Authors

David Lippman received his master's degree in mathematics from Western Washington University and has been teaching at Pierce College since Fall 2000.

Melonie Rasmussen also received her master's degree in mathematics from Western Washington University and has been teaching at Pierce College since Fall 2002. Prior to this Melonie taught for the Puyallup School district for 6 years after receiving her teaching credentials from Pacific Lutheran University.

We have both been long time advocates of open learning, open materials, and basically any idea that will reduce the cost of education for students. It started by supporting the college's calculator rental program, and running a book loan scholarship program. Eventually the frustration with the escalating costs of commercial text books and the online homework systems that charged for access led them to take action.

First, David developed IMathAS, open source online math homework software that runs WAMAP.org and MyOpenMath.com. Through this platform, we became integral parts of a vibrant sharing and learning community of teachers from around Washington State that support and contribute to WAMAP. Our pioneering efforts, supported by dozens of other dedicated faculty and financial support from the WA-SBCTC, have led to a system used by thousands of students every quarter, saving hundreds of thousands of dollars over comparable commercial offerings.

David continued further and wrote his first open textbook, *Math in Society*, a math for liberal arts majors book, after being frustrated by students having to pay $100+ for a textbook for a terminal course. Together, frustrated by both cost and the style of commercial texts, we began writing *PreCalculus: An Investigation of Functions* in 2010. Since then, David has contributed to several other open texts.

Acknowledgements

We would like to thank the following for their generous support and feedback.

- The community of WAMAP users and developers for creating a majority of the homework content used in our online homework sets.

- Pierce College students in our Fall 2010 - Summer 2011 Math 141 and Math 142 classes for helping correct typos, identifying videos related to the homework, and being our willing test subjects.

- The Open Course Library Project for providing the support needed to produce a full course package for these courses.

- Mike Kenyon, Chris Willett, Tophe Anderson, and Vauhn Foster-Grahler for reviewing the course and giving feedback and suggestions.

- Our Pierce College colleagues for providing their suggestions.

- Tophe Anderson, James Gray, and Lawrence Morales for their feedback and suggestions in content and examples.

- Jeff Eldridge for extensive proofreading and suggestions for clarification.

- James Sousa for developing videos associated with the online homework.

- Kevin Dimond for his work on indexing the book and creating PowerPoint slides.

- Faculty at Green River Community College and the Maricopa College District for their feedback and suggestions.

- Lara Michaels for contributing the basis for a conics chapter.

- The dozens of instructors who have sent us typos or suggestions over the years.

Preface

Over the years, when reviewing books we found that many had been mainstreamed by the publishers in an effort to appeal to everyone, leaving them with very little character. There were only a handful of books that had the conceptual and application driven focus we liked, and most of those were lacking in other aspects we cared about, like providing students sufficient examples and practice of basic skills. The largest frustration, however, was the never ending escalation of cost and being forced into new editions every three years. We began researching open textbooks, however the ability for those books to be adapted, remixed, or printed were often limited by the types of licenses, or didn't approach the material the way we wanted.

This book is available online for free, in both Word and PDF format. You are free to change the wording, add materials and sections or take them away. We welcome feedback, comments and suggestions for future development at precalc@opentextbookstore.com. Additionally, if you add a section, chapter or problems, we would love to hear from you and possibly add your materials so everyone can benefit.

In writing this book, our focus was on the story of functions. We begin with function notation, a basic toolkit of functions, and the basic operation with functions: composition and transformation. Building from these basic functions, as each new family of functions is introduced we explore the important features of the function: its graph, domain and range, intercepts, and asymptotes. The exploration then moves to evaluating and solving equations involving the function, finding inverses, and culminates with modeling using the function.

The "rule of four" is integrated throughout - looking at the functions verbally, graphically, numerically, as well as algebraically. We feel that using the "rule of four" gives students the tools they need to approach new problems from various angles. Often the "story problems of life" do not always come packaged in a neat equation. Being able to think critically, see the parts and build a table or graph a trend, helps us change the words into meaningful and measurable functions that model the world around us.

There is nothing we hate more than a chapter on exponential equations that begins "Exponential functions are functions that have the form $f(x)=a^x$." As each family of functions is introduced, we motivate the topic by looking at how the function arises from life scenarios or from modeling. Also, we feel it is important that precalculus be the bridge in level of thinking between algebra and calculus. In algebra, it is common to see numerous examples with very similar homework exercises, encouraging the student to mimic the examples. Precalculus provides a link that takes students from the basic plug & chug of formulaic calculations towards building an understanding that equations and formulas have deeper meaning and purpose. While you will find examples and similar exercises for the basic skills in this book, you will also find examples of multistep problem solving along with exercises in multistep problem solving. Often times these exercises will not exactly mimic the exercises, forcing the students to employ their critical thinking skills and apply the skills they've learned to new situations. By developing students' critical thinking and problem solving skills this course prepares students for the rigors of Calculus.

While we followed a fairly standard ordering of material in the first half of the book, we took some liberties in the trig portion of the book. It is our opinion that there is no need to separate unit circle trig from triangle trig, and instead integrated them in the first chapter. Identities are introduced in the first chapter, and revisited throughout. Likewise, solving is introduced in the second chapter and revisited more extensively in the third chapter. As with the first part of the book, an emphasis is placed on motivating the concepts and on modeling and interpretation.

About the Second Edition

About 4 years and several minor typo revisions after the original release of this book, we started contemplating creating a second edition. We didn't want to change much; we've always found it very annoying when new editions change things just for the sake of making it seem different. However, in talking with instructors from around the country, we knew there were a few topics that we had left out that other schools need. We didn't want to suffer the same "content bloat" that many commercial books do, but we also wanted to make it easier for more schools to adopt open resources.

We put our plans for a new revision on hold after OpenStax started working on a precalculus book, using the first edition of this text as a base. After the final product came out, though, we felt it had strayed a bit far from our original vision. We had written this text, not to be an encyclopedic reference text, but to be a concise, easy-to-read, student-friendly approach to precalculus. We valued contextual motivation and conceptual understanding over procedural skills. Our book took, in places, a non-traditional approach to topics and content ordering. Ultimately, we decided to go ahead with this second edition.

The primary changes in the second edition are:

- New, higher resolution graphs throughout
- New sections added to Chapter 3:
    - 3.4 Factor theorem (includes long division of polynomials)
    - 3.5 Real zeros of polynomials (using rational roots theorem)
    - 3.6 Complex zeros of polynomials
- Coverage of oblique asymptotes added to the rational equations section (now 3.7)
- A new section 8.5 on dot product of vectors
- A new chapter 9 on conic sections

There were many additional refinements, some new examples added, and Try it Now answers expanded, but most of the book remains unchanged.

Instructor Resources

As part of the Washington Open Course Library project, we developed a full course package to accompany this text. The course shell was built for the IMathAS online homework platform, and is available for Washington State faculty at www.wamap.org and mirrored for others at www.myopenmath.com. It contains:

- Online homework for each section (algorithmically generated, free response), most with video help associated.
- Video lessons for each section. The videos were mostly created and selected by James Sousa, of Mathispower4u.
- A selection of printable class worksheets, activities, and handouts
- Support materials for an example course (does not include all sections):
    - Suggested syllabus and Day by day course guide
    - Instructor guide with lecture outlines and examples
    - Discussion forums
    - Diagnostic review
    - Chapter review problems
    - Sample quizzes and sample chapter exams

The course shell was designed to follow Quality Matters (QM) guidelines, but has not yet been formally reviewed.

Getting Started

To get started using this textbook and the online supplementary materials,

- Request an instructor account on WAMAP (in Washington) or MyOpenMath (outside Washington).
- Review the table of contents of the text, and compare it to your course outcomes or student learning objectives. Determine which sections you will need to cover, and which to omit. If there are topics in your outcomes that are not in the text, explore other sources like the Stitz/Zeager Precalc or OpenStax Precalc to supplement from. Also check the book's website, as we may offer additional online-only topics.
- Once your instructor account is approved, log in, and click **Add New Course**
- From the "Use content from a template course", select "Precalculus – Lippman/Rasmussen 2nd Ed". Note that you might also see two half-book templates, one covering chapters 1 – 4, and the other covering chapters 5 – 9.
- Once you have copied the course, go through and remove any sections you don't need for your course. Refer to the Training Course Quickstart videos in MyOpenMath and WAMAP for more details on how to make those changes.

How To Be Successful In This Course

This is not a high school math course, although for some of you the content may seem familiar. There are key differences to what you will learn here, how quickly you will be required to learn it and how much work will be required of you.

You will no longer be shown a technique and be asked to mimic it repetitively as the only way to prove learning. Not only will you be required to master the technique, but you will also be required to extend that knowledge to new situations and build bridges between the material at hand and the next topic, making the course highly cumulative.

As a rule of thumb, for each hour you spend in class, you should expect this course will require an average of 2 hours of out-of-class focused study. This means that some of you with a stronger background in mathematics may take less, but if you have a weaker background or any math anxiety it will take you more.

Notice how this is the equivalent of having a part time job, and if you are taking a fulltime load of courses as many college students do, this equates to more than a full time job. If you must work, raise a family and take a full load of courses all at the same time, we recommend that you get a head start & get organized as soon as possible. We also recommend that you spread out your learning into daily chunks and avoid trying to cram or learn material quickly before an exam.

To be prepared, read through the material before it is covered in class and note or highlight the material that is new or confusing. The instructor's lecture and activities should not be the first exposure to the material. As you read, test your understanding with the Try it Now problems in the book. If you can't figure one out, try again after class, and ask for help if you still can't get it.

As soon as possible after the class session recap the day's lecture or activities into a meaningful format to provide a third exposure to the material. You could summarize your notes into a list of key points, or reread your notes and try to work examples done in class without referring back to your notes. Next, begin any assigned homework. The next day, if the instructor provides the opportunity to clarify topics or ask questions, do not be afraid to ask. If you are afraid to ask, then you are not getting your money's worth! If the instructor does not provide this opportunity, be prepared to go to a tutoring center or build a peer study group. Put in quality effort and time and you can get quality results.

Lastly, if you feel like you do not understand a topic. Don't wait, ASK FOR HELP!

ASK: **A**sk a teacher or tutor, **S**earch for ancillaries, **K**eep a detailed list of questions
FOR: **F**ind additional resources, **O**rganize the material, **R**esearch other learning options
HELP: **H**ave a support network, **E**xamine your weaknesses, **L**ist specific examples & **P**ractice

Best of luck learning! We hope you like the course & love the price.
David & Melonie

Table of Contents

NOTE: This printed text only contains Chapters 5-9 of the book. The first half of the book can be read online at http://www.opentextbookstore.com/precalc/ or purchased as a separated printed text.

Chapter 5: Trigonometric Functions of Angles

In the previous chapters, we have explored a variety of functions which could be combined to form a variety of shapes. In this discussion, one common shape has been missing: the circle. We already know certain things about the circle, like how to find area and circumference, and the relationship between radius and diameter, but now, in this chapter, we explore the circle and its unique features that lead us into the rich world of trigonometry.

Section 5.1 Circles

To begin, we need to find distances. Starting with the Pythagorean Theorem, which relates the sides of a right triangle, we can find the distance between two points.

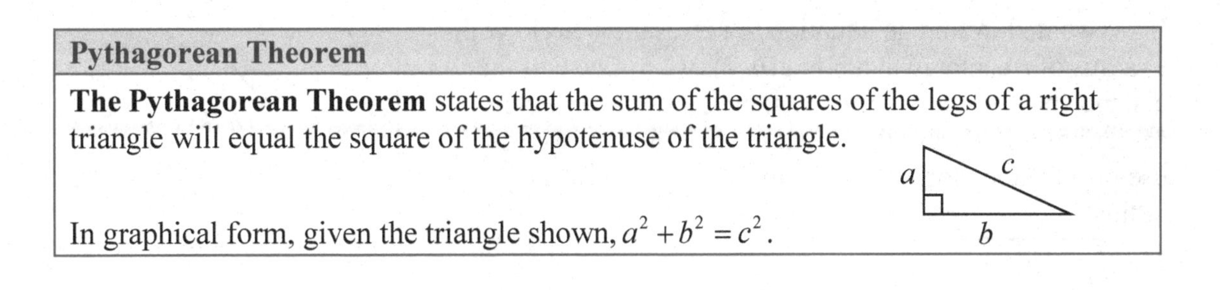

Pythagorean Theorem

The Pythagorean Theorem states that the sum of the squares of the legs of a right triangle will equal the square of the hypotenuse of the triangle.

In graphical form, given the triangle shown, $a^2 + b^2 = c^2$.

We can use the Pythagorean Theorem to find the distance between two points on a graph.

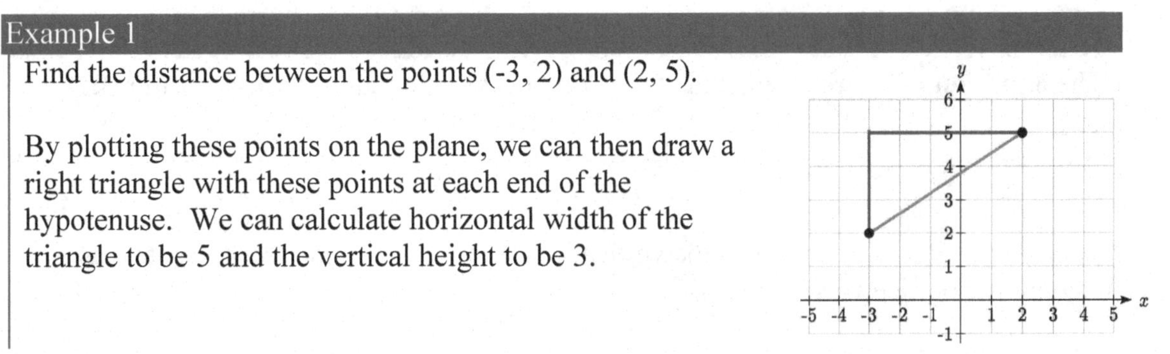

Example 1

Find the distance between the points (-3, 2) and (2, 5).

By plotting these points on the plane, we can then draw a right triangle with these points at each end of the hypotenuse. We can calculate horizontal width of the triangle to be 5 and the vertical height to be 3.

From these we can find the distance between the points using the Pythagorean Theorem:

$dist^2 = 5^2 + 3^2 = 34$

$dist = \sqrt{34}$

Notice that the width of the triangle was calculated using the difference between the x (input) values of the two points, and the height of the triangle was found using the difference between the y (output) values of the two points. Generalizing this process gives us the distance formula.

Distance Formula
The distance between two points (x_1, y_1) and (x_2, y_2) can be calculated as $dist = \sqrt{(x_2 - x_1)^2 + (y_2 - y_1)^2}$

Try it Now

1. Find the distance between the points (1, 6) and (3, -5).

Circles

If we wanted to find an equation to represent a circle with a radius of r centered at a point (h, k), we notice that the distance between any point (x, y) on the circle and the center point is always the same: r. Noting this, we can use our distance formula to write an equation for the radius:

$r = \sqrt{(x - h)^2 + (y - k)^2}$

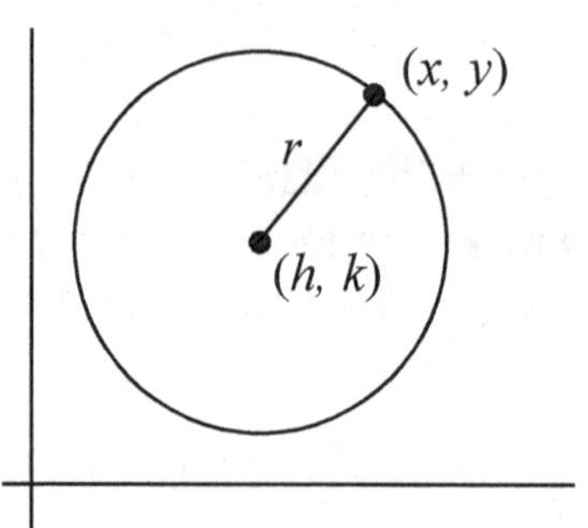

Squaring both sides of the equation gives us the standard equation for a circle.

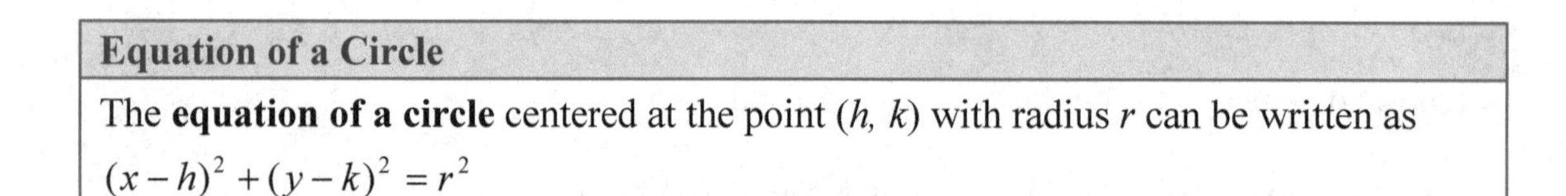

Equation of a Circle
The **equation of a circle** centered at the point (h, k) with radius r can be written as $(x - h)^2 + (y - k)^2 = r^2$

Notice that a circle does not pass the vertical line test. It is not possible to write y as a function of x or vice versa.

Example 2

Write an equation for a circle centered at the point (-3, 2) with radius 4.

Using the equation from above, h = -3, k = 2, and the radius r = 4. Using these in our formula,

$(x-(-3))^2+(y-2)^2=4^2$ simplified, this gives

$(x+3)^2+(y-2)^2=16$

Example 3

Write an equation for the circle graphed here.

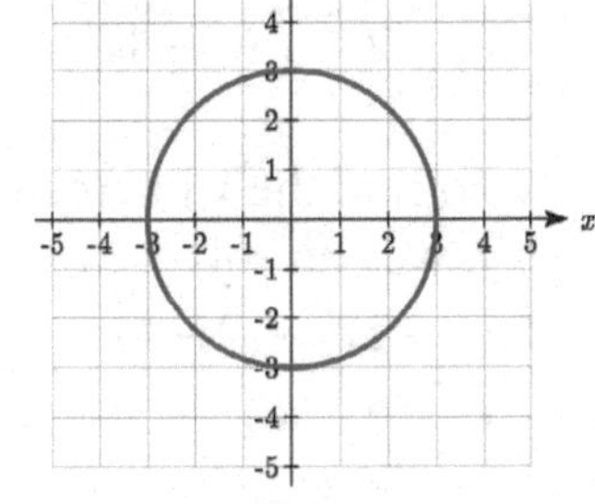

This circle is centered at the origin, the point (0, 0). By measuring horizontally or vertically from the center out to the circle, we can see the radius is 3. Using this information in our formula gives:

$(x-0)^2+(y-0)^2=3^2$ simplified, this gives

$x^2+y^2=9$

Try it Now

2. Write an equation for a circle centered at (4, -2) with radius 6.

Notice that, relative to a circle centered at the origin, horizontal and vertical shifts of the circle are revealed in the values of h and k, which are the coordinates for the center of the circle.

Points on a Circle

As noted earlier, an equation for a circle cannot be written so that y is a function of x or vice versa. To find coordinates on the circle given only the x or y value, we must solve algebraically for the unknown values.

Example 4

Find the points on a circle of radius 5 centered at the origin with an x value of 3.

We begin by writing an equation for the circle centered at the origin with a radius of 5.

$x^2+y^2=25$

Substituting in the desired x value of 3 gives an equation we can solve for y.

$3^2 + y^2 = 25$
$y^2 = 25 - 9 = 16$
$y = \pm\sqrt{16} = \pm 4$

There are two points on the circle with an x value of 3: (3, 4) and (3, -4).

Example 5

Find the x intercepts of a circle with radius 6 centered at the point (2, 4).

We can start by writing an equation for the circle.
$(x-2)^2 + (y-4)^2 = 36$

To find the x intercepts, we need to find the points where $y = 0$. Substituting in zero for y, we can solve for x.
$(x-2)^2 + (0-4)^2 = 36$
$(x-2)^2 + 16 = 36$
$(x-2)^2 = 20$
$x - 2 = \pm\sqrt{20}$
$x = 2 \pm \sqrt{20} = 2 \pm 2\sqrt{5}$

The x intercepts of the circle are $\left(2+2\sqrt{5},0\right)$ and $\left(2-2\sqrt{5},0\right)$

Example 6

In a town, Main Street runs east to west, and Meridian Road runs north to south. A pizza store is located on Meridian 2 miles south of the intersection of Main and Meridian. If the store advertises that it delivers within a 3-mile radius, how much of Main Street do they deliver to?

This type of question is one in which introducing a coordinate system and drawing a picture can help us solve the problem. We could either place the origin at the intersection of the two streets, or place the origin at the pizza store itself. It is often easier to work with circles centered at the origin, so we'll place the origin at the pizza store, though either approach would work fine.

Placing the origin at the pizza store, the delivery area with radius 3 miles can be described as the region inside the circle described by $x^2 + y^2 = 9$.

Main Street, located 2 miles north of the pizza store and running east to west, can be described by the equation $y = 2$.

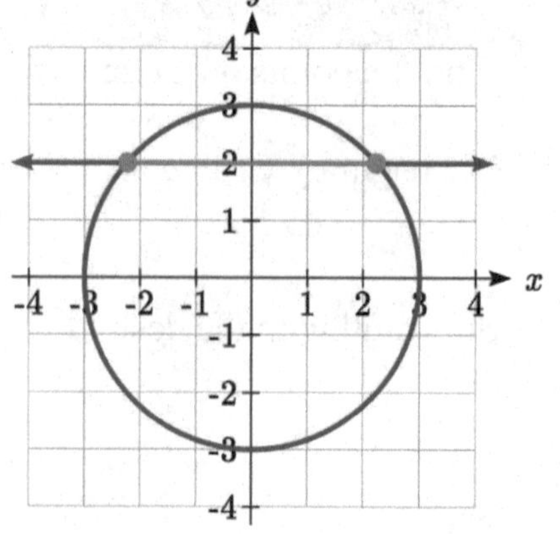

To find the portion of Main Street the store will deliver to, we first find the boundary of their delivery region by looking for where the delivery circle intersects Main Street. To find the intersection, we look for the points on the circle where $y = 2$. Substituting $y = 2$ into the circle equation lets us solve for the corresponding x values.

$$x^2 + 2^2 = 9$$
$$x^2 = 9 - 4 = 5$$
$$x = \pm\sqrt{5} \approx \pm 2.236$$

This means the pizza store will deliver 2.236 miles down Main Street east of Meridian and 2.236 miles down Main Street west of Meridian. We can conclude that the pizza store delivers to a 4.472 mile long segment of Main St.

In addition to finding where a vertical or horizontal line intersects the circle, we can also find where an arbitrary line intersects a circle.

Example 7

Find where the line $f(x) = 4x$ intersects the circle $(x-2)^2 + y^2 = 16$.

Normally, to find an intersection of two functions *f(x)* and *g(x)* we would solve for the *x* value that would make the functions equal by solving the equation *f(x)* = *g(x)*. In the case of a circle, it isn't possible to represent the equation as a function, but we can utilize the same idea.

The output value of the line determines the *y* value: $y = f(x) = 4x$. We want the *y* value of the circle to equal the *y* value of the line, which is the output value of the function. To do this, we can substitute the expression for *y* from the line into the circle equation.

$(x-2)^2 + y^2 = 16$	replace *y* with the line formula: $y = 4x$
$(x-2)^2 + (4x)^2 = 16$	expand
$x^2 - 4x + 4 + 16x^2 = 16$	simplify
$17x^2 - 4x + 4 = 16$	since this equation is quadratic, we arrange one side to be 0
$17x^2 - 4x - 12 = 0$	

Since this quadratic doesn't appear to be easily factorable, we can use the quadratic formula to solve for *x*:

$$x = \frac{-(-4) \pm \sqrt{(-4)^2 - 4(17)(-12)}}{2(17)} = \frac{4 \pm \sqrt{832}}{34}$$, or approximately $x \approx 0.966$ or -0.731

From these *x* values we can use either equation to find the corresponding *y* values.

Since the line equation is easier to evaluate, we might choose to use it:
$y = f(0.966) = 4(0.966) = 3.864$
$y = f(-0.731) = 4(-0.731) = -2.923$

The line intersects the circle at the points (0.966, 3.864) and (-0.731, -2.923).

Try it Now

3. A small radio transmitter broadcasts in a 50 mile radius. If you drive along a straight line from a city 60 miles north of the transmitter to a second city 70 miles east of the transmitter, during how much of the drive will you pick up a signal from the transmitter?

Important Topics of This Section
Distance formula
Equation of a Circle
Finding the *x* coordinate of a point on the circle given the *y* coordinate or vice versa
Finding the intersection of a circle and a line

Try it Now Answers

1. $5\sqrt{5}$
2. $(x-4)^2 + (y+2)^2 = 36$
3. The circle can be represented by $x^2 + y^2 = 50^2$.

   Finding a line from (0,60) to (70,0) gives $y = 60 - \frac{60}{70}x$.

   Substituting the line equation into the circle gives $x^2 + \left(60 - \frac{60}{70}x\right)^2 = 50^2$.

   Solving this equation, we find $x = 14$ or $x = 45.29$, corresponding to points (14, 48) and (45.29, 21.18).
   The distance between these points is 41.21 miles.

Section 5.1 Exercises

1. Find the distance between the points (5,3) and (-1,-5).
2. Find the distance between the points (3,3) and (-3,-2).
3. Write an equation of the circle centered at (8 , -10) with radius 8.
4. Write an equation of the circle centered at (-9, 9) with radius 16.
5. Write an equation of the circle centered at (7, -2) that passes through (-10, 0).
6. Write an equation of the circle centered at (3, -7) that passes through (15, 13).
7. Write an equation for a circle where the points (2, 6) and (8, 10) lie along a diameter.
8. Write an equation for a circle where the points (-3, 3) and (5, 7) lie along a diameter.
9. Sketch a graph of $(x-2)^2+(y+3)^2=9$.
10. Sketch a graph of $(x+1)^2+(y-2)^2=16$.
11. Find the y intercept(s) of the circle with center (2, 3) with radius 3.
12. Find the x intercept(s) of the circle with center (2, 3) with radius 4.
13. At what point in the first quadrant does the line with equation $y=2x+5$ intersect a circle with radius 3 and center (0, 5)?
14. At what point in the first quadrant does the line with equation $y=x+2$ intersect the circle with radius 6 and center (0, 2)?
15. At what point in the second quadrant does the line with equation $y=2x+5$ intersect a circle with radius 3 and center (-2, 0)?
16. At what point in the first quadrant does the line with equation $y=x+2$ intersect the circle with radius 6 and center (-1,0)?
17. A small radio transmitter broadcasts in a 53 mile radius. If you drive along a straight line from a city 70 miles north of the transmitter to a second city 74 miles east of the transmitter, during how much of the drive will you pick up a signal from the transmitter?
18. A small radio transmitter broadcasts in a 44 mile radius. If you drive along a straight line from a city 56 miles south of the transmitter to a second city 53 miles west of the transmitter, during how much of the drive will you pick up a signal from the transmitter?

19. A tunnel connecting two portions of a space station has a circular cross-section of radius 15 feet. Two walkway decks are constructed in the tunnel. Deck A is along a horizontal diameter and another parallel Deck B is 2 feet below Deck A. Because the space station is in a weightless environment, you can walk vertically upright along Deck A, or vertically upside down along Deck B. You have been assigned to paint "safety stripes" on each deck level, so that a 6 foot person can safely walk upright along either deck. Determine the width of the "safe walk zone" on each deck. [UW]

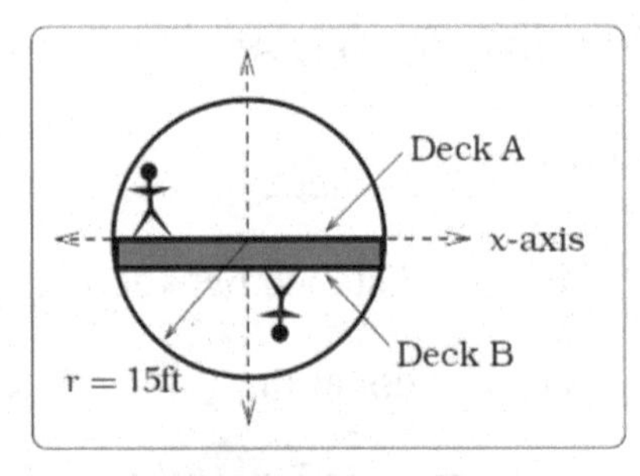

(a) Cross-section of tunnel.

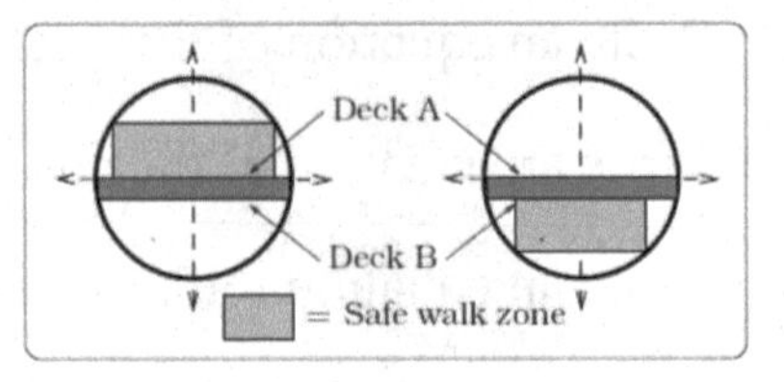

(b) Walk zones.

20. A crawling tractor sprinkler is located as pictured here, 100 feet south of a sidewalk. Once the water is turned on, the sprinkler waters a circular disc of radius 20 feet and moves north along the hose at the rate of ½ inch/second. The hose is perpendicular to the 10 ft. wide sidewalk. Assume there is grass on both sides of the sidewalk. [UW]

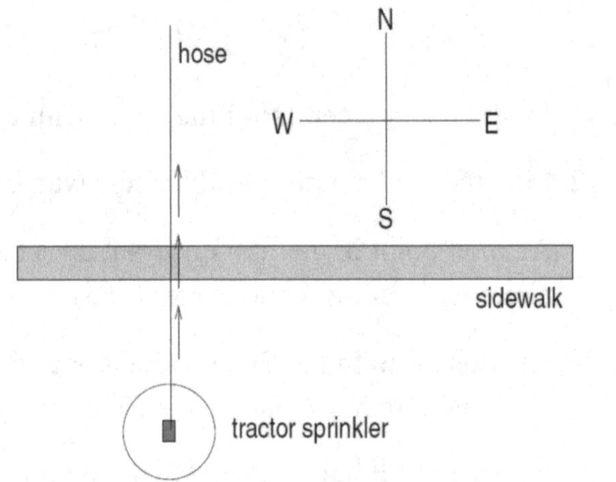

a) Impose a coordinate system. Describe the initial coordinates of the sprinkler and find equations of the lines forming and find equations of the lines forming the north and south boundaries of the sidewalk.
b) When will the water first strike the sidewalk?
c) When will the water from the sprinkler fall completely north of the sidewalk?
d) Find the total amount of time water from the sprinkler falls on the sidewalk.
e) Sketch a picture of the situation after 33 minutes. Draw an accurate picture of the watered portion of the sidewalk.
f) Find the area of grass watered after one hour.

21. Erik's disabled sailboat is floating anchored 3 miles East and 2 miles north of Kingsford. A ferry leaves Kingsford heading toward Eaglerock at 12 mph. Eaglerock is 6 miles due east of Kingsford. After 20 minutes the ferry turns, heading due south. Bander is 8 miles south and 1 mile west of Eaglerock. Impose coordinates with Bander as the origin. [UW]

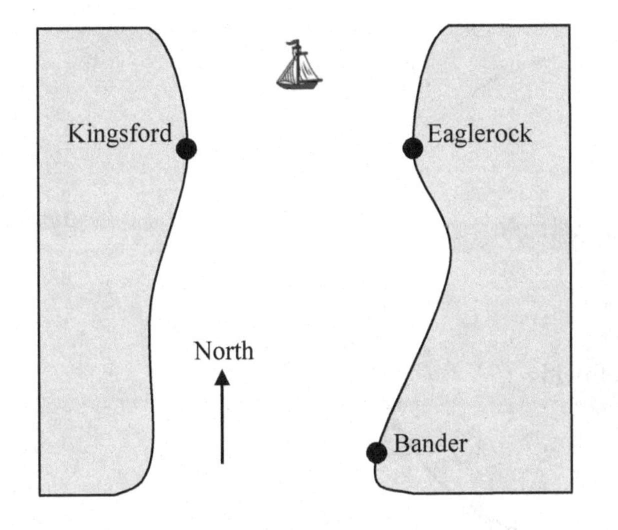

a) Find equations for the lines along which the ferry is moving and draw in these lines.
b) The sailboat has a radar scope that will detect any object within 3 miles of the sailboat. Looking down from above, as in the picture, the radar region looks like a circular disk. The boundary is the "edge" or circle around this disk, the interior is everything inside of the circle, and the exterior is everything outside of the circle. Give the mathematical description (an equation or inequality) of the boundary, interior and exterior of the radar zone. Sketch an accurate picture of the radar zone by determining where the line connecting Kingsford and Eaglerock would cross the radar zone.
c) When does the ferry enter the radar zone?
d) Where and when does the ferry exit the radar zone?
e) How long does the ferry spend inside the radar zone?

22. Nora spends part of her summer driving a combine during the wheat harvest. Assume she starts at the indicated position heading east at 10 ft/sec toward a circular wheat field of radius 200 ft. The combine cuts a swath 20 feet wide and begins when the corner of the machine labeled "a" is 60 feet north and 60 feet west of the western-most edge of the field. [UW]

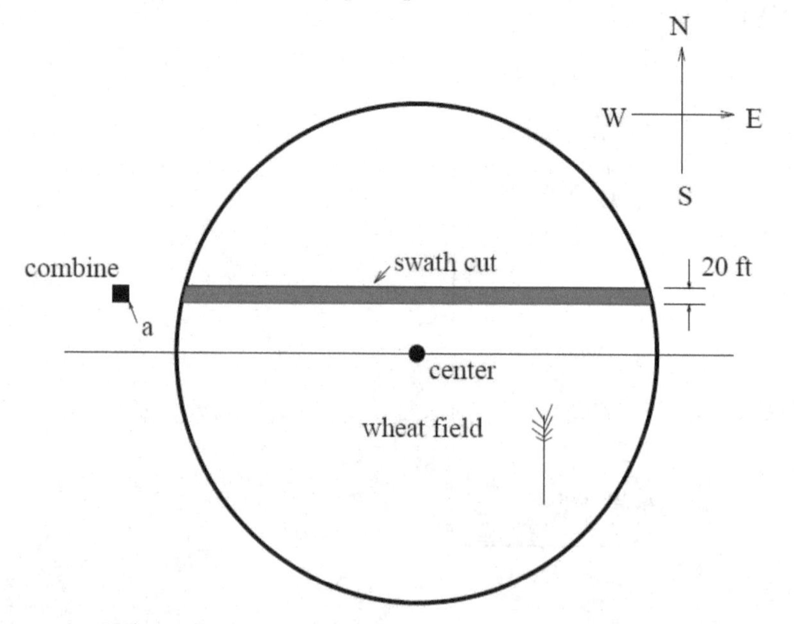

a) When does Nora's combine first start cutting the wheat?
b) When does Nora's combine first start cutting a swath 20 feet wide?
c) Find the total amount of time wheat is being cut during this pass across the field.
d) Estimate the area of the swath cut during this pass across the field.

23. The vertical cross-section of a drainage ditch is pictured to the right. Here, R indicates in each case the radius of a circle with $R = 10$ feet, where all of the indicated circle centers lie along a horizontal line 10 feet above and parallel to the ditch bottom. Assume that water is flowing into the ditch so that the level above the bottom is rising at a rate of 2 inches per minute. [UW]

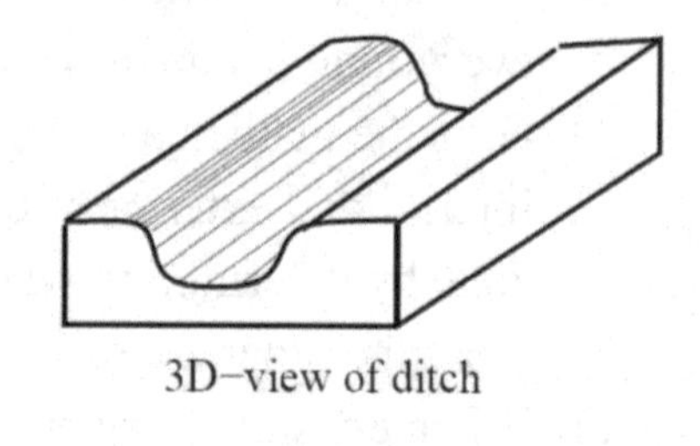

3D–view of ditch

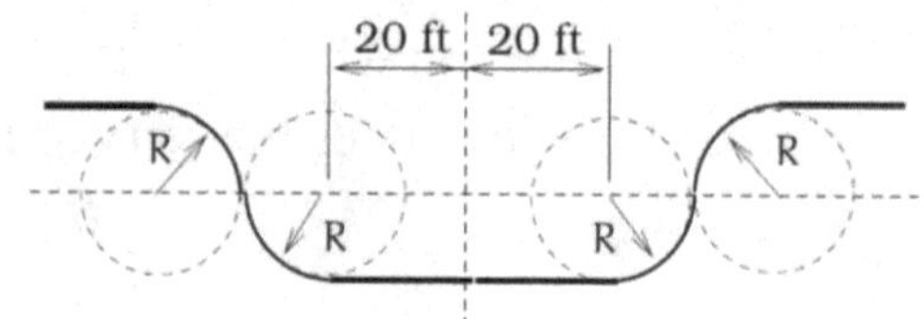

vertical cross-section

a) When will the ditch be completely full?
b) Find a piecewise defined function that models the vertical cross-section of the ditch.
c) What is the width of the filled portion of the ditch after 1 hour and 18 minutes?
d) When will the filled portion of the ditch be 42 feet wide? 50 feet wide? 73 feet wide?

Section 5.2 Angles

Because many applications involving circles also involve a rotation of the circle, it is natural to introduce a measure for the rotation, or angle, between two rays (line segments) emanating from the center of a circle. The angle measurement you are most likely familiar with is degrees, so we'll begin there.

Measure of an Angle

The **measure of an angle** is a measurement between two intersecting lines, line segments or rays, starting at the **initial side** and ending at the **terminal side**. It is a rotational measure, not a linear measure.

terminal side
angle
initial side

Measuring Angles

Degrees

A **degree** is a measurement of angle. One full rotation around the circle is equal to 360 degrees, so one degree is 1/360 of a circle.

An angle measured in degrees should always include the unit "degrees" after the number, or include the degree symbol °. For example, 90 degrees = $90°$.

Standard Position

When measuring angles on a circle, unless otherwise directed, we measure angles in **standard position**: starting at the positive horizontal axis and with counter-clockwise rotation.

Example 1

Give the degree measure of the angle shown on the circle.

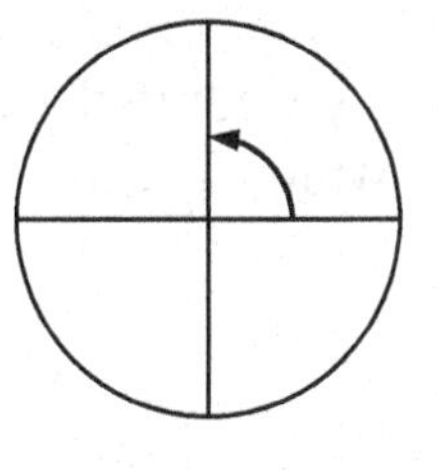

The vertical and horizontal lines divide the circle into quarters. Since one full rotation is 360 degrees= $360°$, each quarter rotation is 360/4 = $90°$ or 90 degrees.

Example 2

Show an angle of $30°$ on the circle.

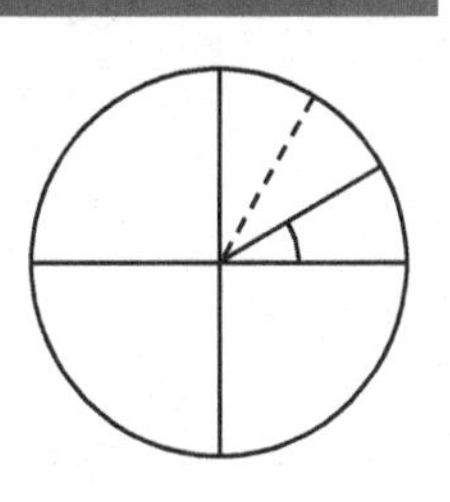

An angle of $30°$ is 1/3 of $90°$, so by dividing a quarter rotation into thirds, we can sketch a line at $30°$.

Going Greek

When representing angles using variables, it is traditional to use Greek letters. Here is a list of commonly encountered Greek letters.

θ	φ or ϕ	α	β	γ
theta	phi	alpha	beta	gamma

Working with Angles in Degrees

Notice that since there are 360 degrees in one rotation, an angle greater than 360 degrees would indicate more than 1 full rotation. Shown on a circle, the resulting direction in which this angle's terminal side points would be the same as for another angle between 0 and 360 degrees. These angles would be called **coterminal.**

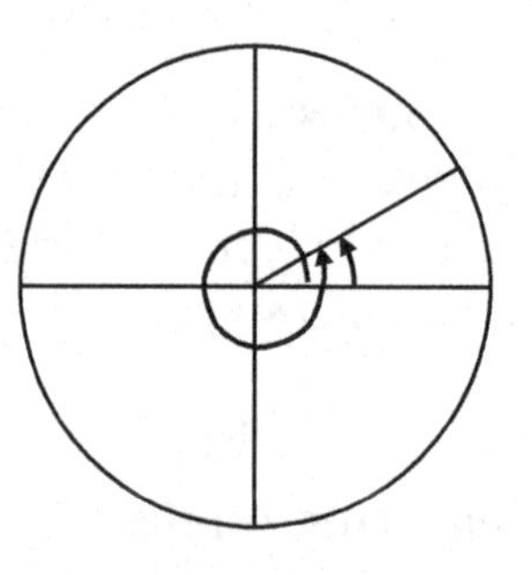

Coterminal Angles

After completing their full rotation based on the given angle, two angles are **coterminal** if they terminate in the same position, so their terminal sides coincide (point in the same direction).

Example 3

Find an angle θ that is coterminal with $800°$, where $0° \le \theta < 360°$

Since adding or subtracting a full rotation, 360 degrees, would result in an angle with terminal side pointing in the same direction, we can find coterminal angles by adding or subtracting 360 degrees. An angle of 800 degrees is coterminal with an angle of 800-360 = 440 degrees. It would also be coterminal with an angle of 440-360 = 80 degrees.

The angle $\theta = 80°$ is coterminal with $800°$.

By finding the coterminal angle between 0 and 360 degrees, it can be easier to see which direction the terminal side of an angle points in.

Try it Now

1. Find an angle α that is coterminal with $870°$, where $0° \le \alpha < 360°$.

On a number line a positive number is measured to the right and a negative number is measured in the opposite direction (to the left). Similarly a positive angle is measured counterclockwise and a negative angle is measured in the opposite direction (clockwise).

Example 4

Show the angle $-45°$ on the circle and find a positive angle α that is coterminal and $0° \leq \alpha < 360°$.

Since 45 degrees is half of 90 degrees, we can start at the positive horizontal axis and measure clockwise half of a 90 degree angle.

Since we can find coterminal angles by adding or subtracting a full rotation of 360 degrees, we can find a positive coterminal angle here by adding 360 degrees:
$-45° + 360° = 315°$

Try it Now

2. Find an angle β coterminal with $-300°$ where $0° \leq \beta < 360°$.

It can be helpful to have a familiarity with the frequently encountered angles in one rotation of a circle. It is common to encounter multiples of 30, 45, 60, and 90 degrees. These values are shown to the right. Memorizing these angles and understanding their properties will be very useful as we study the properties associated with angles

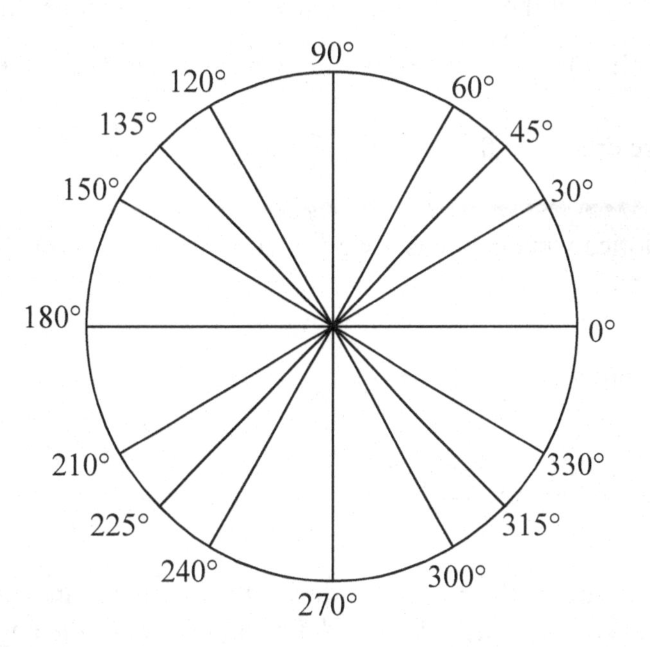

Angles in Radians

While measuring angles in degrees may be familiar, doing so often complicates matters since the units of measure can get in the way of calculations. For this reason, another measure of angles is commonly used. This measure is based on the distance around a circle.

Arclength
Arclength is the length of an arc, *s*, along a circle of radius *r* subtended (drawn out) by an angle θ. It is the portion of the circumference between the initial and terminal sides of the angle.

The length of the arc around an entire circle is called the circumference of a circle. The circumference of a circle is $C = 2\pi r$. The ratio of the circumference to the radius, produces the constant 2π. Regardless of the radius, this ratio is always the same, just as how the degree measure of an angle is independent of the radius.

To elaborate on this idea, consider two circles, one with radius 2 and one with radius 3. Recall the circumference (perimeter) of a circle is $C = 2\pi r$, where *r* is the radius of the circle. The smaller circle then has circumference $2\pi(2) = 4\pi$ and the larger has circumference $2\pi(3) = 6\pi$.

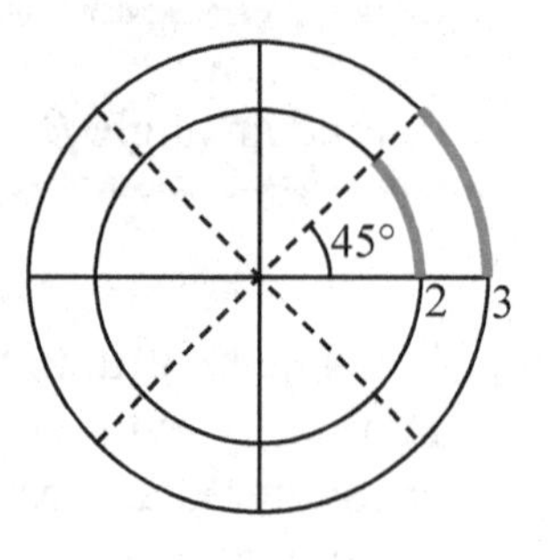

Drawing a 45 degree angle on the two circles, we might be interested in the length of the arc of the circle that the angle indicates. In both cases, the 45 degree angle draws out an arc that is 1/8th of the full circumference, so for the smaller circle, the arclength $= \frac{1}{8}(4\pi) = \frac{1}{2}\pi$, and for the larger circle, the length of the arc or arclength $= \frac{1}{8}(6\pi) = \frac{3}{4}\pi$.

Notice what happens if we find the *ratio* of the arclength divided by the radius of the circle:

Smaller circle: $\dfrac{\frac{1}{2}\pi}{2} = \frac{1}{4}\pi$

Larger circle: $\dfrac{\frac{3}{4}\pi}{3} = \frac{1}{4}\pi$

The ratio is the same regardless of the radius of the circle – it only depends on the angle. This property allows us to define a measure of the angle based on arclength.

Radians

The **radian measure** of an angle is the ratio of the length of the circular arc subtended by the angle to the radius of the circle.

In other words, if s is the length of an arc of a circle, and r is the radius of the circle, then

$$\text{radian measure} = \frac{s}{r}$$

If the circle has radius 1, then the radian measure corresponds to the length of the arc.

Because radian measure is the ratio of two lengths, it is a **unitless measure**. It is not necessary to write the label "radians" after a radian measure, and if you see an angle that is not labeled with "degrees" or the degree symbol, you should assume that it is a radian measure.

Considering the most basic case, the unit circle (a circle with radius 1), we know that 1 rotation equals 360 degrees, $360°$. We can also track one rotation around a circle by finding the circumference, $C = 2\pi r$, and for the unit circle $C = 2\pi$. These two different ways to rotate around a circle give us a way to convert from degrees to radians.

1 rotation = $360° = 2\pi$ radians
½ rotation = $180° = \pi$ radians
¼ rotation = $90° = \frac{\pi}{2}$ radians

Example 5

Find the radian measure of one third of a full rotation.

For any circle, the arclength along such a rotation would be one third of the circumference, $C = \frac{1}{3}(2\pi r) = \frac{2\pi r}{3}$. The radian measure would be the arclength divided by the radius:

$$\text{Radian measure} = \frac{2\pi r}{3r} = \frac{2\pi}{3}.$$

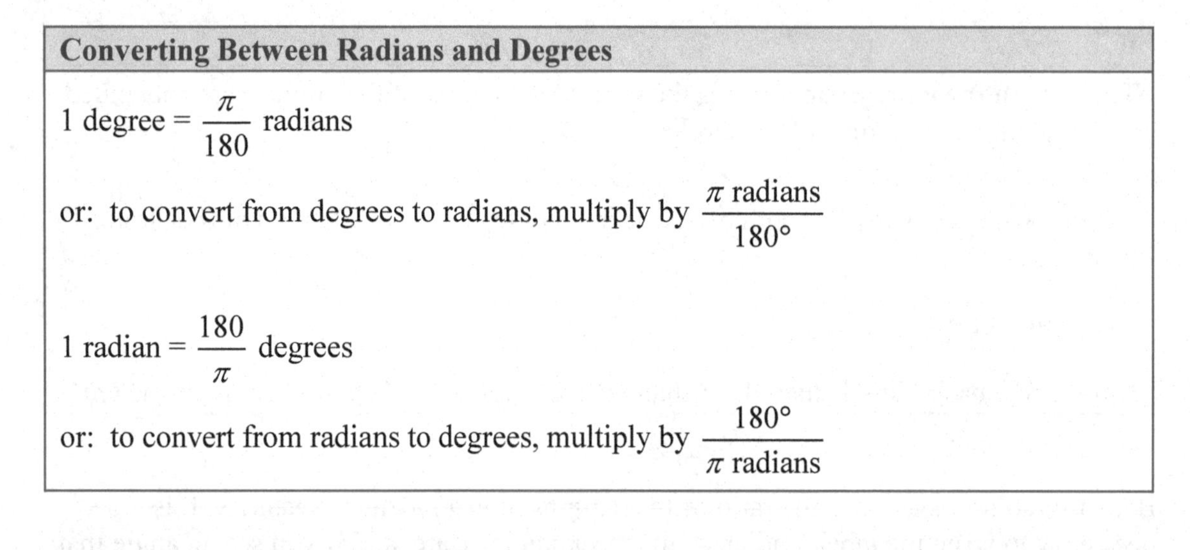

Converting Between Radians and Degrees

1 degree $= \frac{\pi}{180}$ radians

or: to convert from degrees to radians, multiply by $\frac{\pi \text{ radians}}{180°}$

1 radian $= \frac{180}{\pi}$ degrees

or: to convert from radians to degrees, multiply by $\frac{180°}{\pi \text{ radians}}$

Example 6

Convert $\frac{\pi}{6}$ radians to degrees.

Since we are given a problem in radians and we want degrees, we multiply by $\frac{180°}{\pi}$.
Remember radians are a unitless measure, so we don't need to write "radians."

$$\frac{\pi}{6} \text{ radians} = \frac{\pi}{6} \cdot \frac{180°}{\pi} = 30 \text{ degrees.}$$

Example 7

Convert 15 degrees to radians.

In this example, we start with degrees and want radians so we use the other conversion $\frac{\pi}{180°}$ so that the degree units cancel and we are left with the unitless measure of radians.

$$15 \text{ degrees} = 15° \cdot \frac{\pi}{180°} = \frac{\pi}{12}$$

Try it Now

3. Convert $\frac{7\pi}{10}$ radians to degrees.

Just as we listed all the common angles in degrees on a circle, we should also list the corresponding radian values for the common measures of a circle corresponding to degree multiples of 30, 45, 60, and 90 degrees. As with the degree measurements, it would be advisable to commit these to memory.

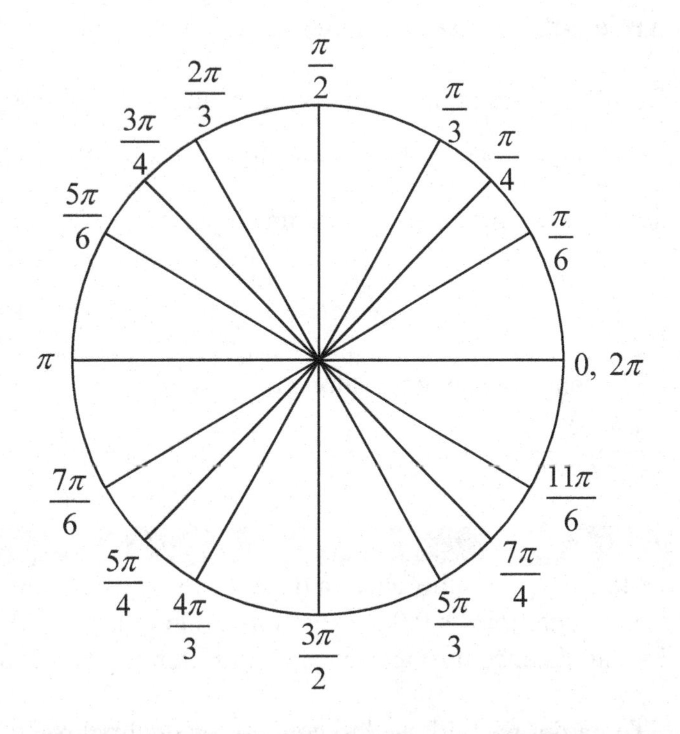

We can work with the radian measures of an angle the same way we work with degrees.

Example 8

Find an angle β that is coterminal with $\frac{19\pi}{4}$, where $0 \le \beta < 2\pi$.

When working in degrees, we found coterminal angles by adding or subtracting 360 degrees, a full rotation. Likewise, in radians, we can find coterminal angles by adding or subtracting full rotations of 2π radians.

$$\frac{19\pi}{4} - 2\pi = \frac{19\pi}{4} - \frac{8\pi}{4} = \frac{11\pi}{4}$$

The angle $\frac{11\pi}{4}$ is coterminal, but not less than 2π, so we subtract another rotation.

$$\frac{11\pi}{4} - 2\pi = \frac{11\pi}{4} - \frac{8\pi}{4} = \frac{3\pi}{4}$$

The angle $\frac{3\pi}{4}$ is coterminal with $\frac{19\pi}{4}$.

Try it Now

4. Find an angle ϕ that is coterminal with $-\frac{17\pi}{6}$ where $0 \le \phi < 2\pi$.

Arclength and Area of a Sector

Recall that the radian measure of an angle was defined as the ratio of the arclength of a circular arc to the radius of the circle, $\theta = \frac{s}{r}$. From this relationship, we can find arclength along a circle given an angle.

Arclength on a Circle

The length of an arc, s, along a circle of radius r subtended by angle θ in radians is
$s = r\theta$

Example 9

Mercury orbits the sun at a distance of approximately 36 million miles. In one Earth day, it completes 0.0114 rotation around the sun. If the orbit was perfectly circular, what distance through space would Mercury travel in one Earth day?

To begin, we will need to convert the decimal rotation value to a radian measure.

Since one rotation = 2π radians,
0.0114 rotation = $2\pi(0.0114) = 0.0716$ radians.

Combining this with the given radius of 36 million miles, we can find the arclength:
$s = (36)(0.0716) = 2.578$ million miles travelled through space.

Try it Now

5. Find the arclength along a circle of radius 10 subtended by an angle of 215 degrees.

In addition to arclength, we can also use angles to find the area of a sector of a circle. A sector is a portion of a circle contained between two lines from the center, like a slice of pizza or pie.

Recall that the area of a circle with radius r can be found using the formula $A = \pi r^2$. If a sector is cut out by an angle of θ, measured in radians, then the fraction of full circle that angle has cut out is $\frac{\theta}{2\pi}$, since 2π is one full rotation. Thus, the area of the sector would be this fraction of the whole area:

$$\text{Area of sector} = \left(\frac{\theta}{2\pi}\right)\pi r^2 = \frac{\theta \pi r^2}{2\pi} = \frac{1}{2}\theta r^2$$

Area of a Sector

The **area of a sector** of a circle with radius r subtended by an angle θ, measured in radians, is

$$\text{Area of sector} = \frac{1}{2}\theta r^2$$

Example 10

An automatic lawn sprinkler sprays a distance of 20 feet while rotating 30 degrees. What is the area of the sector of grass the sprinkler waters?

First, we need to convert the angle measure into radians. Since 30 degrees is one of our common angles, you ideally should already know the equivalent radian measure, but if not we can convert:

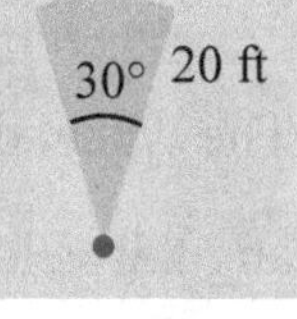

$$30 \text{ degrees} = 30 \cdot \frac{\pi}{180} = \frac{\pi}{6} \text{ radians.}$$

The area of the sector is then Area $= \frac{1}{2}\left(\frac{\pi}{6}\right)(20)^2 = 104.72 \text{ ft}^2$

Try it Now

6. In central pivot irrigation, a large irrigation pipe on wheels rotates around a center point, as pictured here[1]. A farmer has a central pivot system with a radius of 400 meters. If water restrictions only allow her to water 150 thousand square meters a day, what angle should she set the system to cover?

Linear and Angular Velocity

When your car drives down a road, it makes sense to describe its speed in terms of miles per hour or meters per second. These are measures of speed along a line, also called linear velocity. When a point on a circle rotates, we would describe its angular velocity, or rotational speed, in radians per second, rotations per minute, or degrees per hour.

[1] http://commons.wikimedia.org/wiki/File:Pivot_otech_002.JPG CC-BY-SA

Angular and Linear Velocity

As a point moves along a circle of radius r, its **angular velocity,** ω, can be found as the angular rotation θ per unit time, t.

$$\omega = \frac{\theta}{t}$$

The **linear velocity**, v, of the point can be found as the distance travelled, arclength s, per unit time, t.

$$v = \frac{s}{t}$$

Example 11

A water wheel completes 1 rotation every 5 seconds. Find the angular velocity in radians per second.[2]

The wheel completes 1 rotation = 2π radians in 5 seconds, so the angular velocity would be $\omega = \frac{2\pi}{5} \approx 1.257$ radians per second.

Combining the definitions above with the arclength equation, $s = r\theta$, we can find a relationship between angular and linear velocities. The angular velocity equation can be solved for θ, giving $\theta = \omega t$. Substituting this into the arclength equation gives $s = r\theta = r\omega t$.

Substituting this into the linear velocity equation gives

$$v = \frac{s}{t} = \frac{r\omega t}{t} = r\omega$$

Relationship Between Linear and Angular Velocity

When the angular velocity is measured in radians per unit time, linear velocity and angular velocity are related by the equation

$$v = r\omega$$

[2] http://en.wikipedia.org/wiki/File:R%C3%B6mische_S%C3%A4gem%C3%BChle.svg CC-BY

Example 12

A bicycle has wheels 28 inches in diameter. A tachometer determines the wheels are rotating at 180 RPM (revolutions per minute). Find the speed the bicycle is travelling down the road.

Here we have an angular velocity and need to find the corresponding linear velocity, since the linear speed of the outside of the tires is the speed at which the bicycle travels down the road.

We begin by converting from rotations per minute to radians per minute. It can be helpful to utilize the units to make this conversion

$$180\frac{\text{rotations}}{\text{minute}}\cdot\frac{2\pi\text{ radians}}{\text{rotation}}=360\pi\frac{\text{radians}}{\text{minute}}$$

Using the formula from above along with the radius of the wheels, we can find the linear velocity

$$v=(14\text{ inches})\left(360\pi\frac{\text{radians}}{\text{minute}}\right)=5040\pi\frac{\text{inches}}{\text{minute}}$$

You may be wondering where the "radians" went in this last equation. Remember that radians are a unitless measure, so it is not necessary to include them.

Finally, we may wish to convert this linear velocity into a more familiar measurement, like miles per hour.

$$5040\pi\frac{\text{inches}}{\text{minute}}\cdot\frac{1\text{ feet}}{12\text{ inches}}\cdot\frac{1\text{ mile}}{5280\text{ feet}}\cdot\frac{60\text{ minutes}}{1\text{ hour}}=14.99\text{ miles per hour (mph).}$$

Try it Now

7. A satellite is rotating around the earth at 27,934 kilometers per minute at an altitude of 242 km above the earth. If the radius of the earth is 6378 kilometers, find the angular velocity of the satellite.

Important Topics of This Section
Degree measure of angle
Radian measure of angle
Conversion between degrees and radians
Common angles in degrees and radians
Coterminal angles
Arclength
Area of a sector
Linear and angular velocity

Try it Now Answers

1. $\alpha = 870 - 360 - 360 = 150°$

2. $\beta = -300 + 360 = 60°$

3. $\frac{7\pi}{10} \cdot \frac{180°}{\pi} = 126°$

4. $-\frac{17\pi}{6} + 2\pi + 2\pi = -\frac{17\pi}{6} + \frac{12\pi}{6} + \frac{12\pi}{6} = \frac{7\pi}{6}$

5. $215° = \frac{215\pi}{180}$ radians. $s = 10 \cdot \frac{215\pi}{180} = \frac{215\pi}{18} \approx 37.525$

6. $\frac{1}{2}\theta(400)^2 = 150{,}000$. $\theta = 1.875$, or $107.43°$

7. $v = 27934$. $r = 6378+242=6620$. $\omega = \frac{v}{r} = \frac{27934}{6620} = 4.2196$ radians per hour.

Section 5.2 Exercises

1. Indicate each angle on a circle: 30°, 300°, -135°, 70°, $\frac{2\pi}{3}$, $\frac{7\pi}{4}$

2. Indicate each angle on a circle: 30°, 315°, -135°, 80°, $\frac{7\pi}{6}$, $\frac{3\pi}{4}$

3. Convert the angle 180° to radians.

4. Convert the angle 30° to radians.

5. Convert the angle $\frac{5\pi}{6}$ from radians to degrees.

6. Convert the angle $\frac{11\pi}{6}$ from radians to degrees.

7. Find the angle between 0° and 360° that is coterminal with a 685° angle.

8. Find the angle between 0° and 360° that is coterminal with a 451° angle.

9. Find the angle between 0° and 360° that is coterminal with a -1746° angle.

10. Find the angle between 0° and 360° that is coterminal with a -1400° angle.

11. Find the angle between 0 and 2π in radians that is coterminal with the angle $\frac{26\pi}{9}$.

12. Find the angle between 0 and 2π in radians that is coterminal with the angle $\frac{17\pi}{3}$.

13. Find the angle between 0 and 2π in radians that is coterminal with the angle $-\frac{3\pi}{2}$.

14. Find the angle between 0 and 2π in radians that is coterminal with the angle $-\frac{7\pi}{6}$.

15. On a circle of radius 7 miles, find the length of the arc that subtends a central angle of 5 radians.

16. On a circle of radius 6 feet, find the length of the arc that subtends a central angle of 1 radian.

17. On a circle of radius 12 cm, find the length of the arc that subtends a central angle of 120 degrees.

18. On a circle of radius 9 miles, find the length of the arc that subtends a central angle of 800 degrees.

19. Find the distance along an arc on the surface of the Earth that subtends a central angle of 5 minutes (1 minute = 1/60 degree). The radius of the Earth is 3960 miles.

20. Find the distance along an arc on the surface of the Earth that subtends a central angle of 7 minutes (1 minute = 1/60 degree). The radius of the Earth is 3960 miles.

21. On a circle of radius 6 feet, what angle in degrees would subtend an arc of length 3 feet?

22. On a circle of radius 5 feet, what angle in degrees would subtend an arc of length 2 feet?

23. A sector of a circle has a central angle of 45°. Find the area of the sector if the radius of the circle is 6 cm.

24. A sector of a circle has a central angle of 30°. Find the area of the sector if the radius of the circle is 20 cm.

25. A truck with 32-in.-diameter wheels is traveling at 60 mi/h. Find the angular speed of the wheels in rad/min. How many revolutions per minute do the wheels make?

26. A bicycle with 24-in.-diameter wheels is traveling at 15 mi/h. Find the angular speed of the wheels in rad/min. How many revolutions per minute do the wheels make?

27. A wheel of radius 8 in. is rotating 15°/sec. What is the linear speed v, the angular speed in RPM, and the angular speed in rad/sec?

28. A wheel of radius 14 in. is rotating 0.5 rad/sec. What is the linear speed v, the angular speed in RPM, and the angular speed in deg/sec?

29. A CD has diameter of 120 millimeters. When playing audio, the angular speed varies to keep the linear speed constant where the disc is being read. When reading along the outer edge of the disc, the angular speed is about 200 RPM (revolutions per minute). Find the linear speed.

30. When being burned in a writable CD-R drive, the angular speed of a CD is often much faster than when playing audio, but the angular speed still varies to keep the linear speed constant where the disc is being written. When writing along the outer edge of the disc, the angular speed of one drive is about 4800 RPM (revolutions per minute). Find the linear speed.

31. You are standing on the equator of the Earth (radius 3960 miles). What is your linear and angular speed?

32. The restaurant in the Space Needle in Seattle rotates at the rate of one revolution every 47 minutes. [UW]
    a) Through how many radians does it turn in 100 minutes?
    b) How long does it take the restaurant to rotate through 4 radians?
    c) How far does a person sitting by the window move in 100 minutes if the radius of the restaurant is 21 meters?

Section 5.3 Points on Circles Using Sine and Cosine

While it is convenient to describe the location of a point on a circle using an angle or a distance along the circle, relating this information to the x and y coordinates and the circle equation we explored in Section 5.1 is an important application of trigonometry.

A distress signal is sent from a sailboat during a storm, but the transmission is unclear and the rescue boat sitting at the marina cannot determine the sailboat's location. Using high powered radar, they determine the distress signal is coming from a distance of 20 miles at an angle of 225 degrees from the marina. How many miles east/west and north/south of the rescue boat is the stranded sailboat?

In a general sense, to investigate this, we begin by drawing a circle centered at the origin with radius r, and marking the point on the circle indicated by some angle θ. This point has coordinates (x, y).

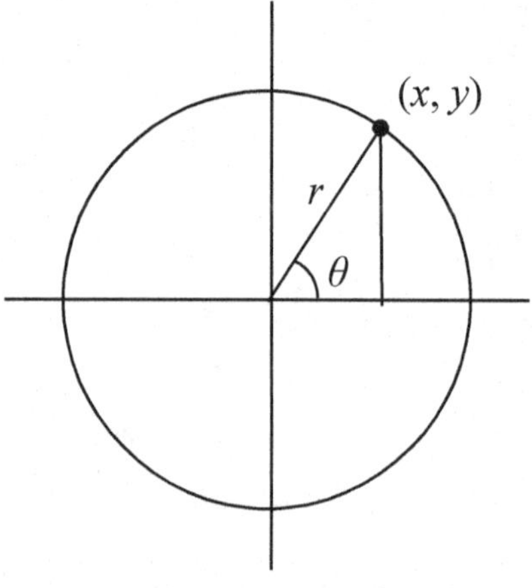

If we drop a line segment vertically down from this point to the x axis, we would form a right triangle inside of the circle.

No matter which quadrant our angle θ puts us in we can draw a triangle by dropping a perpendicular line segment to the x axis, keeping in mind that the values of x and y may be positive or negative, depending on the quadrant.

Additionally, if the angle θ puts us on an axis, we simply measure the radius as the x or y with the other value being 0, again ensuring we have appropriate signs on the coordinates based on the quadrant.

Triangles obtained from different radii will all be similar triangles, meaning corresponding sides scale proportionally. While the lengths of the sides may change, as we saw in the last section, the *ratios* of the side lengths will always remain constant for any given angle.

$$\frac{y_1}{r_1} = \frac{y_2}{r_2}$$

$$\frac{x_1}{r_1} = \frac{x_2}{r_2}$$

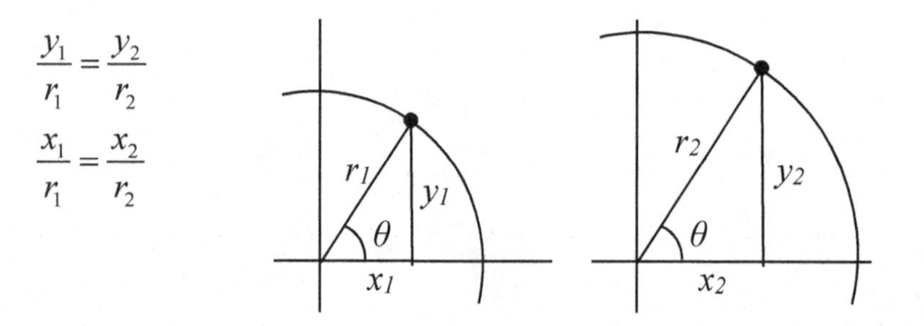

To be able to refer to these ratios more easily, we will give them names. Since the ratios depend on the angle, we will write them as functions of the angle θ.

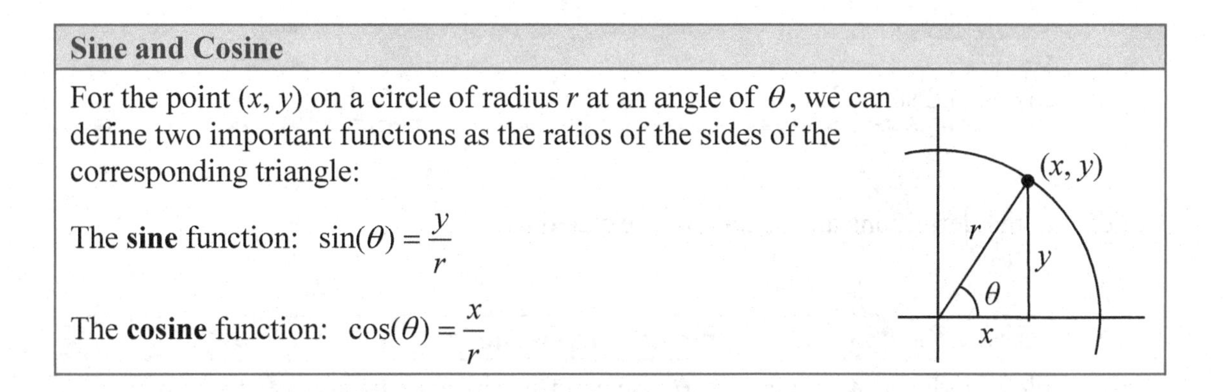

Sine and Cosine

For the point (x, y) on a circle of radius r at an angle of θ, we can define two important functions as the ratios of the sides of the corresponding triangle:

The **sine** function: $\sin(\theta) = \frac{y}{r}$

The **cosine** function: $\cos(\theta) = \frac{x}{r}$

In this chapter, we will explore these functions using both circles and right triangles. In the next chapter, we will take a closer look at the behavior and characteristics of the sine and cosine functions.

Example 1

The point (3, 4) is on the circle of radius 5 at some angle θ. Find $\cos(\theta)$ and $\sin(\theta)$.

Knowing the radius of the circle and coordinates of the point, we can evaluate the cosine and sine functions as the ratio of the sides.

$\cos(\theta) = \frac{x}{r} = \frac{3}{5}$ $\qquad$ $\sin(\theta) = \frac{y}{r} = \frac{4}{5}$

There are a few cosine and sine values which we can determine fairly easily because the corresponding point on the circle falls on the x or y axis.

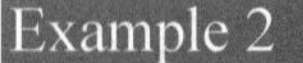

Example 2

Find $\cos(90°)$ and $\sin(90°)$

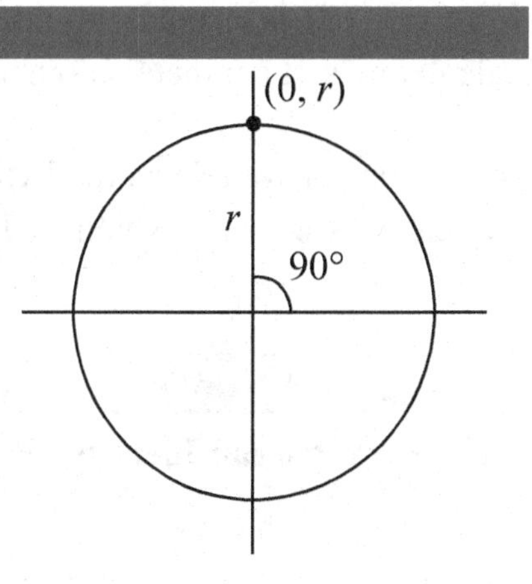

On any circle, the terminal side of a 90 degree angle points straight up, so the coordinates of the corresponding point on the circle would be $(0, r)$. Using our definitions of cosine and sine,

$$\cos(90°) = \frac{x}{r} = \frac{0}{r} = 0$$

$$\sin(90°) = \frac{y}{r} = \frac{r}{r} = 1$$

Try it Now

1. Find cosine and sine of the angle π.

Notice that the definitions above can also be stated as:

Coordinates of the Point on a Circle at a Given Angle

On a circle of radius r at an angle of θ, we can find the **coordinates of the point (x, y)** at that angle using

$x = r\cos(\theta)$

$y = r\sin(\theta)$

On a unit circle, a circle with radius 1, $x = \cos(\theta)$ and $y = \sin(\theta)$.

Utilizing the basic equation for a circle centered at the origin, $x^2 + y^2 = r^2$, combined with the relationships above, we can establish a new identity.

$x^2 + y^2 = r^2$	substituting the relations above,
$(r\cos(\theta))^2 + (r\sin(\theta))^2 = r^2$	simplifying,
$r^2(\cos(\theta))^2 + r^2(\sin(\theta))^2 = r^2$	dividing by r^2
$(\cos(\theta))^2 + (\sin(\theta))^2 = 1$	or using shorthand notation
$\cos^2(\theta) + \sin^2(\theta) = 1$	

Here $\cos^2(\theta)$ is a commonly used shorthand notation for $(\cos(\theta))^2$. Be aware that many calculators and computers do not understand the shorthand notation.

In Section 5.1 we related the Pythagorean Theorem $a^2 + b^2 = c^2$ to the basic equation of a circle $x^2 + y^2 = r^2$, which we have now used to arrive at the Pythagorean Identity.

Pythagorean Identity

The **Pythagorean Identity**. For any angle θ, $\cos^2(\theta) + \sin^2(\theta) = 1$.

One use of this identity is that it helps us to find a cosine value of an angle if we know the sine value of that angle or vice versa. However, since the equation will yield two possible values, we will need to utilize additional knowledge of the angle to help us find the desired value.

Example 3

If $\sin(\theta) = \frac{3}{7}$ and θ is in the second quadrant, find $\cos(\theta)$.

Substituting the known value for sine into the Pythagorean identity,

$$\cos^2(\theta) + \left(\frac{3}{7}\right)^2 = 1$$

$$\cos^2(\theta) + \frac{9}{49} = 1$$

$$\cos^2(\theta) = \frac{40}{49}$$

$$\cos(\theta) = \pm\sqrt{\frac{40}{49}} = \pm\frac{\sqrt{40}}{7} = \pm\frac{2\sqrt{10}}{7}$$

Since the angle is in the second quadrant, we know the x value of the point would be negative, so the cosine value should also be negative. Using this additional information, we can conclude that $\cos(\theta) = -\frac{2\sqrt{10}}{7}$.

Values for Sine and Cosine

At this point, you may have noticed that we haven't found any cosine or sine values from angles not on an axis. To do this, we will need to utilize our knowledge of triangles.

First, consider a point on a circle at an angle of 45 degrees, or $\frac{\pi}{4}$. At this angle, the x and y coordinates of the corresponding point on the circle will be equal because 45 degrees divides the first quadrant in half. Since the x and y values will be the same, the sine and cosine values will also be equal. Utilizing the Pythagorean Identity,

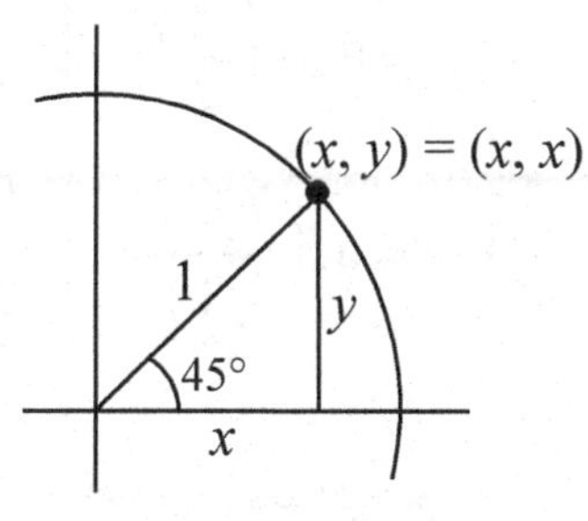

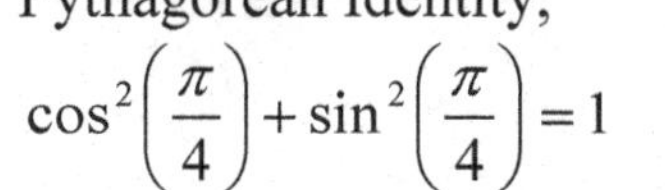

$$\cos^2\left(\frac{\pi}{4}\right) + \sin^2\left(\frac{\pi}{4}\right) = 1$$ since the sine and cosine are equal, we can substitute sine with cosine

$$\cos^2\left(\frac{\pi}{4}\right) + \cos^2\left(\frac{\pi}{4}\right) = 1$$ add like terms

$$2\cos^2\left(\frac{\pi}{4}\right) = 1$$ divide

$$\cos^2\left(\frac{\pi}{4}\right) = \frac{1}{2}$$ since the x value is positive, we'll keep the positive root

$\cos\left(\frac{\pi}{4}\right) = \sqrt{\frac{1}{2}}$ often this value is written with a rationalized denominator

Remember, to rationalize the denominator we multiply by a term equivalent to 1 to get rid of the radical in the denominator.

$$\cos\left(\frac{\pi}{4}\right) = \sqrt{\frac{1}{2}}\sqrt{\frac{2}{2}} = \sqrt{\frac{2}{4}} = \frac{\sqrt{2}}{2}$$

Since the sine and cosine are equal, $\sin\left(\frac{\pi}{4}\right) = \frac{\sqrt{2}}{2}$ as well. The (x, y) coordinates for a point on a circle of radius 1 at an angle of 45 degrees are $\left(\frac{\sqrt{2}}{2}, \frac{\sqrt{2}}{2}\right)$.

Example 4

Find the coordinates of the point on a circle of radius 6 at an angle of $\frac{\pi}{4}$.

Using our new knowledge that $\sin\left(\frac{\pi}{4}\right) = \frac{\sqrt{2}}{2}$ and $\cos\left(\frac{\pi}{4}\right) = \frac{\sqrt{2}}{2}$, along with our relationships that stated $x = r\cos(\theta)$ and $y = r\sin(\theta)$, we can find the coordinates of the point desired:

$$x = 6\cos\left(\frac{\pi}{4}\right) = 6\left(\frac{\sqrt{2}}{2}\right) = 3\sqrt{2}$$

$$y = 6\sin\left(\frac{\pi}{4}\right) = 6\left(\frac{\sqrt{2}}{2}\right) = 3\sqrt{2}$$

Try it Now

2. Find the coordinates of the point on a circle of radius 3 at an angle of $90°$.

Next, we will find the cosine and sine at an angle of 30 degrees, or $\frac{\pi}{6}$. To do this, we will first draw a triangle inside a circle with one side at an angle of 30 degrees, and another at an angle of -30 degrees. If the resulting two right triangles are combined into one

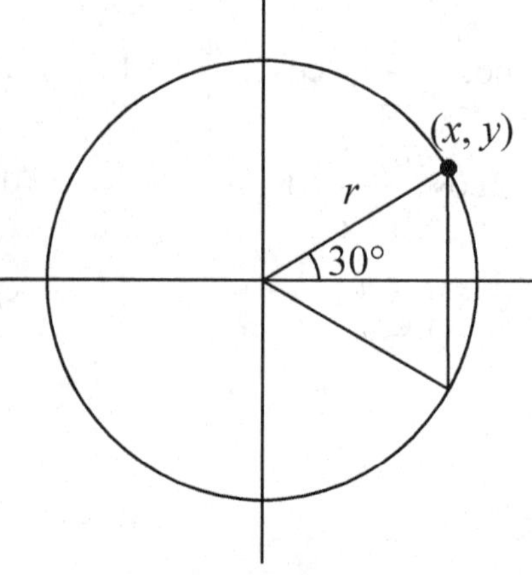

large triangle, notice that all three angles of this larger triangle will be 60 degrees. Since all the angles are equal, the sides will all be equal as well. The vertical line has length $2y$, and since the sides are all equal we can conclude that $2y = r$, or $y = \frac{r}{2}$. Using this, we can find the sine value:

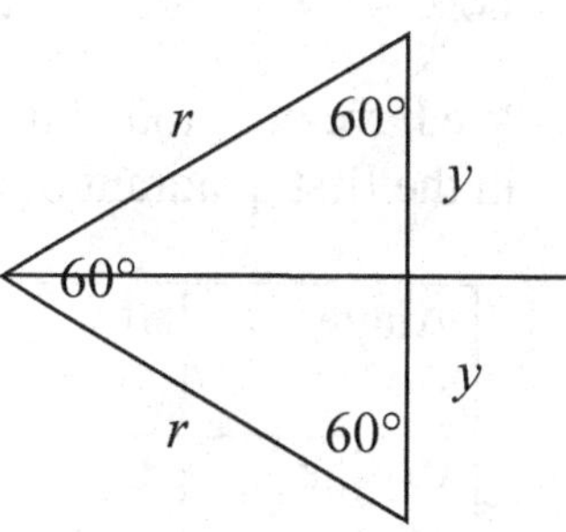

$$\sin\left(\frac{\pi}{6}\right) = \frac{y}{r} = \frac{r/2}{r} = \frac{r}{2} \cdot \frac{1}{r} = \frac{1}{2}$$

Using the Pythagorean Identity, we can find the cosine value:

$$\cos^2\left(\frac{\pi}{6}\right) + \sin^2\left(\frac{\pi}{6}\right) = 1$$

$$\cos^2\left(\frac{\pi}{6}\right) + \left(\frac{1}{2}\right)^2 = 1$$

$$\cos^2\left(\frac{\pi}{6}\right) = \frac{3}{4}$$ since the x value is positive, we'll keep the positive root

$$\cos\left(\frac{\pi}{6}\right) = \sqrt{\frac{3}{4}} = \frac{\sqrt{3}}{2}$$

The (x, y) coordinates for the point on a circle of radius 1 at an angle of 30 degrees are $\left(\frac{\sqrt{3}}{2}, \frac{1}{2}\right)$.

By drawing a the triangle inside the unit circle with a 30 degree angle and reflecting it over the line $y = x$, we can find the cosine and sine for 60 degrees, or $\frac{\pi}{3}$, without any additional work.

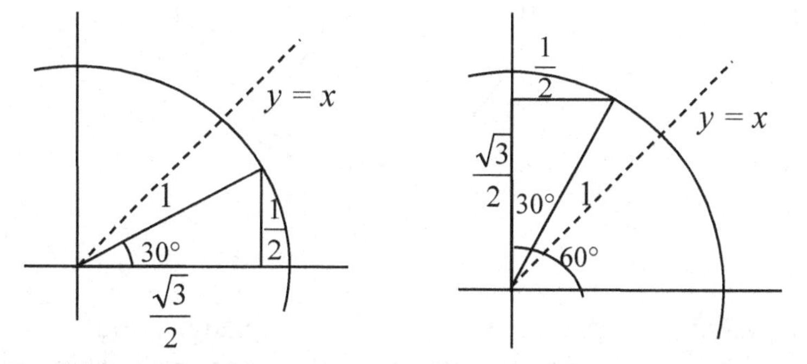

By this symmetry, we can see the coordinates of the point on the unit circle at an angle of 60 degrees will be $\left(\frac{1}{2}, \frac{\sqrt{3}}{2}\right)$, giving

$\cos\left(\frac{\pi}{3}\right) = \frac{1}{2}$ and $\sin\left(\frac{\pi}{3}\right) = \frac{\sqrt{3}}{2}$

We have now found the cosine and sine values for all the commonly encountered angles in the first quadrant of the unit circle.

Angle	0	$\frac{\pi}{6}$, or 30°	$\frac{\pi}{4}$, or 45°	$\frac{\pi}{3}$, or 60°	$\frac{\pi}{2}$, or 90°
Cosine	1	$\frac{\sqrt{3}}{2}$	$\frac{\sqrt{2}}{2}$	$\frac{1}{2}$	0
Sine	0	$\frac{1}{2}$	$\frac{\sqrt{2}}{2}$	$\frac{\sqrt{3}}{2}$	1

For any given angle in the first quadrant, there will be an angle in another quadrant with the same sine value, and yet another angle in yet another quadrant with the same cosine value. Since the sine value is the y coordinate on the unit circle, the other angle with the same sine will share the same y value, but have the opposite x value. Likewise, the angle with the same cosine will share the same x value, but have the opposite y value.

As shown here, angle α has the same sine value as angle θ; the cosine values would be opposites. The angle β has the same cosine value as the angle θ; the sine values would be opposites.

$\sin(\theta) = \sin(\alpha)$ and $\cos(\theta) = -\cos(\alpha)$ $\qquad$ $\sin(\theta) = -\sin(\beta)$ and $\cos(\theta) = \cos(\beta)$

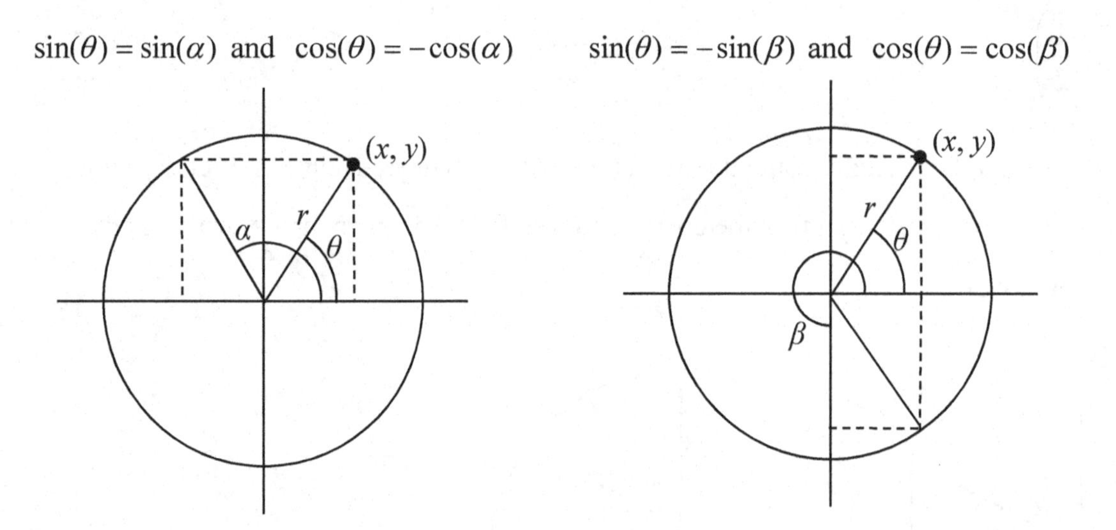

It is important to notice the relationship between the angles. If, from the angle, you measured the smallest angle to the horizontal axis, all would have the same measure in absolute value. We say that all these angles have a **reference angle** of θ.

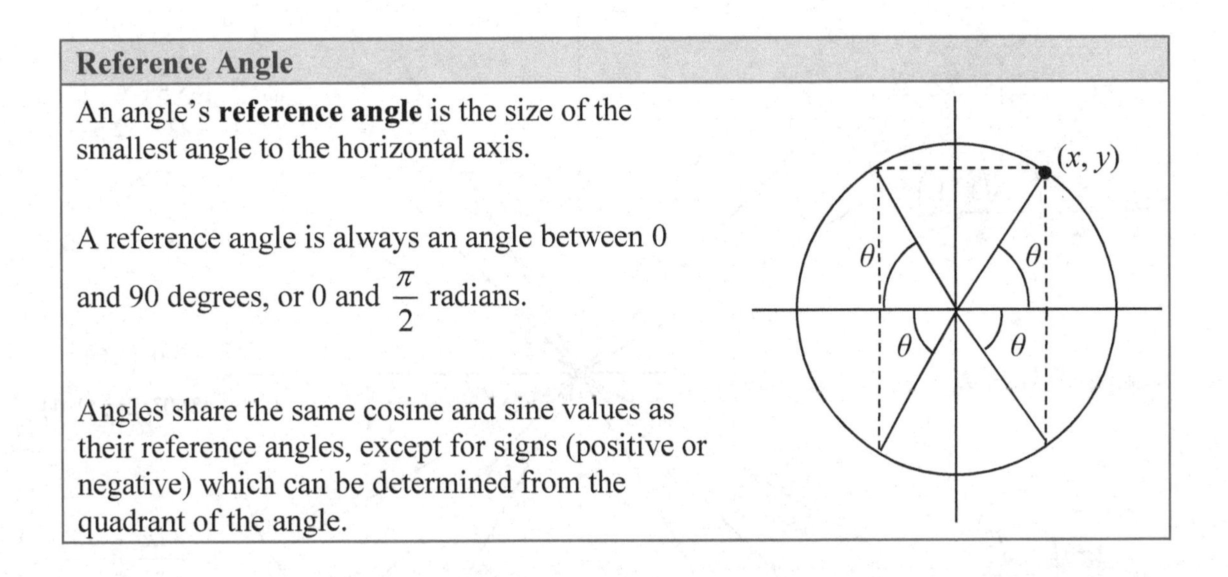

Reference Angle

An angle's **reference angle** is the size of the smallest angle to the horizontal axis.

A reference angle is always an angle between 0 and 90 degrees, or 0 and $\frac{\pi}{2}$ radians.

Angles share the same cosine and sine values as their reference angles, except for signs (positive or negative) which can be determined from the quadrant of the angle.

Example 5

Find the reference angle of 150 degrees. Use it to find $\cos(150°)$ and $\sin(150°)$.

150 degrees is located in the second quadrant. It is 30 degrees short of the horizontal axis at 180 degrees, so the reference angle is 30 degrees.

This tells us that 150 degrees has the same sine and cosine values as 30 degrees, except for sign. We know that $\sin(30°) = \frac{1}{2}$ and $\cos(30°) = \frac{\sqrt{3}}{2}$. Since 150 degrees is in the second quadrant, the x coordinate of the point on the circle would be negative, so the cosine value will be negative. The y coordinate is positive, so the sine value will be positive.

$\sin(150°) = \frac{1}{2}$ and $\cos(150°) = -\frac{\sqrt{3}}{2}$

The (x, y) coordinates for the point on a unit circle at an angle of 150° are $\left(\frac{-\sqrt{3}}{2}, \frac{1}{2}\right)$.

Using symmetry and reference angles, we can fill in cosine and sine values at the rest of the special angles on the unit circle. Take time to learn the (x, y) coordinates of all the major angles in the first quadrant!

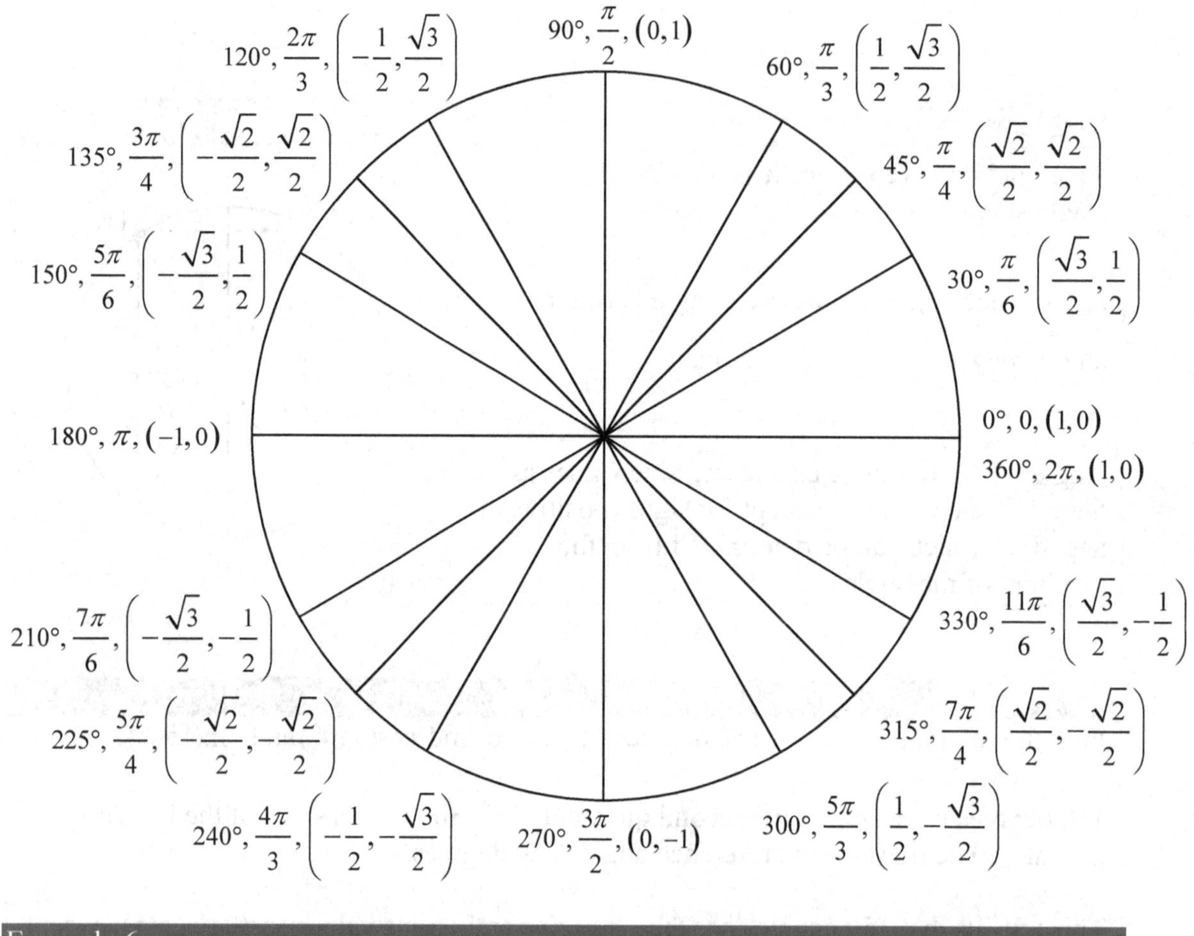

Example 6

Find the coordinates of the point on a circle of radius 12 at an angle of $\frac{7\pi}{6}$.

Note that this angle is in the third quadrant, where both x and y are negative. Keeping this in mind can help you check your signs of the sine and cosine function.

$$x = 12\cos\left(\frac{7\pi}{6}\right) = 12\left(\frac{-\sqrt{3}}{2}\right) = -6\sqrt{3}$$

$$y = 12\sin\left(\frac{7\pi}{6}\right) = 12\left(\frac{-1}{2}\right) = -6$$

The coordinates of the point are $(-6\sqrt{3}, -6)$.

Try it Now

3. Find the coordinates of the point on a circle of radius 5 at an angle of $\frac{5\pi}{3}$.

Example 7

We now have the tools to return to the sailboat question posed at the beginning of this section.

A distress signal is sent from a sailboat during a storm, but the transmission is unclear and the rescue boat sitting at the marina cannot determine the sailboat's location. Using high powered radar, they determine the distress signal is coming from a point 20 miles away at an angle of 225 degrees from the marina. How many miles east/west and north/south of the rescue boat is the stranded sailboat?

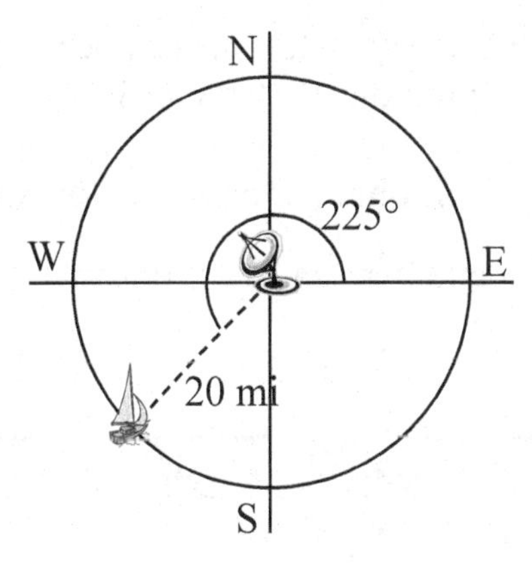

We can now answer the question by finding the coordinates of the point on a circle with a radius of 20 miles at an angle of 225 degrees.

$$x = 20\cos(225°) = 20\left(\frac{-\sqrt{2}}{2}\right) \approx -14.142 \text{ miles}$$

$$y = 20\sin(225°) = 20\left(\frac{-\sqrt{2}}{2}\right) \approx -14.142 \text{ miles}$$

The sailboat is located 14.142 miles west and 14.142 miles south of the marina.

The special values of sine and cosine in the first quadrant are very useful to know, since knowing them allows you to quickly evaluate the sine and cosine of very common angles without needing to look at a reference or use your calculator. However, scenarios do come up where we need to know the sine and cosine of other angles.

To find the cosine and sine of any other angle, we turn to a computer or calculator. **Be aware**: most calculators can be set into "degree" or "radian" mode, which tells the calculator the units for the input value. When you evaluate "cos(30)" on your calculator, it will evaluate it as the cosine of 30 degrees if the calculator is in degree mode, or the cosine of 30 radians if the calculator is in radian mode. Most computer software with cosine and sine functions only operates in radian mode.

Example 8

Evaluate the cosine of 20 degrees using a calculator or computer.

On a calculator that can be put in degree mode, you can evaluate this directly to be approximately 0.939693.

On a computer or calculator without degree mode, you would first need to convert the angle to radians, or equivalently evaluate the expression $\cos\left(20 \cdot \frac{\pi}{180}\right)$.

Important Topics of This Section
The sine function
The cosine function
Pythagorean Identity
Unit Circle values
Reference angles
Using technology to find points on a circle

Try it Now Answers

1. $\cos(\pi) = -1 \quad \sin(\pi) = 0$

2. $$x = 3\cos\left(\frac{\pi}{2}\right) = 3 \cdot 0 = 0$$
$$y = 3\sin\left(\frac{\pi}{2}\right) = 3 \cdot 1 = 3$$

3. $\left(5\cos\left(\frac{5\pi}{3}\right), 5\sin\left(\frac{5\pi}{3}\right)\right) = \left(\frac{5}{2}, \frac{-5\sqrt{3}}{2}\right)$

Section 5.3 Exercises

1. Find the quadrant in which the terminal point determined by t lies if
   a. $\sin(t) < 0$ and $\cos(t) < 0$ b. $\sin(t) > 0$ and $\cos(t) < 0$

2. Find the quadrant in which the terminal point determined by t lies if
   a. $\sin(t) < 0$ and $\cos(t) > 0$ b. $\sin(t) > 0$ and $\cos(t) > 0$

3. The point P is on the unit circle. If the y-coordinate of P is $\frac{3}{5}$, and P is in quadrant II, find the x coordinate.

4. The point P is on the unit circle. If the x-coordinate of P is $\frac{1}{5}$, and P is in quadrant IV, find the y coordinate.

5. If $\cos(\theta) = \frac{1}{7}$ and θ is in the 4th quadrant, find $\sin(\theta)$.

6. If $\cos(\theta) = \frac{2}{9}$ and θ is in the 1st quadrant, find $\sin(\theta)$.

7. If $\sin(\theta) = \frac{3}{8}$ and θ is in the 2nd quadrant, find $\cos(\theta)$.

8. If $\sin(\theta) = -\frac{1}{4}$ and θ is in the 3rd quadrant, find $\cos(\theta)$.

9. For each of the following angles, find the reference angle and which quadrant the angle lies in. Then compute sine and cosine of the angle.
   a. 225° b. 300° c. 135° d. 210°

10. For each of the following angles, find the reference angle and which quadrant the angle lies in. Then compute sine and cosine of the angle.
    a. 120° b. 315° c. 250° d. 150°

11. For each of the following angles, find the reference angle and which quadrant the angle lies in. Then compute sine and cosine of the angle.
    a. $\frac{5\pi}{4}$ b. $\frac{7\pi}{6}$ c. $\frac{5\pi}{3}$ d. $\frac{3\pi}{4}$

12. For each of the following angles, find the reference angle and which quadrant the angle lies in. Then compute sine and cosine of the angle.
    a. $\frac{4\pi}{3}$ b. $\frac{2\pi}{3}$ c. $\frac{5\pi}{6}$ d. $\frac{7\pi}{4}$

13. Give exact values for $\sin(\theta)$ and $\cos(\theta)$ for each of these angles.

a. $-\frac{3\pi}{4}$ b. $\frac{23\pi}{6}$ c. $-\frac{\pi}{2}$ d. 5π

14. Give exact values for $\sin(\theta)$ and $\cos(\theta)$ for each of these angles.

a. $-\frac{2\pi}{3}$ b. $\frac{17\pi}{4}$ c. $-\frac{\pi}{6}$ d. 10π

15. Find an angle θ with $0 < \theta < 360°$ or $0 < \theta < 2\pi$ that has the same sine value as:

a. $\frac{\pi}{3}$ b. $80°$ c. $140°$ d. $\frac{4\pi}{3}$ e. $305°$

16. Find an angle θ with $0 < \theta < 360°$ or $0 < \theta < 2\pi$ that has the same sine value as:

a. $\frac{\pi}{4}$ b. $15°$ c. $160°$ d. $\frac{7\pi}{6}$ e. $340°$

17. Find an angle θ with $0 < \theta < 360°$ or $0 < \theta < 2\pi$ that has the same cosine value as:

a. $\frac{\pi}{3}$ b. $80°$ c. $140°$ d. $\frac{4\pi}{3}$ e. $305°$

18. Find an angle θ with $0 < \theta < 360°$ or $0 < \theta < 2\pi$ that has the same cosine value as:

a. $\frac{\pi}{4}$ b. $15°$ c. $160°$ d. $\frac{7\pi}{6}$ e. $340°$

19. Find the coordinates of the point on a circle with radius 15 corresponding to an angle of 220°.

20. Find the coordinates of the point on a circle with radius 20 corresponding to an angle of 280°.

21. Marla is running clockwise around a circular track. She runs at a constant speed of 3 meters per second. She takes 46 seconds to complete one lap of the track. From her starting point, it takes her 12 seconds to reach the northernmost point of the track. Impose a coordinate system with the center of the track at the origin, and the northernmost point on the positive *y*-axis. [UW]

a) Give Marla's coordinates at her starting point.
b) Give Marla's coordinates when she has been running for 10 seconds.
c) Give Marla's coordinates when she has been running for 901.3 seconds.

Section 5.4 The Other Trigonometric Functions

In the previous section, we defined the sine and cosine functions as ratios of the sides of a right triangle in a circle. Since the triangle has 3 sides there are 6 possible combinations of ratios. While the sine and cosine are the two prominent ratios that can be formed, there are four others, and together they define the 6 trigonometric functions.

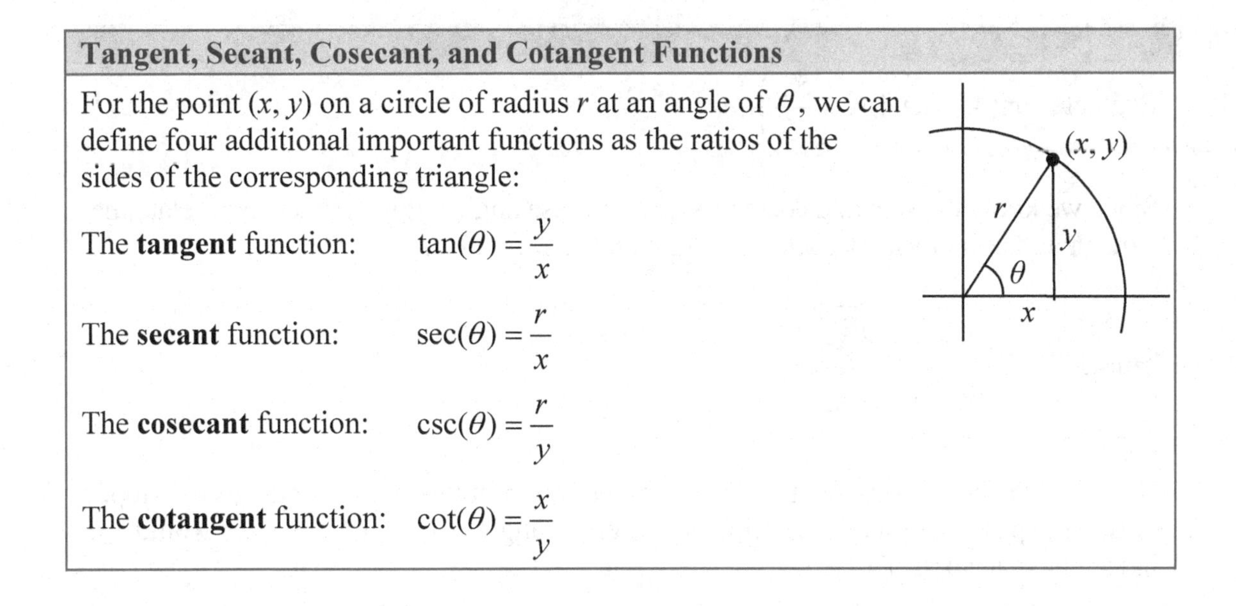

Tangent, Secant, Cosecant, and Cotangent Functions

For the point (x, y) on a circle of radius r at an angle of θ, we can define four additional important functions as the ratios of the sides of the corresponding triangle:

The **tangent** function: $\tan(\theta) = \frac{y}{x}$

The **secant** function: $\sec(\theta) = \frac{r}{x}$

The **cosecant** function: $\csc(\theta) = \frac{r}{y}$

The **cotangent** function: $\cot(\theta) = \frac{x}{y}$

Geometrically, notice that the definition of tangent corresponds with the slope of the line segment between the origin (0, 0) and the point (x, y). This relationship can be very helpful in thinking about tangent values.

You may also notice that the ratios defining the secant, cosecant, and cotangent are the reciprocals of the ratios defining the cosine, sine, and tangent functions, respectively. Additionally, notice that using our results from the last section,

$$\tan(\theta) = \frac{y}{x} = \frac{r\sin(\theta)}{r\cos(\theta)} = \frac{\sin(\theta)}{\cos(\theta)}$$

Applying this concept to the other trig functions we can state the reciprocal identities.

Identities

The other four trigonometric functions can be related back to the sine and cosine functions using these basic relationships:

$$\tan(\theta) = \frac{\sin(\theta)}{\cos(\theta)} \qquad \sec(\theta) = \frac{1}{\cos(\theta)} \qquad \csc(\theta) = \frac{1}{\sin(\theta)} \qquad \cot(\theta) = \frac{1}{\tan(\theta)} = \frac{\cos(\theta)}{\sin(\theta)}$$

These relationships are called **identities**. Identities are statements that are true for all values of the input on which they are defined. Identities are usually something that can be derived from definitions and relationships we already know, similar to how the identities above were derived from the circle relationships of the six trig functions. The Pythagorean Identity we learned earlier was derived from the Pythagorean Theorem and the definitions of sine and cosine. We will discuss the role of identities more after an example.

Example 1

Evaluate $\tan(45°)$ and $\sec\left(\frac{5\pi}{6}\right)$.

Since we know the sine and cosine values for these angles, it makes sense to relate the tangent and secant values back to the sine and cosine values.

$$\tan(45°) = \frac{\sin(45°)}{\cos(45°)} = \frac{\sqrt{2}/2}{\sqrt{2}/2} = 1$$

Notice this result is consistent with our interpretation of the tangent value as the slope of the line passing through the origin at the given angle: a line at 45 degrees would indeed have a slope of 1.

$$\sec\left(\frac{5\pi}{6}\right) = \frac{1}{\cos\left(\frac{5\pi}{6}\right)} = \frac{1}{-\sqrt{3}/2} = \frac{-2}{\sqrt{3}}$$, which could also be written as $\frac{-2\sqrt{3}}{3}$.

Try it Now

1. Evaluate $\csc\left(\frac{7\pi}{6}\right)$.

Just as we often need to simplify algebraic expressions, it is often also necessary or helpful to simplify trigonometric expressions. To do so, we utilize the definitions and identities we have established.

Example 2

Simplify $\frac{\sec(\theta)}{\tan(\theta)}$.

We can simplify this by rewriting both functions in terms of sine and cosine

$$\frac{\sec(\theta)}{\tan(\theta)} = \frac{1/\cos(\theta)}{\sin(\theta)/\cos(\theta)}$$ To divide the fractions we could invert and multiply

$$= \frac{1}{\cos(\theta)}\frac{\cos(\theta)}{\sin(\theta)}$$ cancelling the cosines,

$$= \frac{1}{\sin(\theta)} = \csc(\theta)$$ simplifying and using the identity

By showing that $\frac{\sec(\theta)}{\tan(\theta)}$ can be simplified to $\csc(\theta)$, we have, in fact, established a new identity: that $\frac{\sec(\theta)}{\tan(\theta)} = \csc(\theta)$.

Occasionally a question may ask you to "prove the identity" or "establish the identity." This is the same idea as when an algebra book asks a question like "show that $(x-1)^2 = x^2 - 2x + 1$." In this type of question, we must show the algebraic manipulations that demonstrate that the left and right side of the equation are in fact equal. You can think of a "prove the identity" problem as a simplification problem where you *know the answer*: you know what the end goal of the simplification should be, and just need to show the steps to get there.

To prove an identity, in most cases you will start with the expression on one side of the identity and manipulate it using algebra and trigonometric identities until you have simplified it to the expression on the other side of the equation. **Do not** treat the identity like an equation to solve – it isn't! The proof is establishing *if* the two expressions are equal, so we must take care to work with one side at a time rather than applying an operation simultaneously to both sides of the equation.

Example 3

Prove the identity $\frac{1+\cot(\alpha)}{\csc(\alpha)} = \sin(\alpha) + \cos(\alpha)$.

Since the left side seems a bit more complicated, we will start there and simplify the expression until we obtain the right side. We can use the right side as a guide for what might be good steps to make. In this case, the left side involves a fraction while the right side doesn't, which suggests we should look to see if the fraction can be reduced.

Additionally, since the right side involves sine and cosine and the left does not, it suggests that rewriting the cotangent and cosecant using sine and cosine might be a good idea.

$\dfrac{1+\cot(\alpha)}{\csc(\alpha)}$ Rewriting the cotangent and cosecant

$=\dfrac{1+\dfrac{\cos(\alpha)}{\sin(\alpha)}}{\dfrac{1}{\sin(\alpha)}}$ To divide the fractions, we invert and multiply

$=\left(1+\dfrac{\cos(\alpha)}{\sin(\alpha)}\right)\dfrac{\sin(\alpha)}{1}$ Distributing,

$=1\cdot\dfrac{\sin(\alpha)}{1}+\dfrac{\cos(\alpha)}{\sin(\alpha)}\cdot\dfrac{\sin(\alpha)}{1}$ Simplifying the fractions,

$=\sin(\alpha)+\cos(\alpha)$ Establishing the identity.

Notice that in the second step, we could have combined the 1 and $\dfrac{\cos(\alpha)}{\sin(\alpha)}$ before inverting and multiplying. It is very common when proving or simplifying identities for there to be more than one way to obtain the same result.

We can also utilize identities we have previously learned, like the Pythagorean Identity, while simplifying or proving identities.

Example 4

Establish the identity $\dfrac{\cos^2(\theta)}{1+\sin(\theta)}=1-\sin(\theta)$.

Since the left side of the identity is more complicated, it makes sense to start there. To simplify this, we will have to reduce the fraction, which would require the numerator to have a factor in common with the denominator. Additionally, we notice that the right side only involves sine. Both of these suggest that we need to convert the cosine into something involving sine.

Recall the Pythagorean Identity told us $\cos^2(\theta)+\sin^2(\theta)=1$. By moving one of the trig functions to the other side, we can establish:

$\sin^2(\theta)=1-\cos^2(\theta)$ and $\cos^2(\theta)=1-\sin^2(\theta)$

Utilizing this, we now can establish the identity. We start on one side and manipulate:

$$\frac{\cos^2(\theta)}{1+\sin(\theta)}$$ Utilizing the Pythagorean Identity

$$=\frac{1-\sin^2(\theta)}{1+\sin(\theta)}$$ Factoring the numerator

$$=\frac{(1-\sin(\theta))(1+\sin(\theta))}{1+\sin(\theta)}$$ Cancelling the like factors

$$=1-\sin(\theta)$$ Establishing the identity

We can also build new identities from previously established identities. For example, if we divide both sides of the Pythagorean Identity by cosine squared (which is allowed since we've already shown the identity is true),

$$\frac{\cos^2(\theta)+\sin^2(\theta)}{\cos^2(\theta)}=\frac{1}{\cos^2(\theta)}$$ Splitting the fraction on the left,

$$\frac{\cos^2(\theta)}{\cos^2(\theta)}+\frac{\sin^2(\theta)}{\cos^2(\theta)}=\frac{1}{\cos^2(\theta)}$$ Simplifying and using the definitions of tan and sec

$$1+\tan^2(\theta)=\sec^2(\theta).$$

Try it Now

2. Use a similar approach to establish that $\cot^2(\theta)+1=\csc^2(\theta)$.

Identities

Alternate forms of the Pythagorean Identity

$1+\tan^2(\theta)=\sec^2(\theta)$

$\cot^2(\theta)+1=\csc^2(\theta)$

Example 5

If $\tan(\theta)=\frac{2}{7}$ and θ is in the 3rd quadrant, find $\cos(\theta)$.

There are two approaches to this problem, both of which work equally well.

Approach 1

Since $\tan(\theta) = \frac{y}{x}$ and the angle is in the third quadrant, we can imagine a triangle in a circle of some radius so that the point on the circle is (-7, -2), so $\frac{y}{x} = \frac{-2}{-7} = \frac{2}{7}$.

Using the Pythagorean Theorem, we can find the radius of the circle:
$(-7)^2 + (-2)^2 = r^2$, so $r = \sqrt{53}$.

Now we can find the cosine value:
$\cos(\theta) = \frac{x}{r} = \frac{-7}{\sqrt{53}}$

Approach 2

Using the $1 + \tan^2(\theta) = \sec^2(\theta)$ form of the Pythagorean Identity with the known tangent value,

$$1 + \tan^2(\theta) = \sec^2(\theta)$$

$$1 + \left(\frac{2}{7}\right)^2 = \sec^2(\theta)$$

$$\frac{53}{49} = \sec^2(\theta)$$

$$\sec(\theta) = \pm\sqrt{\frac{53}{49}} = \pm\frac{\sqrt{53}}{7}$$

Since the angle is in the third quadrant, the cosine value will be negative so the secant value will also be negative. Keeping the negative result, and using definition of secant,

$$\sec(\theta) = -\frac{\sqrt{53}}{7}$$

$$\frac{1}{\cos(\theta)} = -\frac{\sqrt{53}}{7} \qquad \text{Inverting both sides}$$

$$\cos(\theta) = -\frac{7}{\sqrt{53}} = -\frac{7\sqrt{53}}{53}$$

Try it Now

3. If $\sec(\phi) = -\frac{7}{3}$ and $\frac{\pi}{2} < \phi < \pi$, find $\tan(\phi)$ and $\sin(\phi)$.

Important Topics of This Section
6 Trigonometric Functions:
Sine
Cosine
Tangent
Cosecant
Secant
Cotangent
Trig identities

Try it Now Answers

1. $\csc\left(\frac{7\pi}{6}\right) = \frac{1}{\sin\left(\frac{7\pi}{6}\right)} = \frac{1}{-1/2} = -2$

2.

$$\frac{\cos^2(\theta) + \sin^2(\theta)}{\sin^2\theta} = 1$$

$$\frac{\cos^2(\theta)}{\sin^2(\theta)} + \frac{\sin^2(\theta)}{\sin^2(\theta)} = \frac{1}{\sin^2(\theta)}$$

$$\cot^2(\theta) + 1 = \csc^2(\theta)$$

3. $\sec(\phi) = -\frac{7}{3}$. By definition, $\frac{1}{\cos(\phi)} = -\frac{7}{3}$, so $\cos(\phi) = -\frac{3}{7}$.

Using Pythagorean Identity with the sec, $1 + \tan^2(\phi) = \left(-\frac{7}{3}\right)^2$. Solving gives

$\tan(\phi) = \frac{\sqrt{40}}{-3}$. We use the negative square root since an angle in the second quadrant would have a negative tangent.

Using Pythagorean Identity with the cos, $\sin^2(\phi) + \left(-\frac{3}{7}\right)^2 = 1$. Solving,

$\sin(\phi) = \frac{\sqrt{40}}{7}$.

Section 5.4 Exercises

1. If $\theta = \frac{\pi}{4}$, find exact values for $\sec(\theta), \csc(\theta), \tan(\theta), \cot(\theta)$.

2. If $\theta = \frac{7\pi}{4}$, find exact values for $\sec(\theta), \csc(\theta), \tan(\theta), \cot(\theta)$.

3. If $\theta = \frac{5\pi}{6}$, find exact values for $\sec(\theta), \csc(\theta), \tan(\theta), \cot(\theta)$.

4. If $\theta = \frac{\pi}{6}$, find exact values for $\sec(\theta), \csc(\theta), \tan(\theta), \cot(\theta)$.

5. If $\theta = \frac{2\pi}{3}$, find exact values for $\sec(\theta), \csc(\theta), \tan(\theta), \cot(\theta)$.

6. If $\theta = \frac{4\pi}{3}$, find exact values for $\sec(\theta), \csc(\theta), \tan(\theta), \cot(\theta)$.

7. Evaluate: a. $\sec(135°)$ b. $\csc(210°)$ c. $\tan(60°)$ d. $\cot(225°)$

8. Evaluate: a. $\sec(30°)$ b. $\csc(315°)$ c. $\tan(135°)$ d. $\cot(150°)$

9. If $\sin(\theta) = \frac{3}{4}$, and θ is in quadrant II, find $\cos(\theta), \sec(\theta), \csc(\theta), \tan(\theta), \cot(\theta)$.

10. If $\sin(\theta) = \frac{2}{7}$, and θ is in quadrant II, find $\cos(\theta), \sec(\theta), \csc(\theta), \tan(\theta), \cot(\theta)$.

11. If $\cos(\theta) = -\frac{1}{3}$, and θ is in quadrant III, find $\sin(\theta), \sec(\theta), \csc(\theta), \tan(\theta), \cot(\theta)$.

12. If $\cos(\theta) = \frac{1}{5}$, and θ is in quadrant I, find $\sin(\theta), \sec(\theta), \csc(\theta), \tan(\theta), \cot(\theta)$.

13. If $\tan(\theta) = \frac{12}{5}$, and $0 \le \theta < \frac{\pi}{2}$, find $\sin(\theta), \cos(\theta), \sec(\theta), \csc(\theta), \cot(\theta)$.

14. If $\tan(\theta) = 4$, and $0 \le \theta < \frac{\pi}{2}$, find $\sin(\theta), \cos(\theta), \sec(\theta), \csc(\theta), \cot(\theta)$.

15. Use a calculator to find sine, cosine, and tangent of the following values:

a. 0.15 b. 4 c. 70° d. 283°

16. Use a calculator to find sine, cosine, and tangent of the following values:

a. 0.5 b. 5.2 c. 10° d. 195°

Simplify each of the following to an expression involving a single trig function with no fractions.

17. $\csc(t)\tan(t)$

18. $\cos(t)\csc(t)$

19. $\dfrac{\sec(t)}{\csc(t)}$

20. $\dfrac{\cot(t)}{\csc(t)}$

21. $\dfrac{\sec(t)-\cos(t)}{\sin(t)}$

22. $\dfrac{\tan(t)}{\sec(t)-\cos(t)}$

23. $\dfrac{1+\cot(t)}{1+\tan(t)}$

24. $\dfrac{1+\sin(t)}{1+\csc(t)}$

25. $\dfrac{\sin^2(t)+\cos^2(t)}{\cos^2(t)}$

26. $\dfrac{1-\sin^2(t)}{\sin^2(t)}$

Prove the identities.

27. $\dfrac{\sin^2(\theta)}{1+\cos(\theta)}=1-\cos(\theta)$

28. $\tan^2(t)=\dfrac{1}{\cos^2(t)}-1$

29. $\sec(a)-\cos(a)=\sin(a)\tan(a)$

30. $\dfrac{1+\tan^2(b)}{\tan^2(b)}=\csc^2(b)$

31. $\dfrac{\csc^2(x)-\sin^2(x)}{\csc(x)+\sin(x)}=\cos(x)\cot(x)$

32. $\dfrac{\sin(\theta)-\cos(\theta)}{\sec(\theta)-\csc(\theta)}=\sin(\theta)\cos(\theta)$

33. $\dfrac{\csc^2(\alpha)-1}{\csc^2(\alpha)-\csc(\alpha)}=1+\sin(\alpha)$

34. $1+\cot(x)=\cos(x)\left(\sec(x)+\csc(x)\right)$

35. $\dfrac{1+\cos(u)}{\sin(u)}=\dfrac{\sin(u)}{1-\cos(u)}$

36. $2\sec^2(t)=\dfrac{1-\sin(t)}{\cos^2(t)}+\dfrac{1}{1-\sin(t)}$

37. $\dfrac{\sin^4(\gamma)-\cos^4(\gamma)}{\sin(\gamma)-\cos(\gamma)}=\sin(\gamma)+\cos(\gamma)$

38. $\dfrac{\left(1+\cos(A)\right)\left(1-\cos(A)\right)}{\sin(A)}=\sin(A)$

Section 5.5 Right Triangle Trigonometry

In section 5.3 we were introduced to the sine and cosine function as ratios of the sides of a triangle drawn inside a circle, and spent the rest of that section discussing the role of those functions in finding points on the circle. In this section, we return to the triangle, and explore the applications of the trigonometric functions to right triangles where circles may not be involved.

Recall that we defined sine and cosine as

$$\sin(\theta) = \frac{y}{r}$$

$$\cos(\theta) = \frac{x}{r}$$

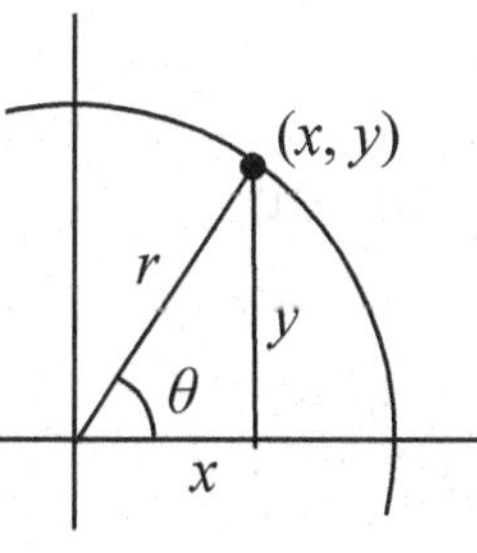

Separating the triangle from the circle, we can make equivalent but more general definitions of the sine, cosine, and tangent on a right triangle. On the right triangle, we will label the hypotenuse as well as the side opposite the angle and the side adjacent (next to) the angle.

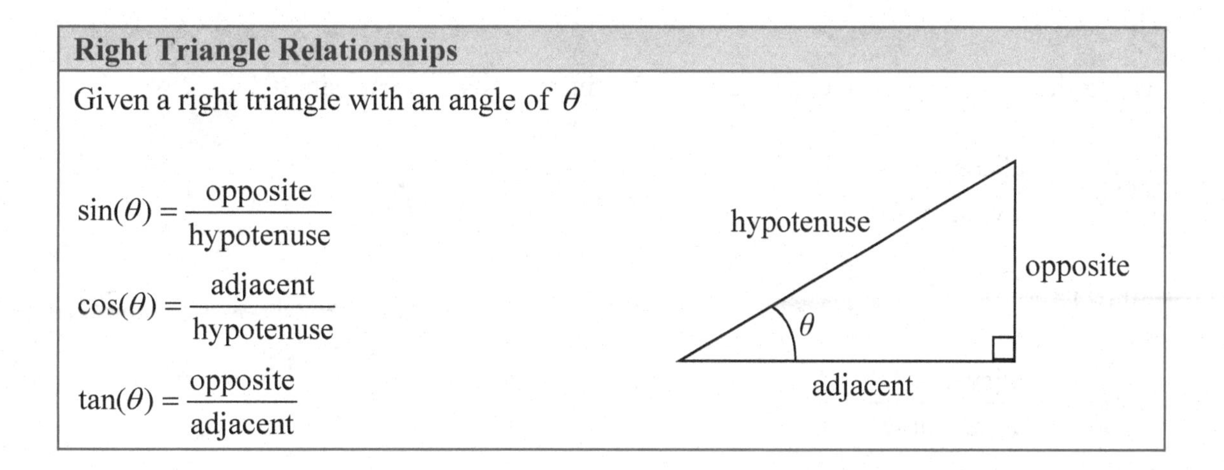

Right Triangle Relationships

Given a right triangle with an angle of θ

$$\sin(\theta) = \frac{\text{opposite}}{\text{hypotenuse}}$$

$$\cos(\theta) = \frac{\text{adjacent}}{\text{hypotenuse}}$$

$$\tan(\theta) = \frac{\text{opposite}}{\text{adjacent}}$$

A common mnemonic for remembering these relationships is SohCahToa, formed from the first letters of "Sine is opposite over hypotenuse, Cosine is adjacent over hypotenuse, Tangent is opposite over adjacent."

Example 1

Given the triangle shown, find the value for $\cos(\alpha)$.

The side adjacent to the angle is 15, and the hypotenuse of the triangle is 17, so

$$\cos(\alpha) = \frac{\text{adjacent}}{\text{hypotenuse}} = \frac{15}{17}$$

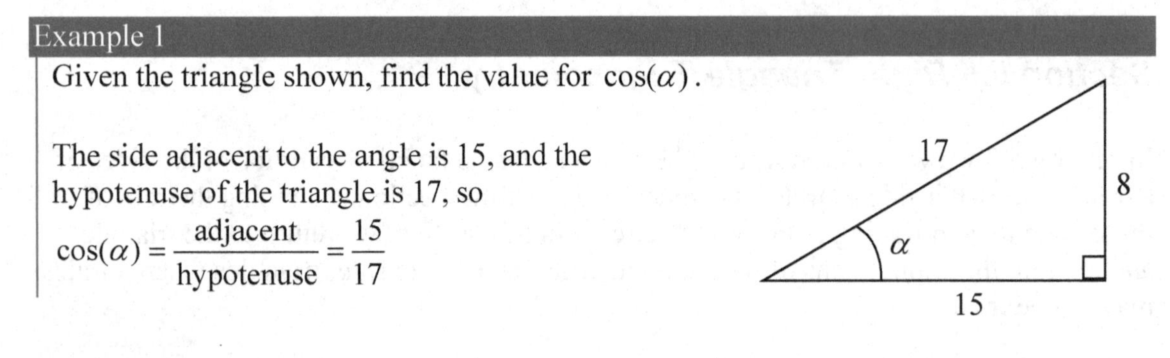

When working with general right triangles, the same rules apply regardless of the orientation of the triangle. In fact, we can evaluate the sine and cosine of either of the two acute angles in the triangle.

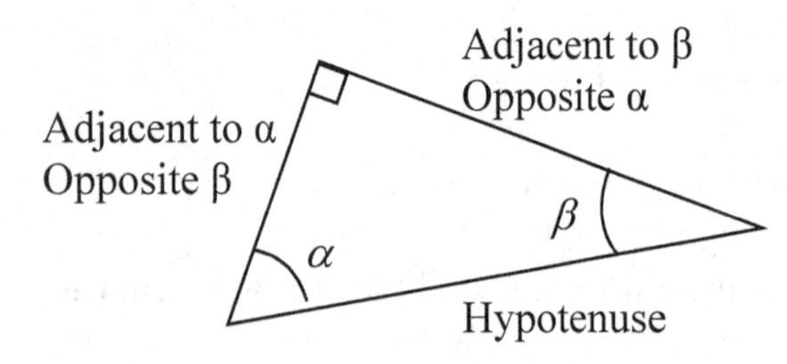

Example 2

Using the triangle shown, evaluate $\cos(\alpha)$, $\sin(\alpha)$, $\cos(\beta)$, and $\sin(\beta)$.

$$\cos(\alpha) = \frac{\text{adjacent to } \alpha}{\text{hypotenuse}} = \frac{3}{5}$$

$$\sin(\alpha) = \frac{\text{opposite } \alpha}{\text{hypotenuse}} = \frac{4}{5}$$

$$\cos(\beta) = \frac{\text{adjacent to } \beta}{\text{hypotenuse}} = \frac{4}{5}$$

$$\sin(\beta) = \frac{\text{opposite } \beta}{\text{hypotenuse}} = \frac{3}{5}$$

Try it Now

1. A right triangle is drawn with angle α opposite a side with length 33, angle β opposite a side with length 56, and hypotenuse 65. Find the sine and cosine of α and β.

You may have noticed that in the above example that $\cos(\alpha) = \sin(\beta)$ and $\cos(\beta) = \sin(\alpha)$. This makes sense since the side opposite α is also adjacent to β. Since the three angles in a triangle need to add to π, or 180 degrees, then the other two angles must add to $\frac{\pi}{2}$, or 90 degrees, so $\beta = \frac{\pi}{2} - \alpha$, and $\alpha = \frac{\pi}{2} - \beta$. Since $\cos(\alpha) = \sin(\beta)$, then $\cos(\alpha) = \sin\left(\frac{\pi}{2} - \alpha\right)$.

Cofunction Identities

The **cofunction identities** for sine and cosine are:

$\cos(\theta) = \sin\left(\frac{\pi}{2} - \theta\right)$ $\qquad$ $\sin(\theta) = \cos\left(\frac{\pi}{2} - \theta\right)$

In the previous examples, we evaluated the sine and cosine on triangles where we knew all three sides of the triangle. Right triangle trigonometry becomes powerful when we start looking at triangles in which we know an angle but don't know all the sides.

Example 3

Find the unknown sides of the triangle pictured here.

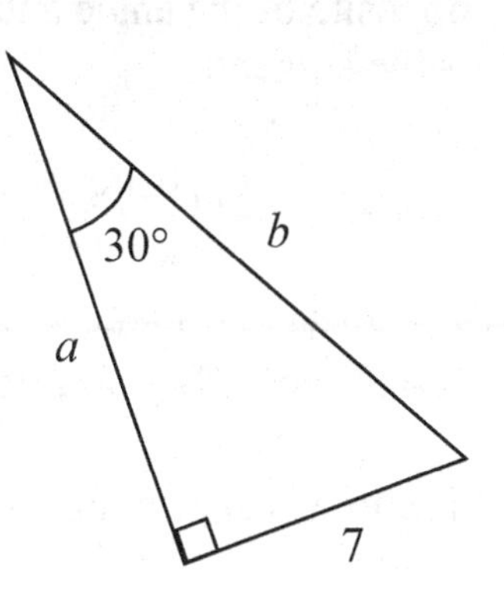

Since $\sin(\theta) = \frac{\text{opposite}}{\text{hypotenuse}}$, $\sin(30°) = \frac{7}{b}$.

From this, we can solve for the side b.

$b\sin(30°) = 7$

$b = \frac{7}{\sin(30°)}$

To obtain a value, we can evaluate the sine and simplify

$b = \frac{7}{1/2} = 14$

To find the value for side a, we could use the cosine, or simply apply the Pythagorean Theorem:

$a^2 + 7^2 = b^2$

$a^2 + 7^2 = 14^2$

$a = \sqrt{147}$

Notice that if we know at least one of the non-right angles of a right triangle and one side, we can find the rest of the sides and angles.

Try it Now

2. A right triangle has one angle of $\frac{\pi}{3}$ and a hypotenuse of 20. Find the unknown sides and angles of the triangle.

Example 4

To find the height of a tree, a person walks to a point 30 feet from the base of the tree, and measures the angle from the ground to the top of the tree to be 57 degrees. Find the height of the tree.

We can introduce a variable, h, to represent the height of the tree. The two sides of the triangle that are most important to us are the side opposite the angle, the height of the tree we are looking for, and the adjacent side, the side we are told is 30 feet long.

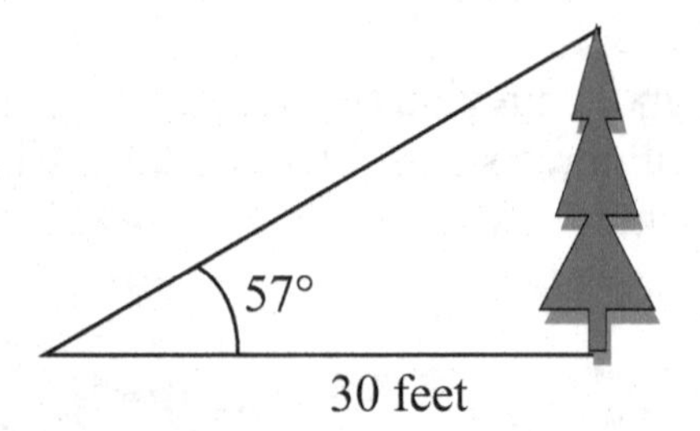

The trigonometric function which relates the side opposite of the angle and the side adjacent to the angle is the tangent.

$\tan(57°) = \frac{\text{opposite}}{\text{adjacent}} = \frac{h}{30}$ Solving for h,

$h = 30\tan(57°)$ Using technology, we can approximate a value

$h = 30\tan(57°) \approx 46.2$ feet

The tree is approximately 46 feet tall.

Example 5

A person standing on the roof of a 100 foot tall building is looking towards a skyscraper a few blocks away, wondering how tall it is. She measures the angle of declination from the roof of the building to the base of the skyscraper to be 20 degrees and the angle of inclination to the top of the skyscraper to be 42 degrees.

To approach this problem, it would be good to start with a picture. Although we are interested in the height, h, of the skyscraper, it can be helpful to also label other unknown quantities in the picture – in this case the horizontal distance x between the buildings and a, the height of the skyscraper above the person.

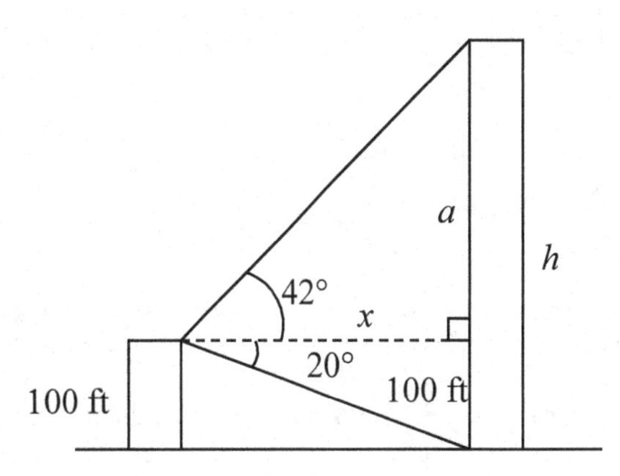

To start solving this problem, notice we have two right triangles. In the top triangle, we know one angle is 42 degrees, but we don't know any of the sides of the triangle, so we don't yet know enough to work with this triangle.

In the lower right triangle, we know one angle is 20 degrees, and we know the vertical height measurement of 100 ft. Since we know these two pieces of information, we can solve for the unknown distance x.

$$\tan(20°) = \frac{\text{opposite}}{\text{adjacent}} = \frac{100}{x} \qquad \text{Solving for } x$$

$$x\tan(20°) = 100$$

$$x = \frac{100}{\tan(20°)}$$

Now that we have found the distance x, we know enough information to solve the top right triangle.

$$\tan(42°) = \frac{\text{opposite}}{\text{adjacent}} = \frac{a}{x} = \frac{a}{100/\tan(20°)}$$

$$\tan(42°) = \frac{a\tan(20°)}{100}$$

$$100\tan(42°) = a\tan(20°)$$

$$\frac{100\tan(42°)}{\tan(20°)} = a$$

Approximating a value,

$$a = \frac{100\tan(42°)}{\tan(20°)} \approx 247.4 \text{ feet}$$

Adding the height of the first building, we determine that the skyscraper is about 347 feet tall.

Important Topics of This Section
SOH CAH TOA
Cofunction identities
Applications with right triangles

Try it Now Answers

1. $\sin(\alpha)=\dfrac{33}{65}$ $\cos(\alpha)=\dfrac{56}{65}$ $\sin(\beta)=\dfrac{56}{65}$ $\cos(\beta)=\dfrac{33}{65}$

2. $\cos\left(\dfrac{\pi}{3}\right)=\dfrac{\text{adjacent}}{\text{hypoteuse}}=\dfrac{\text{Adj}}{20}$ so, adjacent $=20\cos\left(\dfrac{\pi}{3}\right)=20\left(\dfrac{1}{2}\right)=10$

   $\sin\left(\dfrac{\pi}{3}\right)=\dfrac{\text{Opposite}}{\text{hypoteuse}}=\dfrac{\text{Opp}}{20}$ so, opposite $=20\sin\left(\dfrac{\pi}{3}\right)=20\left(\dfrac{\sqrt{3}}{2}\right)=10\sqrt{3}$

   Missing angle = 180-90-60 = 30 degrees or $\pi/6$.

Section 5.5 Exercises

Note: pictures may not be drawn to scale.

In each of the triangles below, find $\sin(A), \cos(A), \tan(A), \sec(A), \csc(A), \cot(A)$.

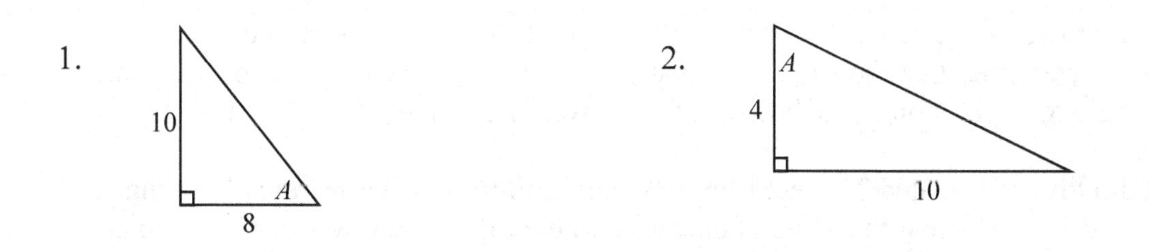

In each of the following triangles, solve for the unknown sides and angles.

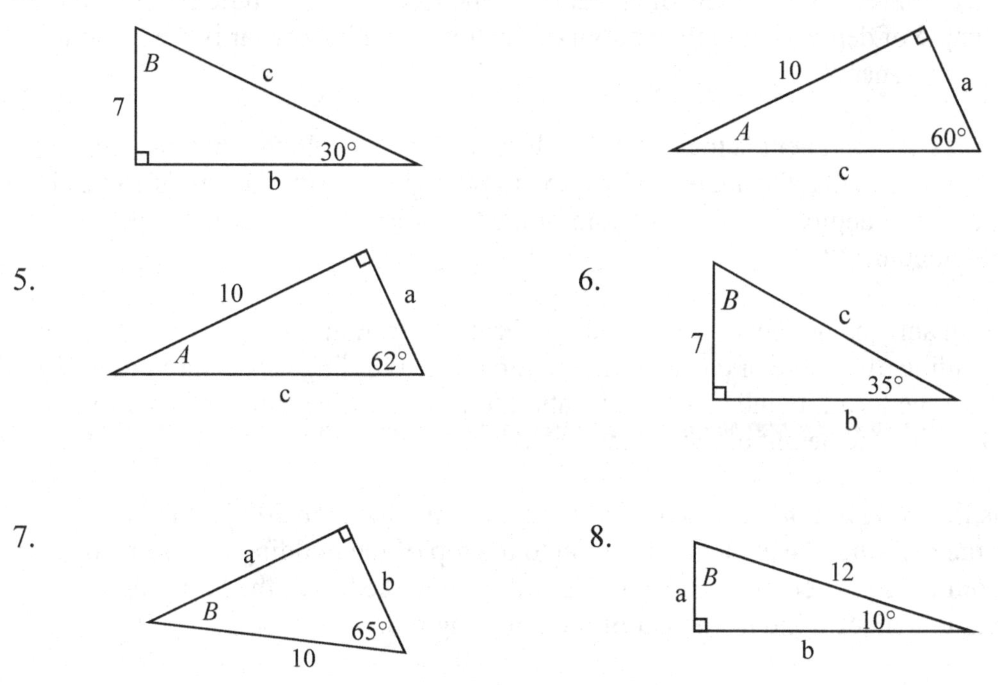

9. A 33-ft ladder leans against a building so that the angle between the ground and the ladder is 80°. How high does the ladder reach up the side of the building?

10. A 23-ft ladder leans against a building so that the angle between the ground and the ladder is 80°. How high does the ladder reach up the side of the building?

11. The angle of elevation to the top of a building in New York is found to be 9 degrees from the ground at a distance of 1 mile from the base of the building. Using this information, find the height of the building.

12. The angle of elevation to the top of a building in Seattle is found to be 2 degrees from the ground at a distance of 2 miles from the base of the building. Using this information, find the height of the building.

13. A radio tower is located 400 feet from a building. From a window in the building, a person determines that the angle of elevation to the top of the tower is 36° and that the angle of depression to the bottom of the tower is 23°. How tall is the tower?

14. A radio tower is located 325 feet from a building. From a window in the building, a person determines that the angle of elevation to the top of the tower is 43° and that the angle of depression to the bottom of the tower is 31°. How tall is the tower?

15. A 200 foot tall monument is located in the distance. From a window in a building, a person determines that the angle of elevation to the top of the monument is 15° and that the angle of depression to the bottom of the tower is 2°. How far is the person from the monument?

16. A 400 foot tall monument is located in the distance. From a window in a building, a person determines that the angle of elevation to the top of the monument is 18° and that the angle of depression to the bottom of the tower is 3°. How far is the person from the monument?

17. There is an antenna on the top of a building. From a location 300 feet from the base of the building, the angle of elevation to the top of the building is measured to be 40°. From the same location, the angle of elevation to the top of the antenna is measured to be 43°. Find the height of the antenna.

18. There is lightning rod on the top of a building. From a location 500 feet from the base of the building, the angle of elevation to the top of the building is measured to be 36°. From the same location, the angle of elevation to the top of the lightning rod is measured to be 38°. Find the height of the lightning rod.

19. Find the length x.

20. Find the length x.

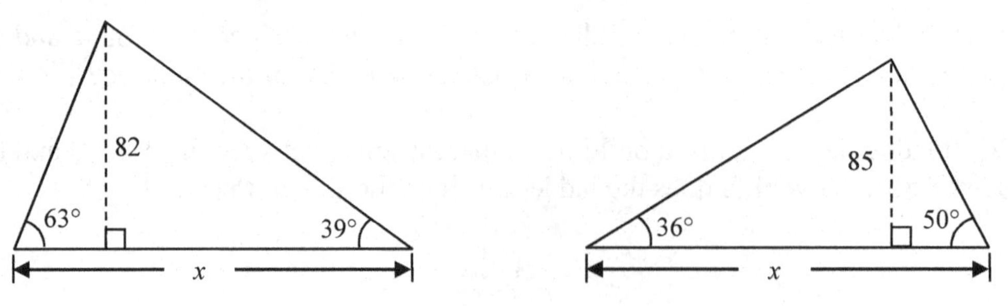

21. Find the length x.

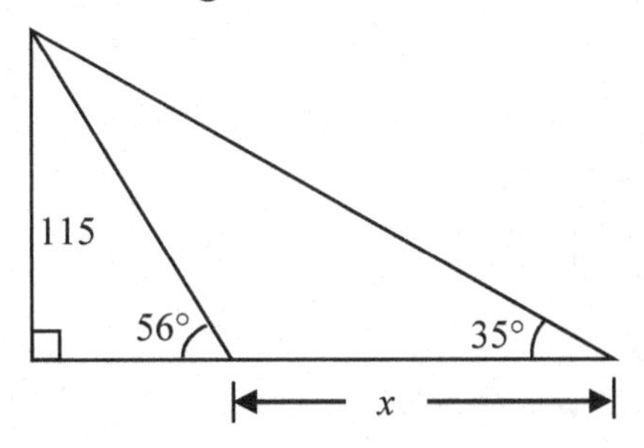

22. Find the length x.

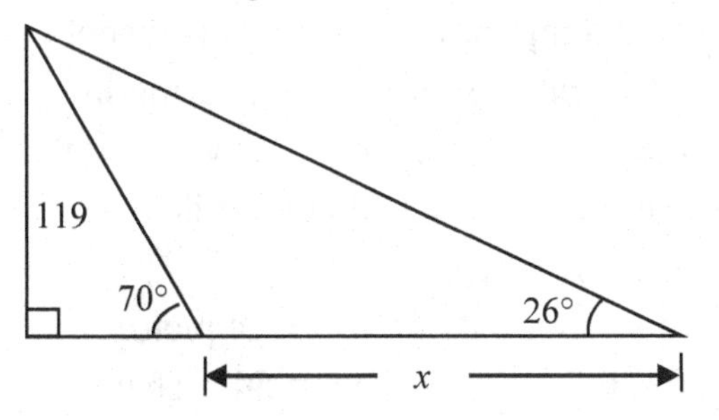

23. A plane is flying 2000 feet above sea level toward a mountain. The pilot observes the top of the mountain to be 18° above the horizontal, then immediately flies the plane at an angle of 20° above horizontal. The airspeed of the plane is 100 mph. After 5 minutes, the plane is directly above the top of the mountain. How high is the plane above the top of the mountain (when it passes over)? What is the height of the mountain? [UW]

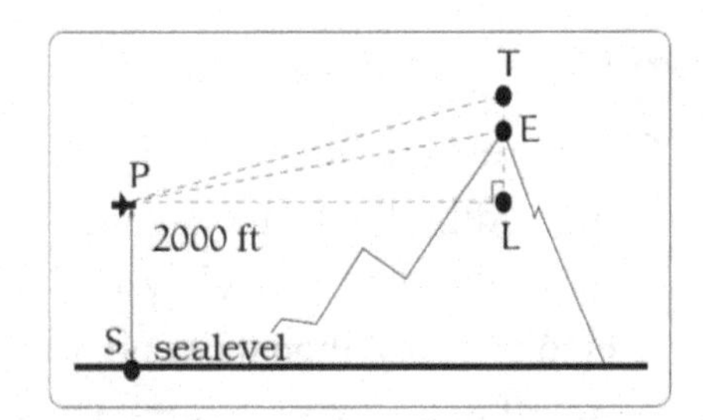

24. Three airplanes depart SeaTac Airport. A United flight is heading in a direction 50° counterclockwise from east, an Alaska flight is heading 115° counterclockwise from east and a Delta flight is heading 20° clockwise from east. [UW]
   a. Find the location of the United flight when it is 20 miles north of SeaTac.
   b. Find the location of the Alaska flight when it is 50 miles west of SeaTac.
   c. Find the location of the Delta flight when it is 30 miles east of SeaTac.

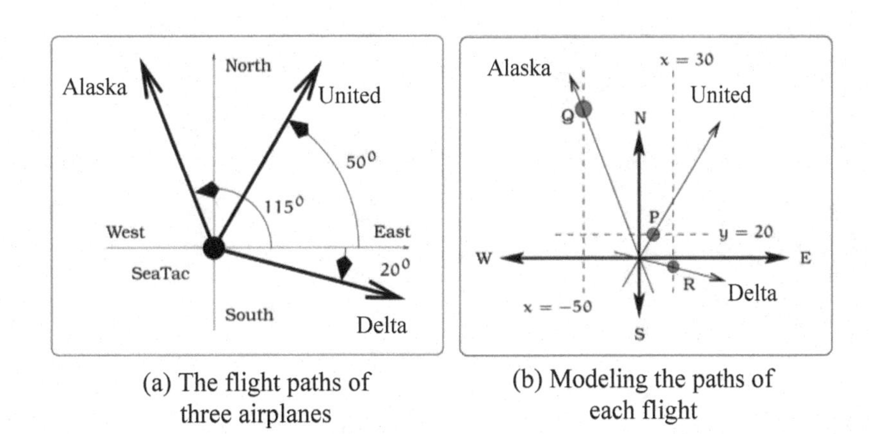

(a) The flight paths of three airplanes

(b) Modeling the paths of each flight

25. The crew of a helicopter needs to land temporarily in a forest and spot a flat piece of ground (a clearing in the forest) as a potential landing site, but are uncertain whether it is wide enough. They make two measurements from A (see picture) finding $\alpha = 25°$ and $\beta = 54°$. They rise vertically 100 feet to B and measure $\gamma = 47°$. Determine the width of the clearing to the nearest foot. [UW]

26. A Forest Service helicopter needs to determine the width of a deep canyon. While hovering, they measure the angle $\gamma = 48°$ at position B (see picture), then descend 400 feet to position A and make two measurements: $\alpha = 13°$ (the measure of $\angle EAD$), $\beta = 53°$ (the measure of $\angle CAD$). Determine the width of the canyon to the nearest foot. [UW]

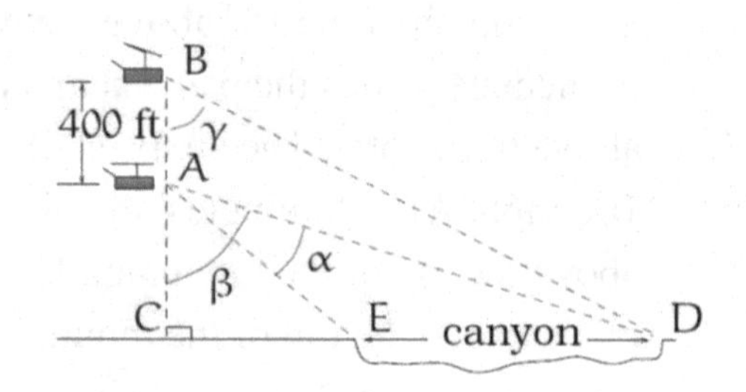

Chapter 6: Periodic Functions

In the previous chapter, the trigonometric functions were introduced as ratios of sides of a right triangle, and related to points on a circle. We noticed how the x and y values of the points did not change with repeated revolutions around the circle by finding coterminal angles. In this chapter, we will take a closer look at the important characteristics and applications of these types of functions, and begin solving equations involving them.

Section 6.1 Sinusoidal Graphs

The London Eye[1] is a huge Ferris wheel 135 meters (394 feet) tall in London, England, which completes one rotation every 30 minutes. When we look at the behavior of this Ferris wheel it is clear that it completes 1 cycle, or 1 revolution, and then repeats this revolution over and over again.

This is an example of a periodic function, because the Ferris wheel repeats its revolution or one cycle every 30 minutes, and so we say it has a period of 30 minutes.

In this section, we will work to sketch a graph of a rider's height above the ground over time and express this height as a function of time.

Periodic Functions
A **periodic function** is a function for which a specific horizontal shift, P, results in the original function: $f(x+P)=f(x)$ for all values of x. When this occurs we call the smallest such horizontal shift with $P>0$ the **period** of the function.

[1] London Eye photo by authors, 2010, CC-BY

You might immediately guess that there is a connection here to finding points on a circle, since the height above ground would correspond to the *y* value of a point on the circle. We can determine the *y* value by using the sine function. To get a better sense of this function's behavior, we can create a table of values we know, and use them to sketch a graph of the sine and cosine functions.

Listing some of the values for sine and cosine on a unit circle,

θ	0	$\frac{\pi}{6}$	$\frac{\pi}{4}$	$\frac{\pi}{3}$	$\frac{\pi}{2}$	$\frac{2\pi}{3}$	$\frac{3\pi}{4}$	$\frac{5\pi}{6}$	π
cos	1	$\frac{\sqrt{3}}{2}$	$\frac{\sqrt{2}}{2}$	$\frac{1}{2}$	0	$-\frac{1}{2}$	$-\frac{\sqrt{2}}{2}$	$-\frac{\sqrt{3}}{2}$	-1
sin	0	$\frac{1}{2}$	$\frac{\sqrt{2}}{2}$	$\frac{\sqrt{3}}{2}$	1	$\frac{\sqrt{3}}{2}$	$\frac{\sqrt{2}}{2}$	$\frac{1}{2}$	0

Here you can see how for each angle, we use the *y* value of the point on the circle to determine the output value of the sine function.

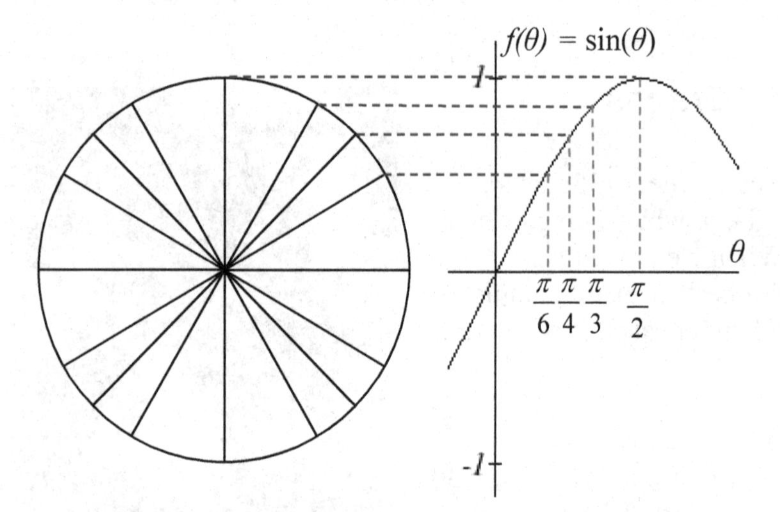

Plotting more points gives the full shape of the sine and cosine functions.

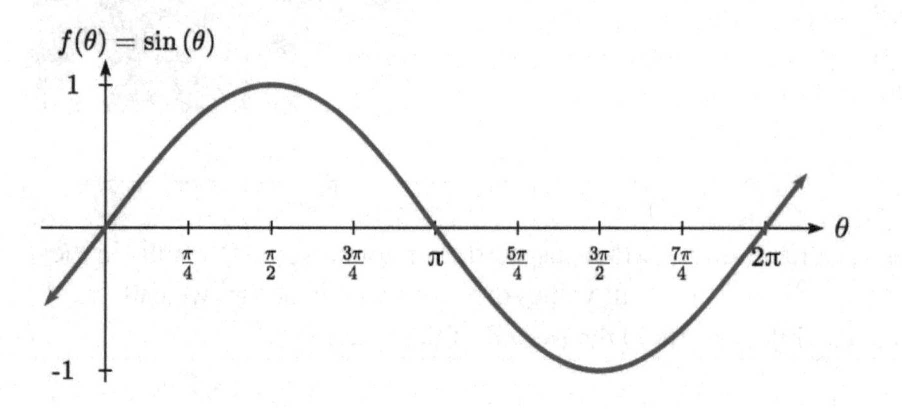

Notice how the sine values are positive between 0 and π, which correspond to the values of sine in quadrants 1 and 2 on the unit circle, and the sine values are negative between π and 2π, corresponding to quadrants 3 and 4.

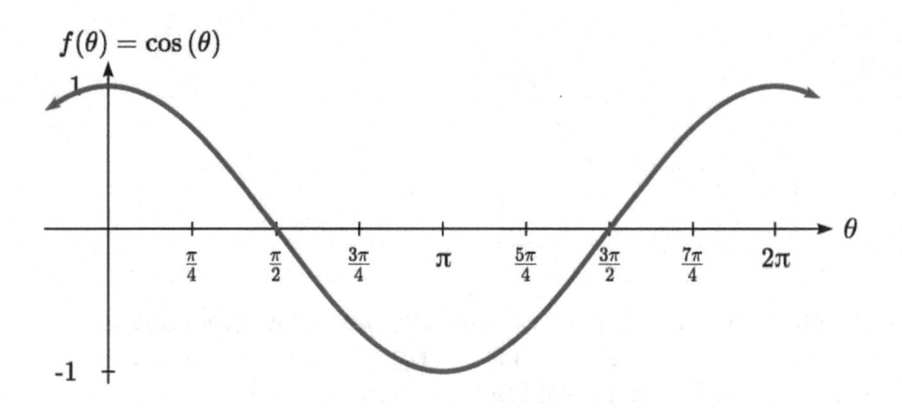

Like the sine function we can track the value of the cosine function through the 4 quadrants of the unit circle as we place it on a graph.

Both of these functions are defined for all real numbers, since we can evaluate the sine and cosine of any angle. By thinking of sine and cosine as coordinates of points on a unit circle, it becomes clear that the range of both functions must be the interval $[-1, 1]$.

Domain and Range of Sine and Cosine
The domain of sine and cosine is all real numbers, $(-\infty, \infty)$.
The range of sine and cosine is the interval [-1, 1].

Both these graphs are called **sinusoidal** graphs.

In both graphs, the shape of the graph begins repeating after 2π. Indeed, since any coterminal angles will have the same sine and cosine values, we could conclude that $\sin(\theta + 2\pi) = \sin(\theta)$ and $\cos(\theta + 2\pi) = \cos(\theta)$.

In other words, if you were to shift either graph horizontally by 2π, the resulting shape would be identical to the original function. Sinusoidal functions are a specific type of periodic function.

Period of Sine and Cosine
The periods of the sine and cosine functions are both 2π.

Looking at these functions on a domain centered at the vertical axis helps reveal symmetries.

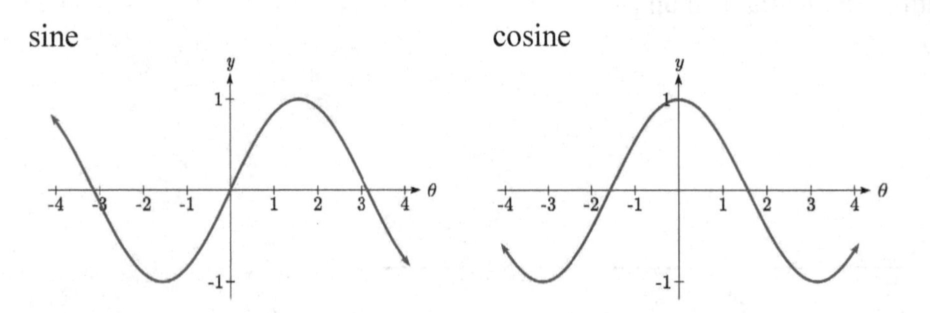

The sine function is symmetric about the origin, the same symmetry the cubic function has, making it an odd function. The cosine function is clearly symmetric about the y axis, the same symmetry as the quadratic function, making it an even function.

Negative Angle Identities
The sine is an odd function, symmetric about the *origin*, so $\sin(-\theta) = -\sin(\theta)$.
The cosine is an even function, symmetric about the y-axis, so $\cos(-\theta) = \cos(\theta)$.

These identities can be used, among other purposes, for helping with simplification and proving identities.

You may recall the cofunction identity from last chapter, $\sin(\theta) = \cos\left(\frac{\pi}{2} - \theta\right)$.

Graphically, this tells us that the sine and cosine graphs are horizontal transformations of each other. We can prove this by using the cofunction identity and the negative angle identity for cosine.

$$\sin(\theta) = \cos\left(\frac{\pi}{2} - \theta\right) = \cos\left(-\theta + \frac{\pi}{2}\right) = \cos\left(-\left(\theta - \frac{\pi}{2}\right)\right) = \cos\left(\theta - \frac{\pi}{2}\right)$$

Now we can clearly see that if we horizontally shift the cosine function to the right by $\pi/2$ we get the sine function.

Remember this shift is not representing the period of the function. It only shows that the cosine and sine function are transformations of each other.

Example 1

Simplify $\dfrac{\sin(-\theta)}{\tan(\theta)}$.

We start by using the negative angle identity for sine.

$\dfrac{-\sin(\theta)}{\tan(\theta)}$ Rewriting the tangent

$\dfrac{-\sin(\theta)}{\sin(\theta)/\cos(\theta)}$ Inverting and multiplying

$-\sin(\theta)\cdot\dfrac{\cos(\theta)}{\sin(\theta)}$ Simplifying we get

$-\cos(\theta)$

Transforming Sine and Cosine

Example 2

A point rotates around a circle of radius 3. Sketch a graph of the y coordinate of the point.

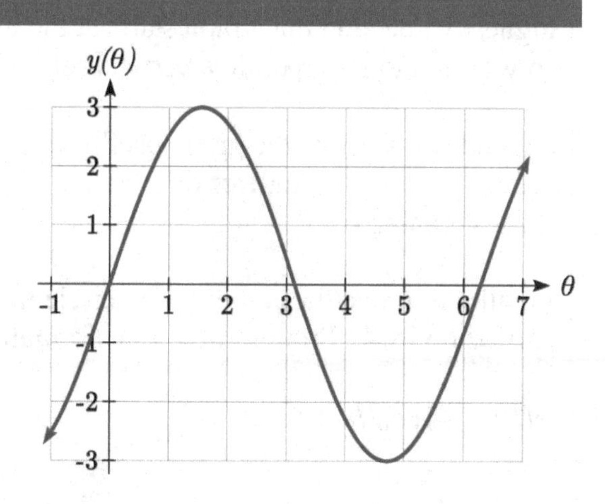

Recall that for a point on a circle of radius r, the y coordinate of the point is $y = r\sin(\theta)$, so in this case, we get the equation $y(\theta) = 3\sin(\theta)$.

The constant 3 causes a vertical stretch of the y values of the function by a factor of 3.

Notice that the period of the function does not change.

Since the outputs of the graph will now oscillate between -3 and 3, we say that the **amplitude** of the sine wave is 3.

Try it Now

1. What is the amplitude of the function $f(\theta) = 7\cos(\theta)$? Sketch a graph of this function.

Example 3

A circle with radius 3 feet is mounted with its center 4 feet off the ground. The point closest to the ground is labeled *P*. Sketch a graph of the height above ground of the point *P* as the circle is rotated, then find a function that gives the height in terms of the angle of rotation.

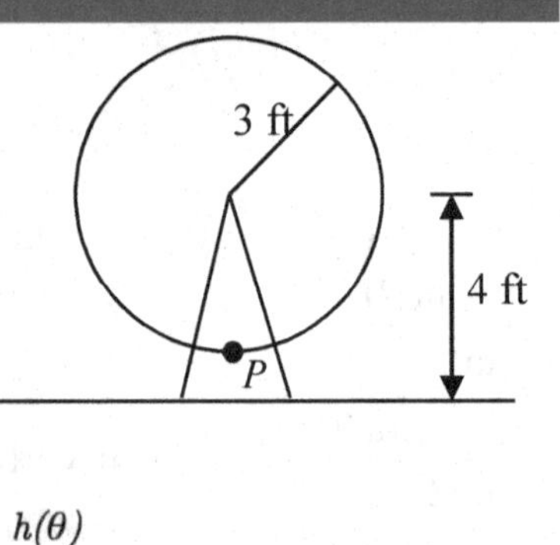

Sketching the height, we note that it will start 1 foot above the ground, then increase up to 7 feet above the ground, and continue to oscillate 3 feet above and below the center value of 4 feet.

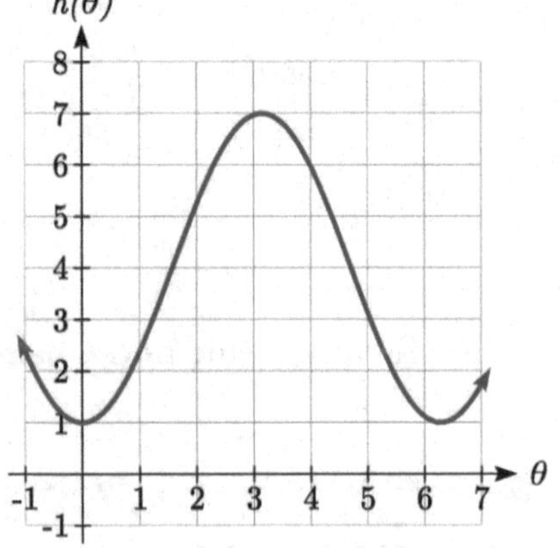

Although we could use a transformation of either the sine or cosine function, we start by looking for characteristics that would make one function easier to use than the other.

We decide to use a cosine function because it starts at the highest or lowest value, while a sine function starts at the middle value. A standard cosine starts at the highest value, and this graph starts at the lowest value, so we need to incorporate a vertical reflection.

Second, we see that the graph oscillates 3 above and below the center, while a basic cosine has an amplitude of one, so this graph has been vertically stretched by 3, as in the last example.

Finally, to move the center of the circle up to a height of 4, the graph has been vertically shifted up by 4. Putting these transformations together,

$$h(\theta) = -3\cos(\theta) + 4$$

Midline

The center value of a sinusoidal function, the value that the function oscillates above and below, is called the **midline** of the function, corresponding to a vertical shift.

The function $f(\theta) = \cos(\theta) + k$ has midline at $y = k$.

Try it Now

2. What is the midline of the function $f(\theta) = 3\cos(\theta) - 4$? Sketch a graph of the function.

To answer the Ferris wheel problem at the beginning of the section, we need to be able to express our sine and cosine functions at inputs of time. To do so, we will utilize composition. Since the sine function takes an input of an angle, we will look for a function that takes time as an input and outputs an angle. If we can find a suitable $\theta(t)$ function, then we can compose this with our $f(\theta) = \cos(\theta)$ function to obtain a sinusoidal function of time: $f(t) = \cos(\theta(t))$.

Example 4

A point completes 1 revolution every 2 minutes around a circle of radius 5. Find the *x* coordinate of the point as a function of time, if it starts at (5, 0).

Normally, we would express the *x* coordinate of a point on a unit circle using $x = r\cos(\theta)$, here we write the function $x(\theta) = 5\cos(\theta)$.

The rotation rate of 1 revolution every 2 minutes is an angular velocity. We can use this rate to find a formula for the angle as a function of time. The point begins at an angle of 0. Since the point rotates 1 revolution = 2π radians every 2 minutes, it rotates π radians every minute. After *t* minutes, it will have rotated:
$\theta(t) = \pi t$ radians

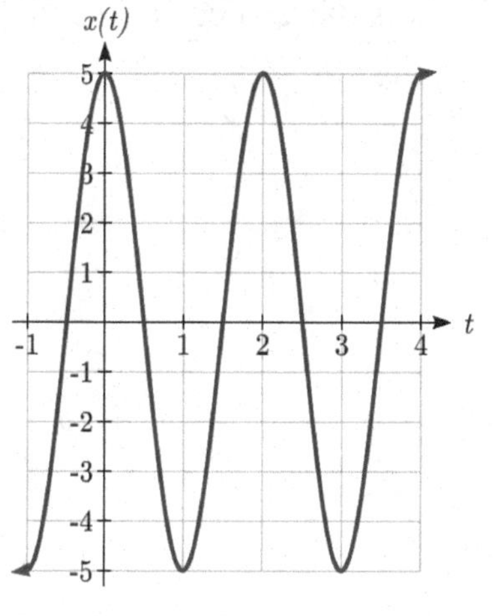

Composing this with the cosine function, we obtain a function of time.
$x(t) = 5\cos(\theta(t)) = 5\cos(\pi t)$

Notice that this composition has the effect of a horizontal compression, changing the period of the function.

To see how the period relates to the stretch or compression coefficient *B* in the equation $f(t) = \sin(Bt)$, note that the period will be the time it takes to complete one full revolution of a circle. If a point takes *P* minutes to complete 1 revolution, then the angular velocity is $\dfrac{2\pi \text{ radians}}{P \text{ minutes}}$. Then $\theta(t) = \dfrac{2\pi}{P}t$. Composing with a sine function,

$$f(t) = \sin(\theta(t)) = \sin\left(\frac{2\pi}{P}t\right)$$

From this, we can determine the relationship between the coefficient *B* and the period:
$B = \dfrac{2\pi}{P}$.

Notice that the stretch or compression coefficient B is a ratio of the "normal period of a sinusoidal function" to the "new period." If we know the stretch or compression coefficient B, we can solve for the "new period": $P = \dfrac{2\pi}{B}$.

Summarizing our transformations so far:

Transformations of Sine and Cosine

Given an equation in the form $f(t) = A\sin(Bt) + k$ or $f(t) = A\cos(Bt) + k$

A is the vertical stretch, and is the **amplitude** of the function.

B is the horizontal stretch/compression, and is related to the **period, P,** by $P = \dfrac{2\pi}{B}$.

k is the vertical shift and determines the **midline** of the function.

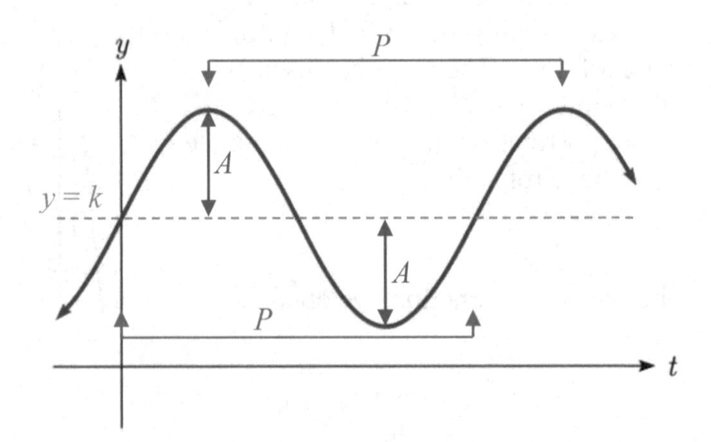

Example 5

What is the period of the function $f(t) = \sin\left(\dfrac{\pi}{6}t\right)$?

Using the relationship above, the stretch/compression factor is $B = \dfrac{\pi}{6}$, so the period will be $P = \dfrac{2\pi}{B} = \dfrac{2\pi}{\frac{\pi}{6}} = 2\pi \cdot \dfrac{6}{\pi} = 12$.

While it is common to compose sine or cosine with functions involving time, the composition can be done so that the input represents any reasonable quantity.

Example 6

A bicycle wheel with radius 14 inches has the bottom-most point on the wheel marked in red. The wheel then begins rolling down the street. Write a formula for the height above ground of the red point after the bicycle has travelled x inches.

The height of the point begins at the lowest value, 0, increases to the highest value of 28 inches, and continues to oscillate above and below a center height of 14 inches. In terms of the angle of rotation, θ:

$h(\theta) = -14\cos(\theta) + 14$

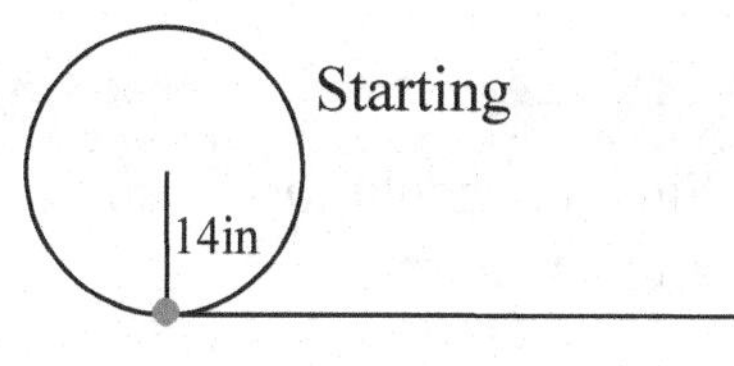

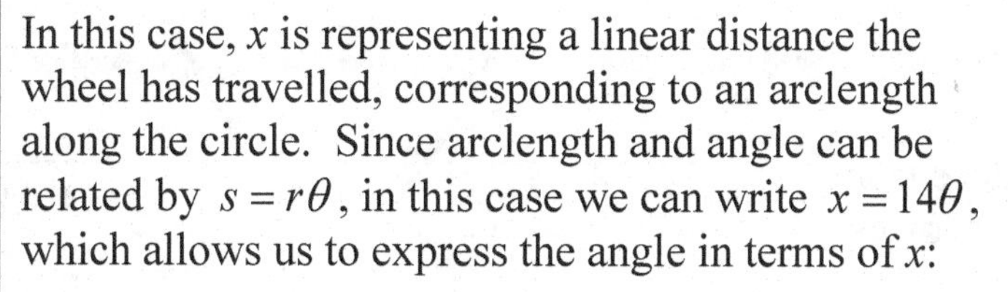

In this case, x is representing a linear distance the wheel has travelled, corresponding to an arclength along the circle. Since arclength and angle can be related by $s = r\theta$, in this case we can write $x = 14\theta$, which allows us to express the angle in terms of x:

$$\theta(x) = \frac{x}{14}$$

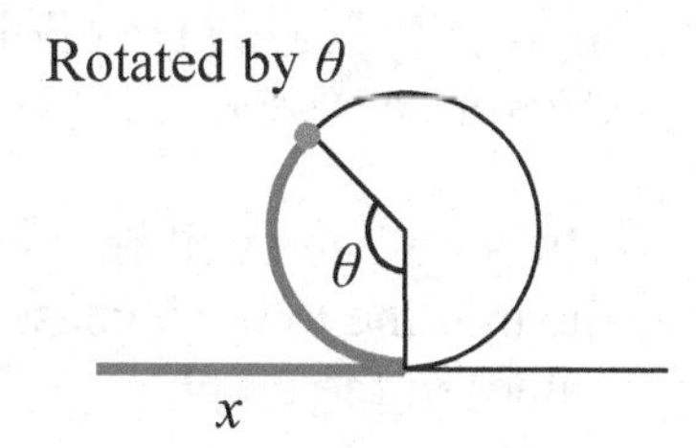

Composing this with our cosine-based function from above,

$$h(x) = h(\theta(x)) = -14\cos\left(\frac{x}{14}\right) + 14 = -14\cos\left(\frac{1}{14}x\right) + 14$$

The period of this function would be $P = \dfrac{2\pi}{B} = \dfrac{2\pi}{\frac{1}{14}} = 2\pi \cdot 14 = 28\pi$, the circumference of the circle. This makes sense – the wheel completes one full revolution after the bicycle has travelled a distance equivalent to the circumference of the wheel.

Example 7

Determine the midline, amplitude, and period of the function $f(t) = 3\sin(2t) + 1$.

The amplitude is 3

The period is $P = \dfrac{2\pi}{B} = \dfrac{2\pi}{2} = \pi$

The midline is at $y = 1$

Amplitude, midline, and period, when combined with vertical flips, allow us to write equations for a variety of sinusoidal situations.

Try it Now

3. If a sinusoidal function starts on the midline at point (0,3), has an amplitude of 2, and a period of 4, write a formula for the function

Example 8

Find a formula for the sinusoidal function graphed here.

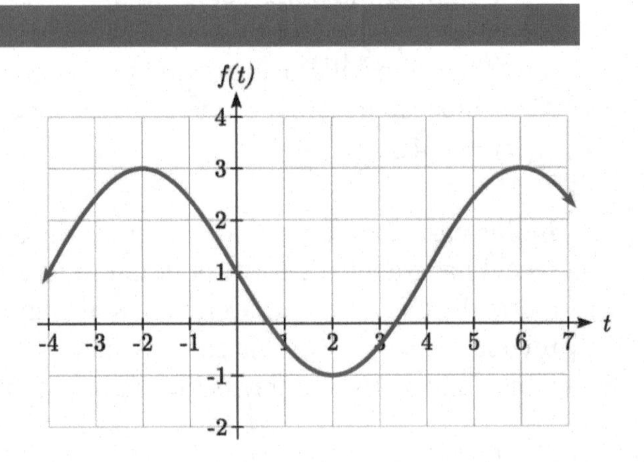

The graph oscillates from a low of -1 to a high of 3, putting the midline at $y = 1$, halfway between.

The amplitude will be 2, the distance from the midline to the highest value (or lowest value) of the graph.

The period of the graph is 8. We can measure this from the first peak at $x = -2$ to the second at $x = 6$. Since the period is 8, the stretch/compression factor we will use will be

$$B = \frac{2\pi}{P} = \frac{2\pi}{8} = \frac{\pi}{4}$$

At $x = 0$, the graph is at the midline value, which tells us the graph can most easily be represented as a sine function. Since the graph then decreases, this must be a vertical reflection of the sine function. Putting this all together,

$$f(t) = -2\sin\left(\frac{\pi}{4}t\right) + 1$$

With these transformations, we are ready to answer the Ferris wheel problem from the beginning of the section.

Example 9

The London Eye is a huge Ferris wheel in London, England, which completes one rotation every 30 minutes. The diameter of the wheel is 120 meters, but the passenger capsules sit outside the wheel. Suppose the diameter at the capsules is 130 meters, and riders board from a platform 5 meters above the ground. Express a rider's height above ground as a function of time in minutes.

It can often help to sketch a graph of the situation before trying to find the equation.

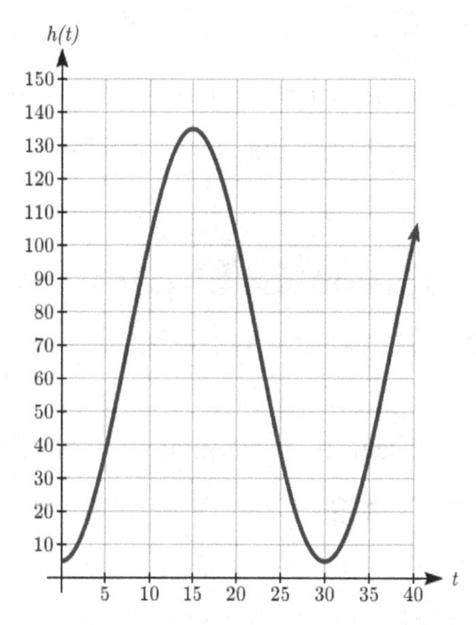

With a diameter of 130 meters, the wheel has a radius of 65 meters. The height will oscillate with amplitude of 65 meters above and below the center.

Passengers board 5 meters above ground level, so the center of the wheel must be located 65 + 5 = 70 meters above ground level. The midline of the oscillation will be at 70 meters.

The wheel takes 30 minutes to complete 1 revolution, so the height will oscillate with period of 30 minutes.

Lastly, since the rider boards at the lowest point, the height will start at the smallest value and increase, following the shape of a flipped cosine curve.
Putting these together:
Amplitude: 65
Midline: 70
Period: 30, so $B = \frac{2\pi}{30} = \frac{\pi}{15}$
Shape: negative cosine

An equation for the rider's height would be
$$h(t) = -65\cos\left(\frac{\pi}{15}t\right) + 70$$

Try it Now

4. The Ferris wheel at the Puyallup Fair[2] has a diameter of about 70 feet and takes 3 minutes to complete a full rotation. Passengers board from a platform 10 feet above the ground. Write an equation for a rider's height above ground over time.

While these transformations are sufficient to represent many situations, occasionally we encounter a sinusoidal function that does not have a vertical intercept at the lowest point, highest point, or midline. In these cases, we need to use horizontal shifts. Since we are combining horizontal shifts with horizontal stretches, we need to be careful. Recall that when the inside of the function is factored, it reveals the horizontal shift.

[2] Photo by photogirl7.1, http://www.flickr.com/photos/kitkaphotogirl/432886205/sizes/z/, CC-BY

Horizontal Shifts of Sine and Cosine

Given an equation in the form $f(t) = A\sin\left(B(t-h)\right)+k$ or $f(t) = A\cos\left(B(t-h)\right)+k$

h is the horizontal shift of the function

Example 10

Sketch a graph of $f(t) = 3\sin\left(\frac{\pi}{4}t - \frac{\pi}{4}\right)$.

To reveal the horizontal shift, we first need to factor inside the function:

$f(t) = 3\sin\left(\frac{\pi}{4}(t-1)\right)$

This graph will have the shape of a sine function, starting at the midline and increasing, with an amplitude of 3. The period of the graph will be $P = \frac{2\pi}{B} = \frac{2\pi}{\frac{\pi}{4}} = 2\pi \cdot \frac{4}{\pi} = 8$.

Finally, the graph will be shifted to the right by 1.

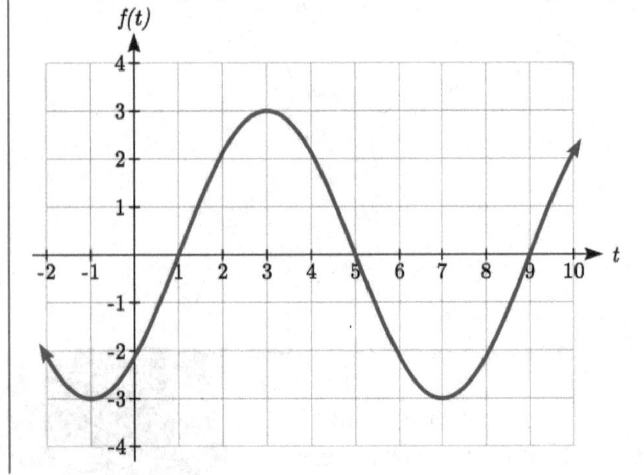

In some physics and mathematics books, you will hear the horizontal shift referred to as **phase shift**. In other physics and mathematics books, they would say the phase shift of the equation above is $\frac{\pi}{4}$, the value in the unfactored form. Because of this ambiguity, we will not use the term phase shift any further, and will only talk about the horizontal shift.

Example 11

Find a formula for the function graphed here.

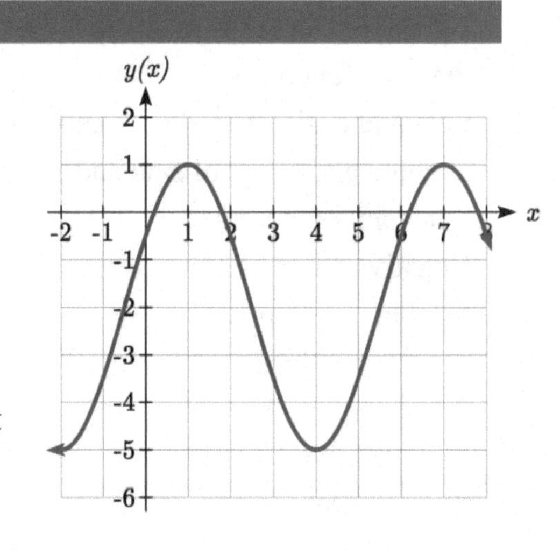

With highest value at 1 and lowest value at -5, the midline will be halfway between at -2.

The distance from the midline to the highest or lowest value gives an amplitude of 3.

The period of the graph is 6, which can be measured from the peak at $x = 1$ to the next peak at $x = 7$, or from the distance between the lowest points. This gives $B = \frac{2\pi}{P} = \frac{2\pi}{6} = \frac{\pi}{3}$.

For the shape and shift, we have more than one option. We could either write this as:
A cosine shifted 1 to the right
A negative cosine shifted 2 to the left
A sine shifted ½ to the left
A negative sine shifted 2.5 to the right

While any of these would be fine, the cosine shifts are easier to work with than the sine shifts in this case, because they involve integer values. Writing these:

$$y(x) = 3\cos\left(\frac{\pi}{3}(x-1)\right) - 2 \qquad \text{or}$$

$$y(x) = -3\cos\left(\frac{\pi}{3}(x+2)\right) - 2$$

Again, these functions are equivalent, so both yield the same graph.

Try it Now

5. Write a formula for the function graphed here.

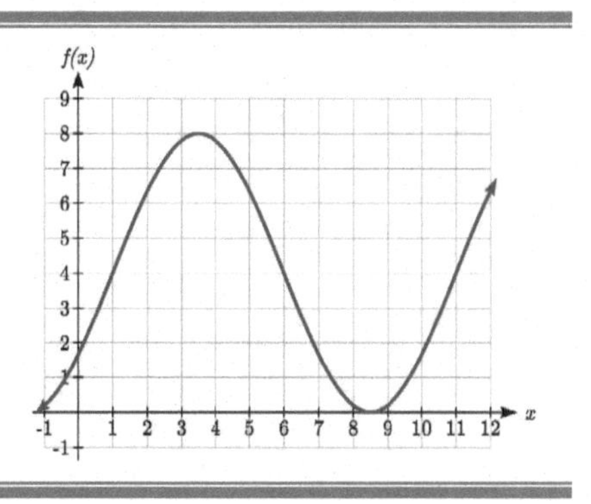

Important Topics of This Section
Periodic functions
Sine and cosine function from the unit circle
Domain and range of sine and cosine functions
Sinusoidal functions
Negative angle identity
Simplifying expressions
Transformations
Amplitude
Midline
Period
Horizontal shifts

Try it Now Answers

1. 7
2. -4
3. $f(x) = 2\sin\left(\frac{\pi}{2}x\right) + 3$
4. $h(t) = -35\cos\left(\frac{2\pi}{3}t\right) + 45$
5. Two possibilities: $f(x) = 4\cos\left(\frac{\pi}{5}(x-3.5)\right) + 4$ or $f(x) = 4\sin\left(\frac{\pi}{5}(x-1)\right) + 4$

Section 6.1 Exercises

1. Sketch a graph of $f(x) = -3\sin(x)$.
2. Sketch a graph of $f(x) = 4\sin(x)$.
3. Sketch a graph of $f(x) = 2\cos(x)$.
4. Sketch a graph of $f(x) = -4\cos(x)$.

For the graphs below, determine the amplitude, midline, and period, then find a formula for the function.

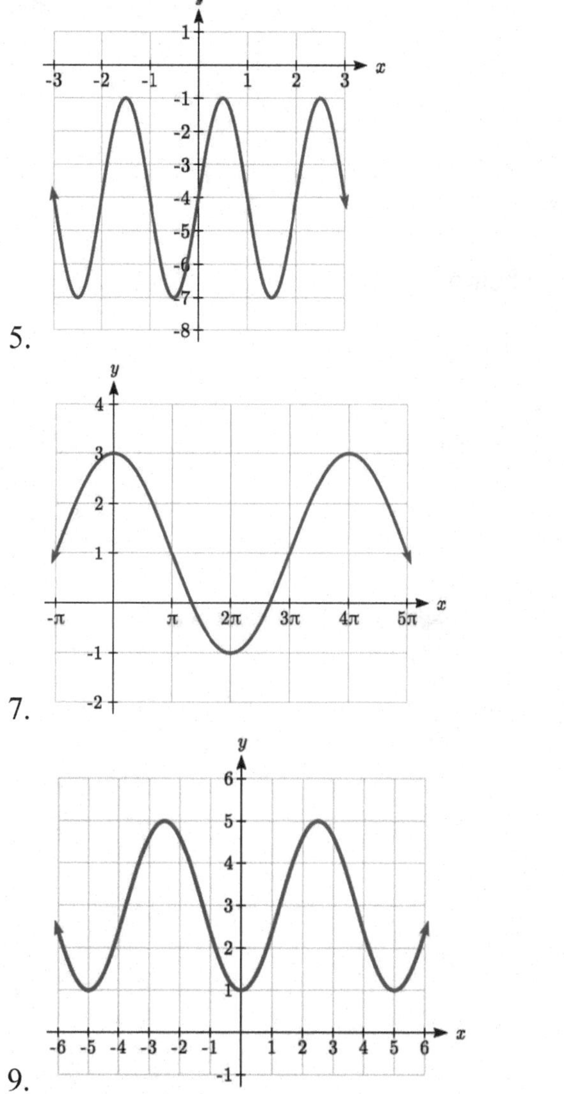

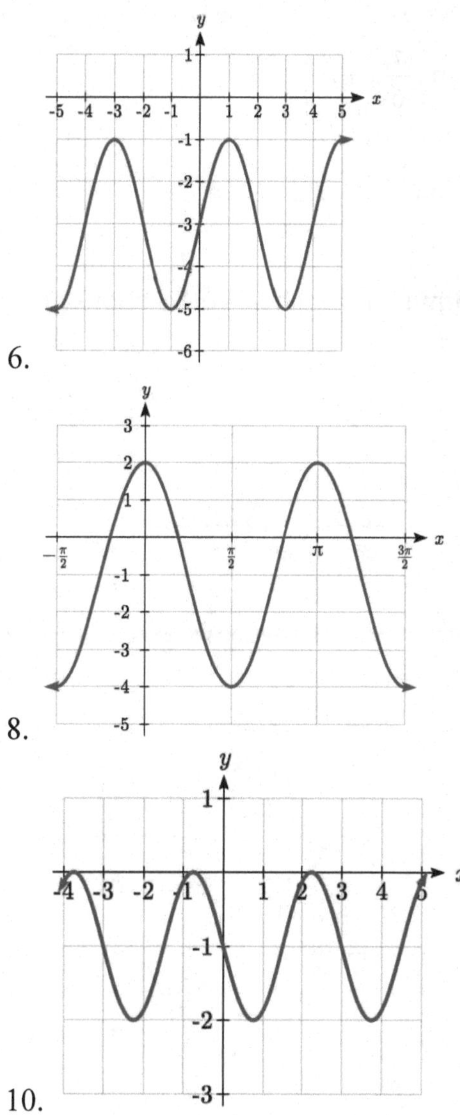

For each of the following equations, find the amplitude, period, horizontal shift, and midline.

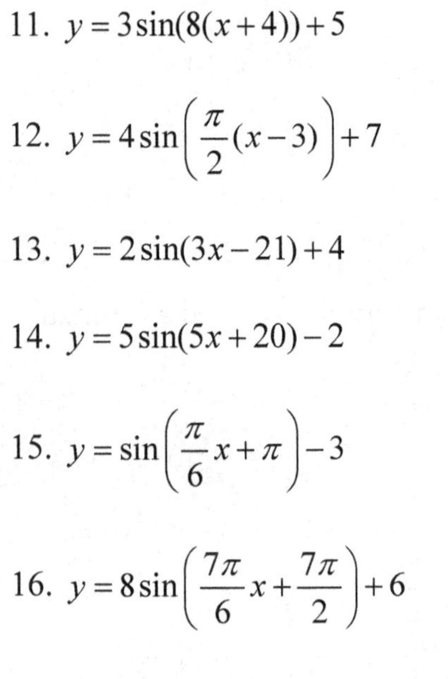

11. $y = 3\sin(8(x+4)) + 5$

12. $y = 4\sin\left(\dfrac{\pi}{2}(x-3)\right) + 7$

13. $y = 2\sin(3x - 21) + 4$

14. $y = 5\sin(5x + 20) - 2$

15. $y = \sin\left(\dfrac{\pi}{6}x + \pi\right) - 3$

16. $y = 8\sin\left(\dfrac{7\pi}{6}x + \dfrac{7\pi}{2}\right) + 6$

Find a formula for each of the functions graphed below.

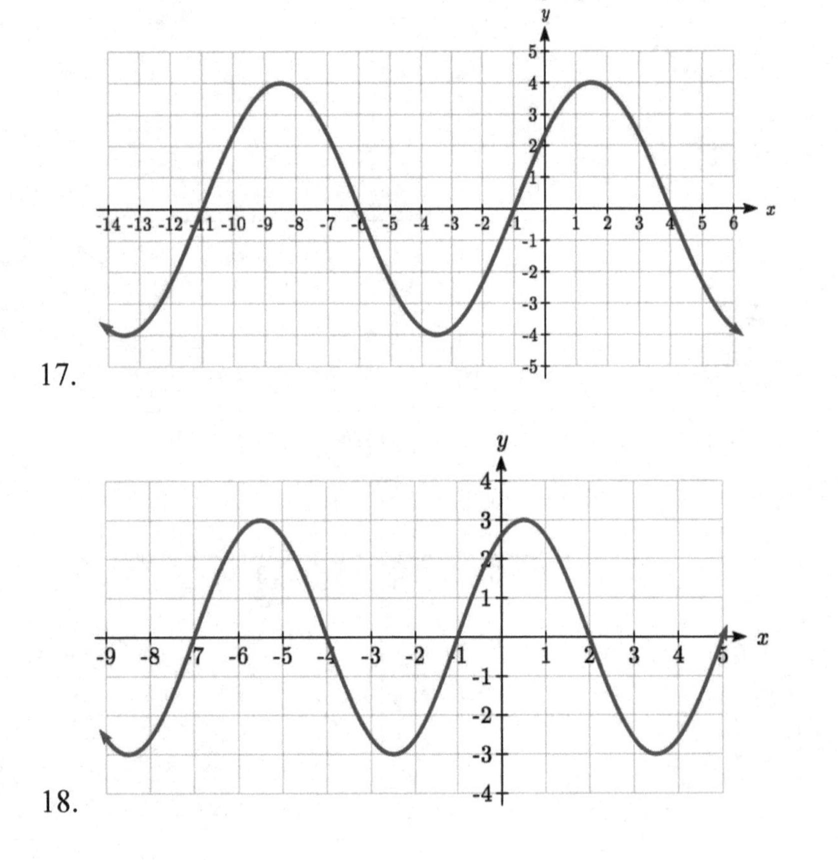

17.

18.

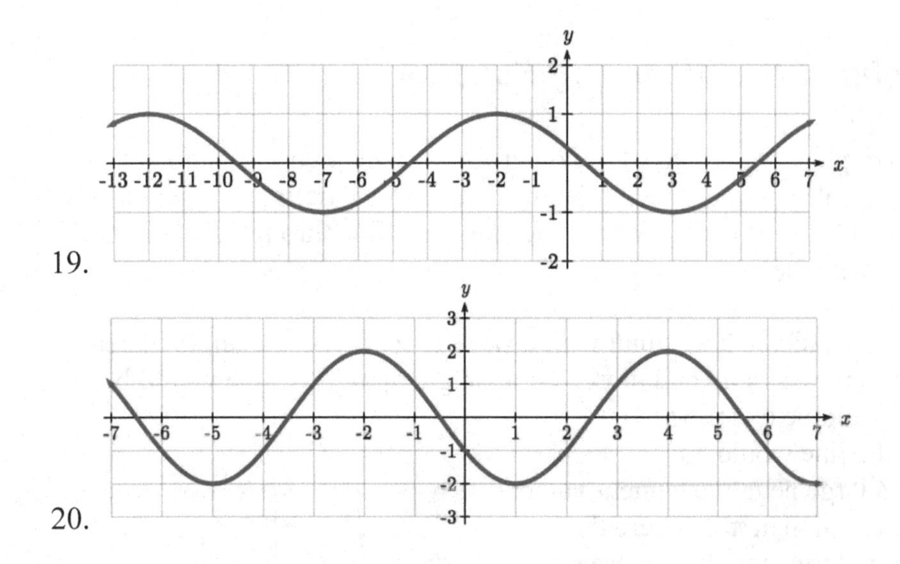

19.

20.

21. Outside temperature over the course of a day can be modeled as a sinusoidal function. Suppose you know the temperature is 50 degrees at midnight and the high and low temperature during the day are 57 and 43 degrees, respectively. Assuming t is the number of hours since midnight, find a function for the temperature, D, in terms of t.

22. Outside temperature over the course of a day can be modeled as a sinusoidal function. Suppose you know the temperature is 68 degrees at midnight and the high and low temperature during the day are 80 and 56 degrees, respectively. Assuming t is the number of hours since midnight, find a function for the temperature, D, in terms of t.

23. A Ferris wheel is 25 meters in diameter and boarded from a platform that is 1 meters above the ground. The six o'clock position on the Ferris wheel is level with the loading platform. The wheel completes 1 full revolution in 10 minutes. The function $h(t)$ gives your height in meters above the ground t minutes after the wheel begins to turn.
    a. Find the amplitude, midline, and period of $h(t)$.
    b. Find a formula for the height function $h(t)$.
    c. How high are you off the ground after 5 minutes?

24. A Ferris wheel is 35 meters in diameter and boarded from a platform that is 3 meters above the ground. The six o'clock position on the Ferris wheel is level with the loading platform. The wheel completes 1 full revolution in 8 minutes. The function $h(t)$ gives your height in meters above the ground t minutes after the wheel begins to turn.
    a. Find the amplitude, midline, and period of $h(t)$.
    b. Find a formula for the height function $h(t)$.
    c. How high are you off the ground after 4 minutes?

Section 6.2 Graphs of the Other Trig Functions

In this section, we will explore the graphs of the other four trigonometric functions. We'll begin with the tangent function. Recall that in Chapter 5 we defined tangent as *y/x* or sine/cosine, so you can think of the tangent as the slope of a line through the origin making the given angle with the positive *x* axis.

At an angle of 0, the line would be horizontal with a slope of zero. As the angle increases towards $\pi/2$, the slope increases more and more. At an angle of $\pi/2$, the line would be vertical and the slope would be undefined. Immediately past $\pi/2$, the line would have a steep negative slope, giving a large negative tangent value. There is a break in the function at $\pi/2$, where the tangent value jumps from large positive to large negative.

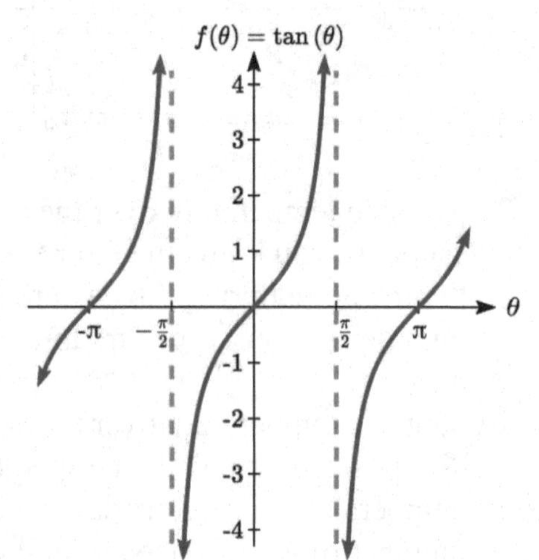

We can use these ideas along with the definition of tangent to sketch a graph. Since tangent is defined as sine/cosine, we can determine that tangent will be zero when sine is zero: at $-\pi$, 0, π, and so on. Likewise, tangent will be undefined when cosine is zero: at $-\pi/2$, $\pi/2$, and so on.

The tangent is positive from 0 to $\pi/2$ and π to $3\pi/2$, corresponding to quadrants 1 and 3 of the unit circle.

Using technology, we can obtain a graph of tangent on a standard grid.

Notice that the graph appears to repeat itself. For any angle on the circle, there is a second angle with the same slope and tangent value halfway around the circle, so the graph repeats itself with a period of π; we can see one continuous cycle from $-\pi/2$ to $\pi/2$, before it jumps and repeats itself.

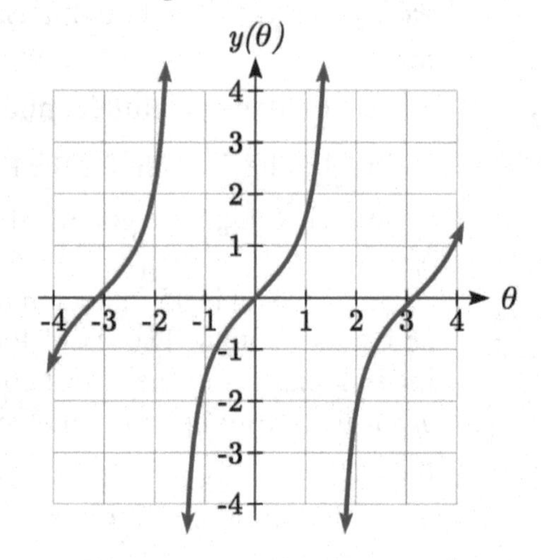

The graph has vertical asymptotes and the tangent is undefined wherever a line at that angle would be vertical: at $\pi/2$, $3\pi/2$, and so on. While the domain of the function is limited in this way, the range of the function is all real numbers.

Features of the Graph of Tangent
The graph of the tangent function $m(\theta) = \tan(\theta)$
The **period** of the tangent function is π
The **domain** of the tangent function is $\theta \neq \frac{\pi}{2} + k\pi$, where k is an integer
The **range** of the tangent function is all real numbers, $(-\infty, \infty)$

With the tangent function, like the sine and cosine functions, horizontal stretches/compressions are distinct from vertical stretches/compressions. The horizontal stretch can typically be determined from the period of the graph. With tangent graphs, it is often necessary to determine a vertical stretch using a point on the graph.

Example 1

Find a formula for the function graphed here.

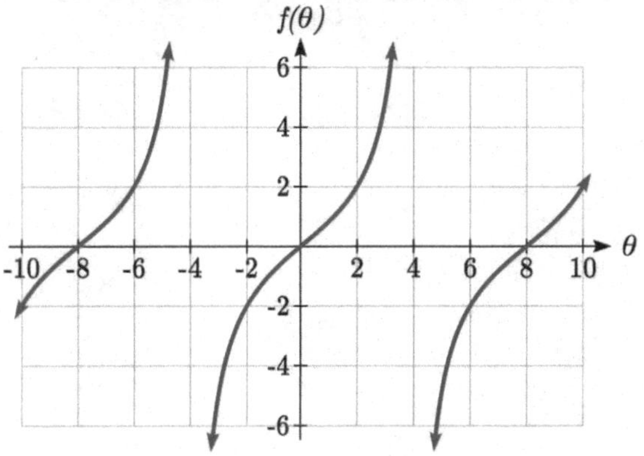

The graph has the shape of a tangent function, however the period appears to be 8. We can see one full continuous cycle from -4 to 4, suggesting a horizontal stretch. To stretch π to 8, the input values would have to be multiplied by $\frac{8}{\pi}$. Since the constant k in $f(\theta) = a\tan(k\theta)$ is the reciprocal of the horizontal stretch $\frac{8}{\pi}$, the equation must have form

$$f(\theta) = a\tan\left(\frac{\pi}{8}\theta\right).$$

We can also think of this the same way we did with sine and cosine. The period of the tangent function is π but it has been transformed and now it is 8; remember the ratio of the "normal period" to the "new period" is $\frac{\pi}{8}$ and so this becomes the value on the inside of the function that tells us how it was horizontally stretched.

To find the vertical stretch a, we can use a point on the graph. Using the point (2, 2)

$$2 = a\tan\left(\frac{\pi}{8}\cdot 2\right) = a\tan\left(\frac{\pi}{4}\right).\quad \text{Since } \tan\left(\frac{\pi}{4}\right) = 1,\ a = 2.$$

This function would have a formula $f(\theta) = 2\tan\left(\frac{\pi}{8}\theta\right)$.

Try it Now

1. Sketch a graph of $f(\theta) = 3\tan\left(\frac{\pi}{6}\theta\right)$.

For the graph of secant, we remember the reciprocal identity where $\sec(\theta) = \dfrac{1}{\cos(\theta)}$.

Notice that the function is undefined when the cosine is 0, leading to a vertical asymptote in the graph at π/2, 3π/2, etc. Since the cosine is always no more than one in absolute value, the secant, being the reciprocal, will always be no less than one in absolute value. Using technology, we can generate the graph. The graph of the cosine is shown dashed so you can see the relationship.

$$f(\theta) = \sec(\theta) = \frac{1}{\cos(\theta)}$$

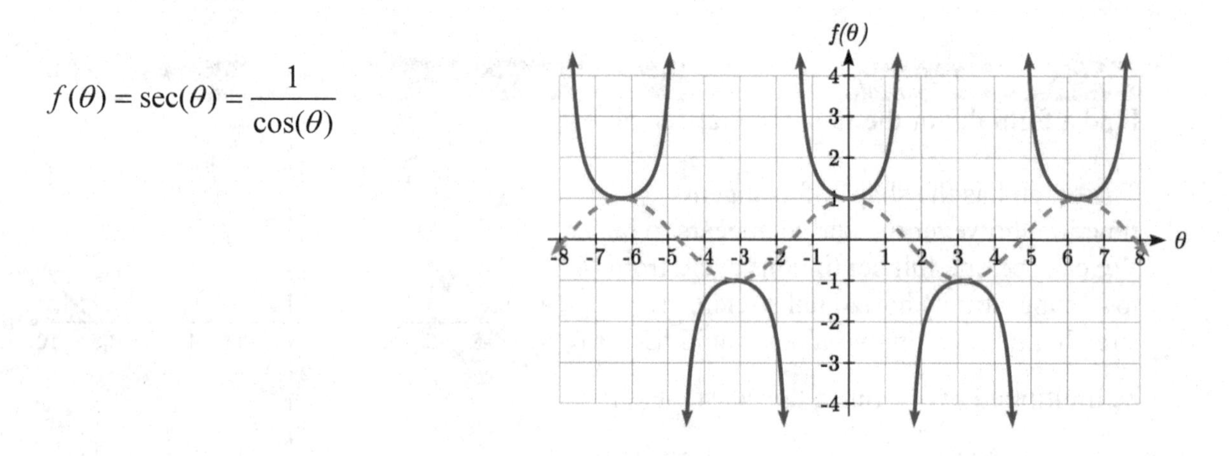

The graph of cosecant is similar. In fact, since $\sin(\theta) = \cos\left(\frac{\pi}{2} - \theta\right)$, it follows that

$\csc(\theta) = \sec\left(\frac{\pi}{2} - \theta\right)$, suggesting the cosecant graph is a horizontal shift of the secant graph. This graph will be undefined where sine is 0. Recall from the unit circle that this occurs at 0, π, 2π, etc. The graph of sine is shown dashed along with the graph of the cosecant.

$$f(\theta) = \csc(\theta) = \frac{1}{\sin(\theta)}$$

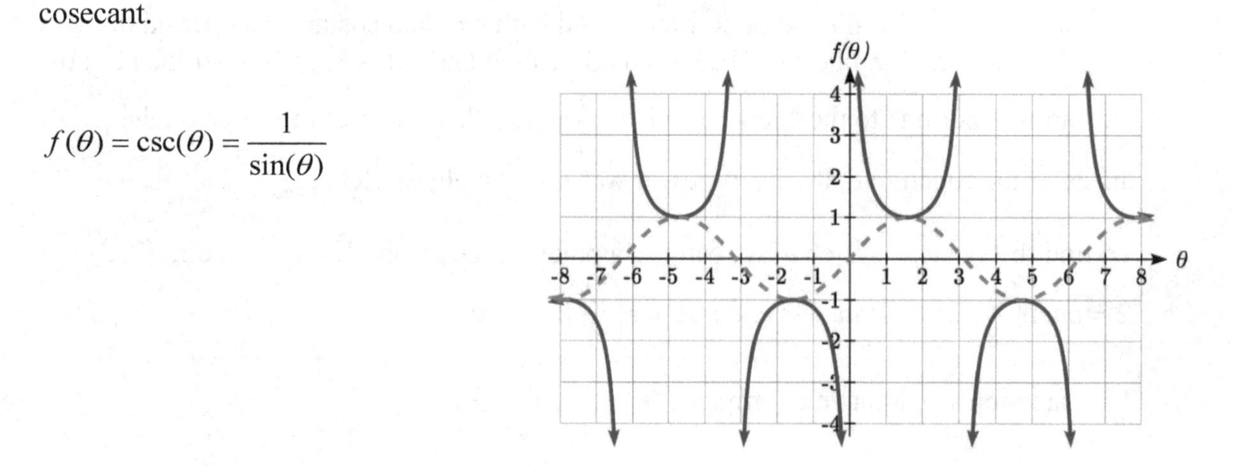

Features of the Graph of Secant and Cosecant
The secant and cosecant graphs have period 2π like the sine and cosine functions.
Secant has domain $\theta \neq \frac{\pi}{2} + k\pi$, where k is an integer
Cosecant has domain $\theta \neq k\pi$, where k is an integer
Both secant and cosecant have range of $(-\infty,-1] \cup [1,\infty)$

Example 2

Sketch a graph of $f(\theta) = 2\csc\left(\frac{\pi}{2}\theta\right) + 1$. What is the domain and range of this function?

The basic cosecant graph has vertical asymptotes at the integer multiples of π. Because of the factor $\frac{\pi}{2}$ inside the cosecant, the graph will be compressed by $\frac{2}{\pi}$, so the vertical asymptotes will be compressed to $\theta = \frac{2}{\pi} \cdot k\pi = 2k$. In other words, the graph will have vertical asymptotes at the integer multiples of 2, and the domain will correspondingly be $\theta \neq 2k$, where k is an integer.

The basic sine graph has a range of [-1, 1]. The vertical stretch by 2 will stretch this to [-2, 2], and the vertical shift up 1 will shift the range of this function to [-1, 3].

The basic cosecant graph has a range of $(-\infty,-1] \cup [1,\infty)$. The vertical stretch by 2 will stretch this to $(-\infty,-2] \cup [2,\infty)$, and the vertical shift up 1 will shift the range of this function to $(-\infty,-1] \cup [3,\infty)$.

The resulting graph is shown to the right.

Notice how the graph of the transformed cosecant relates to the graph of $f(\theta) = 2\sin\left(\frac{\pi}{2}\theta\right) + 1$ shown dashed.

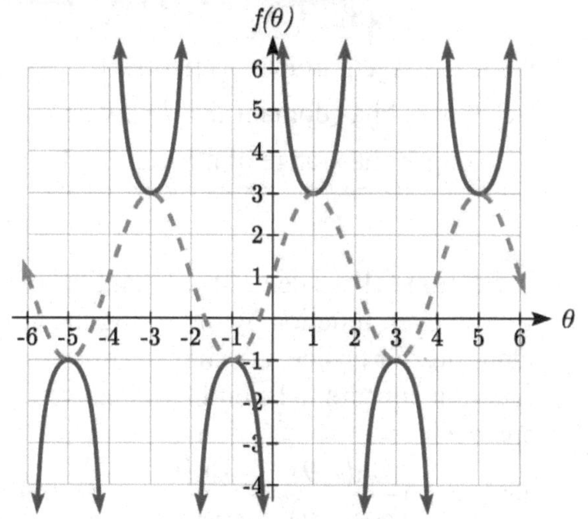

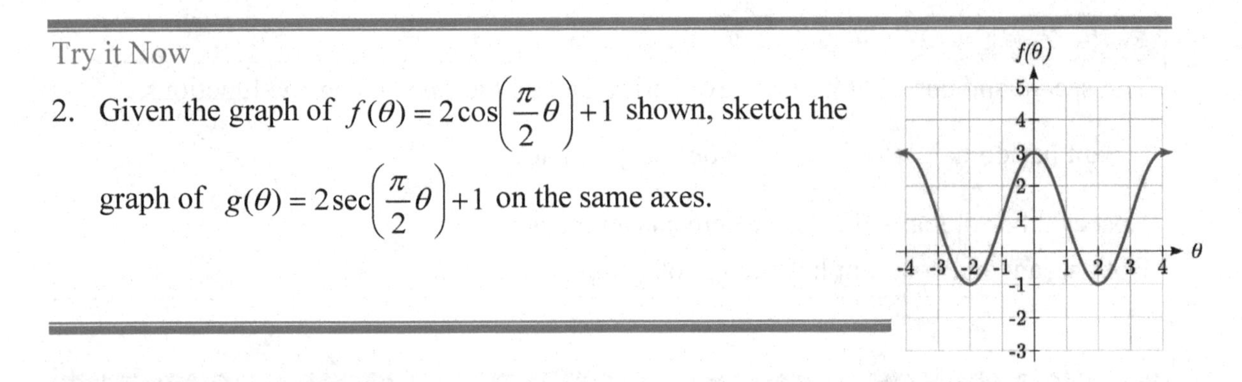

Try it Now

2. Given the graph of $f(\theta)=2\cos\left(\frac{\pi}{2}\theta\right)+1$ shown, sketch the graph of $g(\theta)=2\sec\left(\frac{\pi}{2}\theta\right)+1$ on the same axes.

Finally, we'll look at the graph of cotangent. Based on its definition as the ratio of cosine to sine, it will be undefined when the sine is zero: at at 0, π, 2π, etc. The resulting graph is similar to that of the tangent. In fact, it is a horizontal flip and shift of the tangent function, as we'll see shortly in the next example.

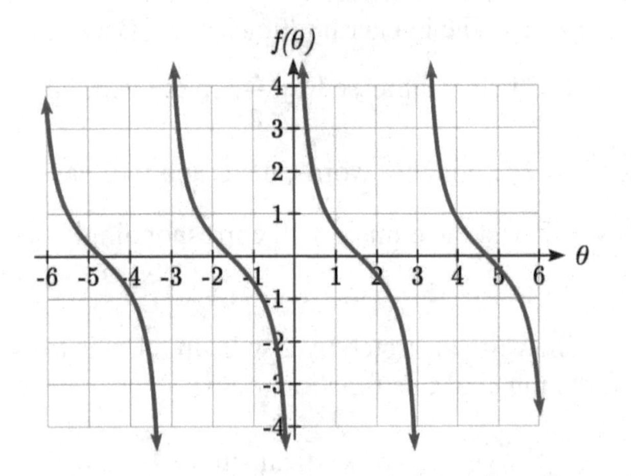

Features of the Graph of Cotangent
The cotangent graph has period π
Cotangent has domain $\theta \neq k\pi$, where k is an integer
Cotangent has range of all real numbers, $(-\infty,\infty)$

In Section 6.1 we determined that the sine function was an odd function and the cosine was an even function by observing the graph and establishing the negative angle identities for cosine and sine. Similarly, you may notice from its graph that the tangent function appears to be odd. We can verify this using the negative angle identities for sine and cosine:

$$\tan(-\theta)=\frac{\sin(-\theta)}{\cos(-\theta)}=\frac{-\sin(\theta)}{\cos(\theta)}=-\tan(\theta)$$

The secant, like the cosine it is based on, is an even function, while the cosecant, like the sine, is an odd function.

Negative Angle Identities Tangent, Cotangent, Secant and Cosecant	
$\tan(-\theta) = -\tan(\theta)$	$\cot(-\theta) = -\cot(\theta)$
$\sec(-\theta) = \sec(\theta)$	$\csc(-\theta) = -\csc(\theta)$

Example 3

Prove that $\tan(\theta) = -\cot\left(\theta - \frac{\pi}{2}\right)$

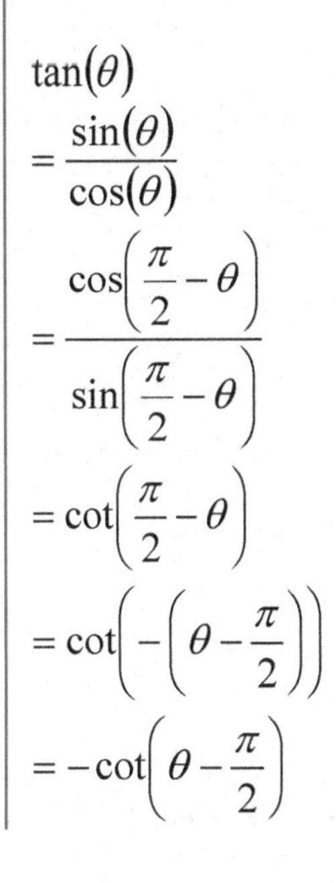

$\tan(\theta)$ — Using the definition of tangent

$= \dfrac{\sin(\theta)}{\cos(\theta)}$ — Using the cofunction identities

$= \dfrac{\cos\left(\frac{\pi}{2} - \theta\right)}{\sin\left(\frac{\pi}{2} - \theta\right)}$ — Using the definition of cotangent

$= \cot\left(\frac{\pi}{2} - \theta\right)$ — Factoring a negative from the inside

$= \cot\left(-\left(\theta - \frac{\pi}{2}\right)\right)$ — Using the negative angle identity for cot

$= -\cot\left(\theta - \frac{\pi}{2}\right)$

Important Topics of This Section
The tangent and cotangent functions
Period
Domain
Range
The secant and cosecant functions
Period
Domain
Range
Transformations
Negative Angle identities

Try it Now Answers

1.

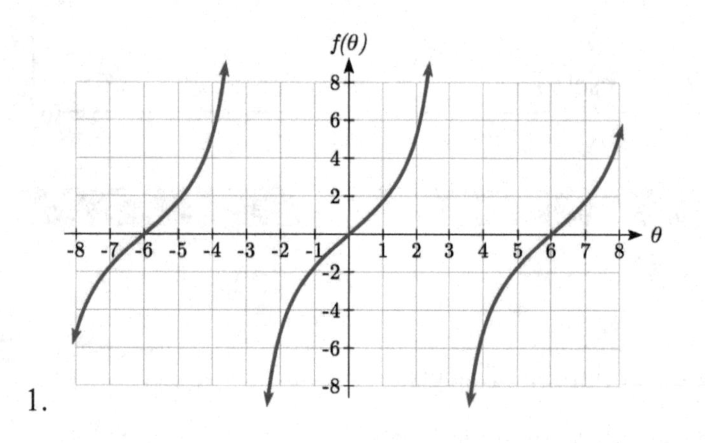

2.

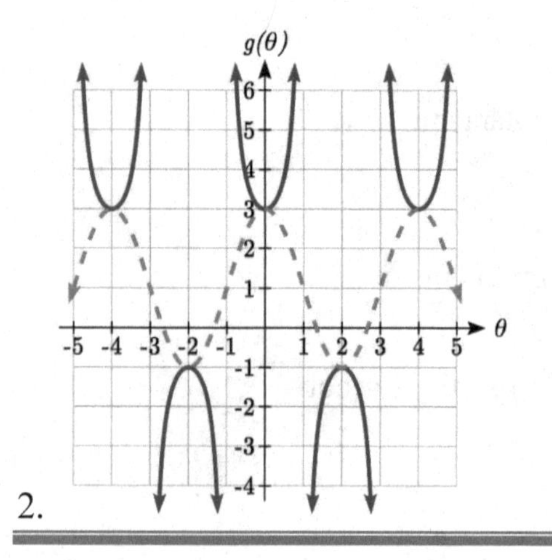

Section 6.2 Exercises

Match each trigonometric function with one of the graphs.

1. $f(x) = \tan(x)$
2. $f(x) = \sec(x)$
3. $f(x) = \csc(x)$
4. $f(x) = \cot(x)$

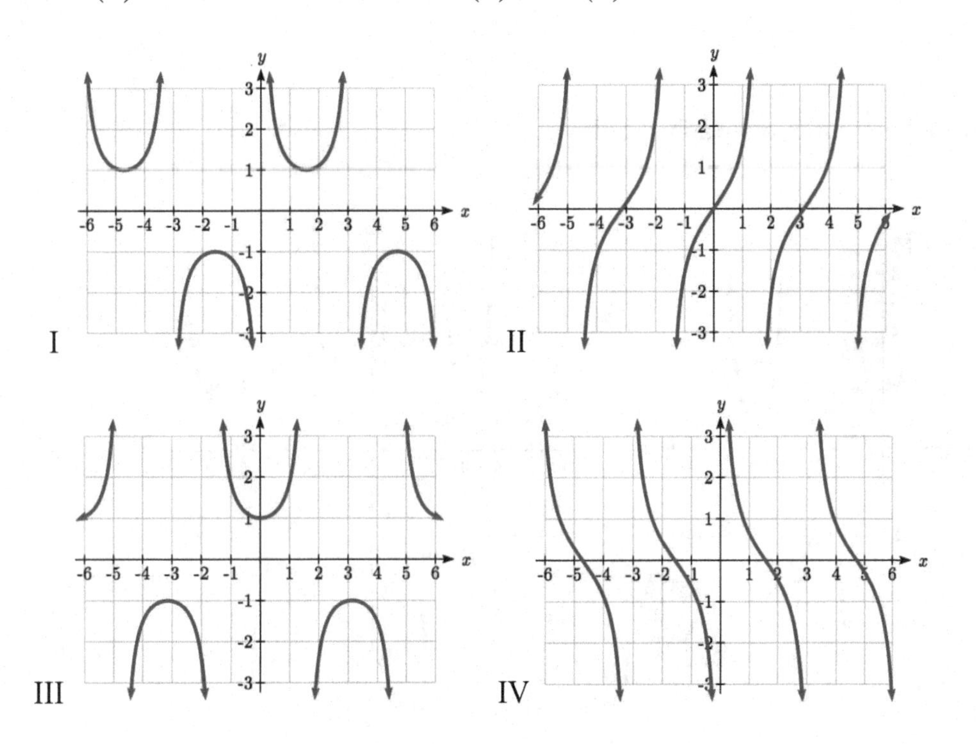

Find the period and horizontal shift of each of the following functions.

5. $f(x) = 2\tan(4x - 32)$
6. $g(x) = 3\tan(6x + 42)$
7. $h(x) = 2\sec\left(\frac{\pi}{4}(x+1)\right)$
8. $k(x) = 3\sec\left(2\left(x + \frac{\pi}{2}\right)\right)$
9. $m(x) = 6\csc\left(\frac{\pi}{3}x + \pi\right)$
10. $n(x) = 4\csc\left(\frac{5\pi}{3}x - \frac{20\pi}{3}\right)$

11. Sketch a graph of #7 above.
12. Sketch a graph of #8 above.
13. Sketch a graph of #9 above.
14. Sketch a graph of #10 above.

15. Sketch a graph of $j(x) = \tan\left(\frac{\pi}{2}x\right)$.

16. Sketch a graph of $p(t) = 2\tan\left(t - \frac{\pi}{2}\right)$.

Find a formula for each function graphed below.

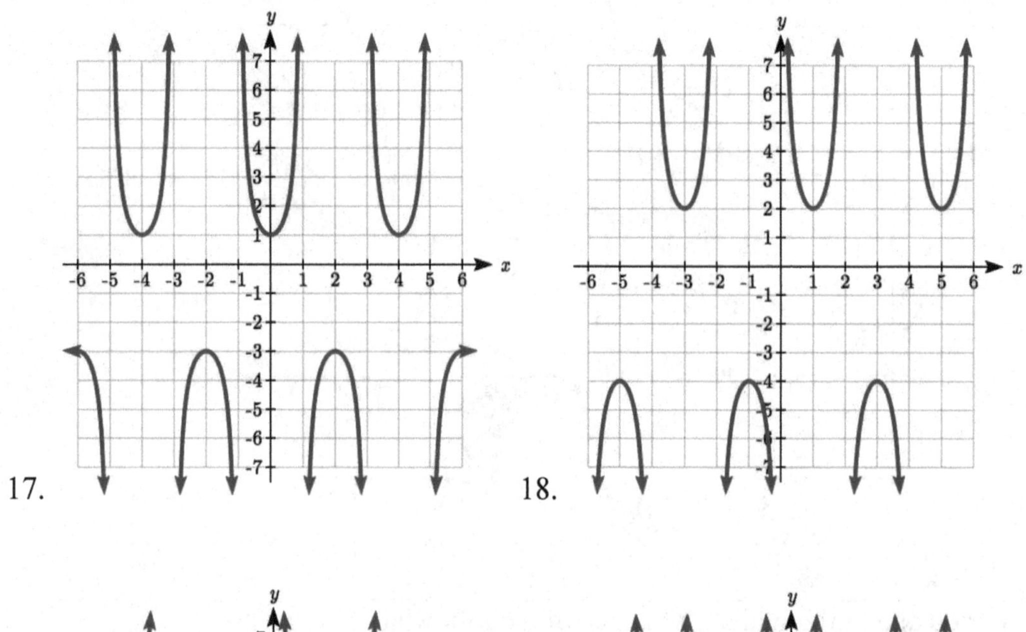

17. 18.

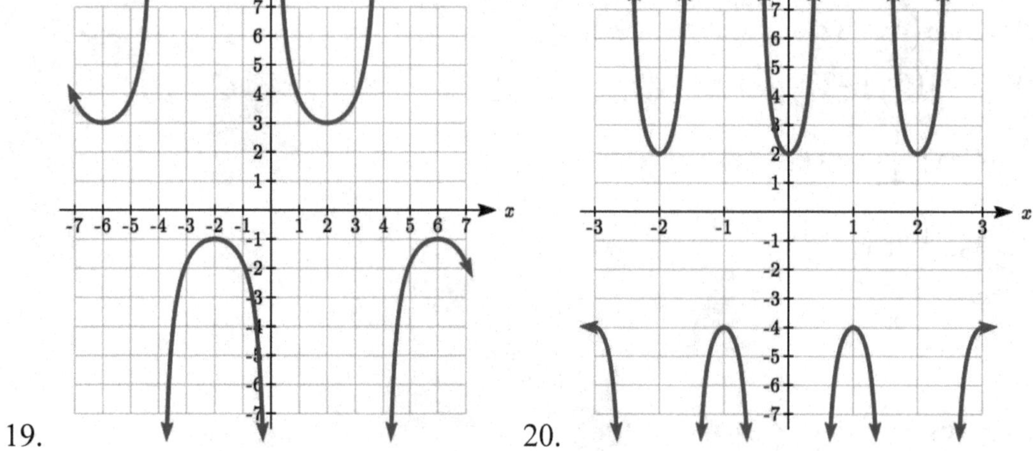

19. 20.

21. If $\tan x = -1.5$, find $\tan(-x)$.
22. If $\tan x = 3$, find $\tan(-x)$.
23. If $\sec x = 2$, find $\sec(-x)$.
24. If $\sec x = -4$, find $\sec(-x)$.
25. If $\csc x = -5$, find $\csc(-x)$.
26. If $\csc x = 2$, find $\csc(-x)$.

Simplify each of the following expressions completely.
27. $\cot(-x)\cos(-x) + \sin(-x)$
28. $\cos(-x) + \tan(-x)\sin(-x)$

Section 6.3 Inverse Trig Functions

In previous sections, we have evaluated the trigonometric functions at various angles, but at times we need to know what angle would yield a specific sine, cosine, or tangent value. For this, we need inverse functions. Recall that for a one-to-one function, if $f(a) = b$, then an inverse function would satisfy $f^{-1}(b) = a$.

You probably are already recognizing an issue – that the sine, cosine, and tangent functions are not one-to-one functions. To define an inverse of these functions, we will need to restrict the domain of these functions to yield a new function that is one-to-one. We choose a domain for each function that includes the angle zero.

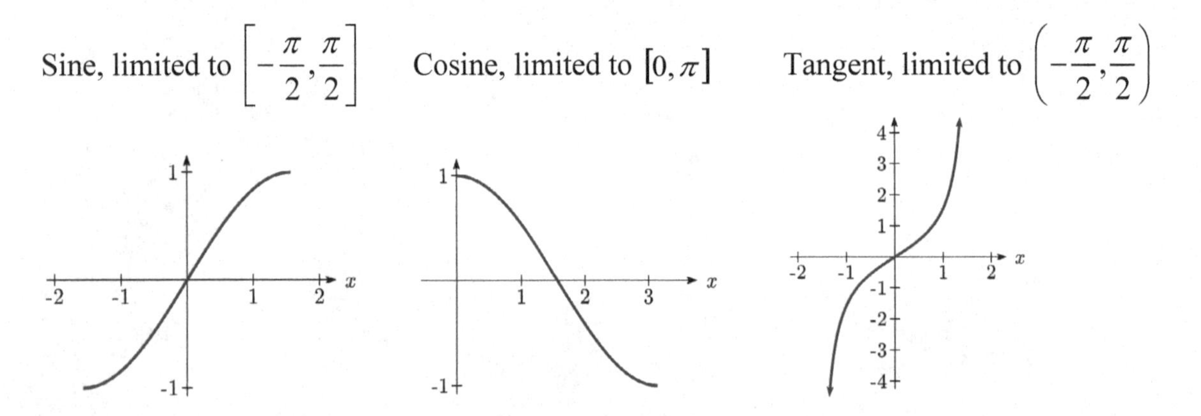

On these restricted domains, we can define the inverse sine, inverse cosine, and inverse tangent functions.

Inverse Sine, Cosine, and Tangent Functions

For angles in the interval $\left[-\frac{\pi}{2},\frac{\pi}{2}\right]$, if $\sin(\theta) = a$, then $\sin^{-1}(a) = \theta$

For angles in the interval $[0,\pi]$, if $\cos(\theta) = a$, then $\cos^{-1}(a) = \theta$

For angles in the interval $\left(-\frac{\pi}{2},\frac{\pi}{2}\right)$, if $\tan(\theta) = a$, then $\tan^{-1}(a) = \theta$

$\sin^{-1}(x)$ has domain [-1, 1] and range $\left[-\frac{\pi}{2},\frac{\pi}{2}\right]$

$\cos^{-1}(x)$ has domain [-1, 1] and range $[0,\pi]$

$\tan^{-1}(x)$ has domain of all real numbers and range $\left(-\frac{\pi}{2},\frac{\pi}{2}\right)$

The $\sin^{-1}(x)$ is sometimes called the **arcsine** function, and notated $\arcsin(a)$.
The $\cos^{-1}(x)$ is sometimes called the **arccosine** function, and notated $\arccos(a)$.
The $\tan^{-1}(x)$ is sometimes called the **arctangent** function, and notated $\arctan(a)$.

The graphs of the inverse functions are shown here:

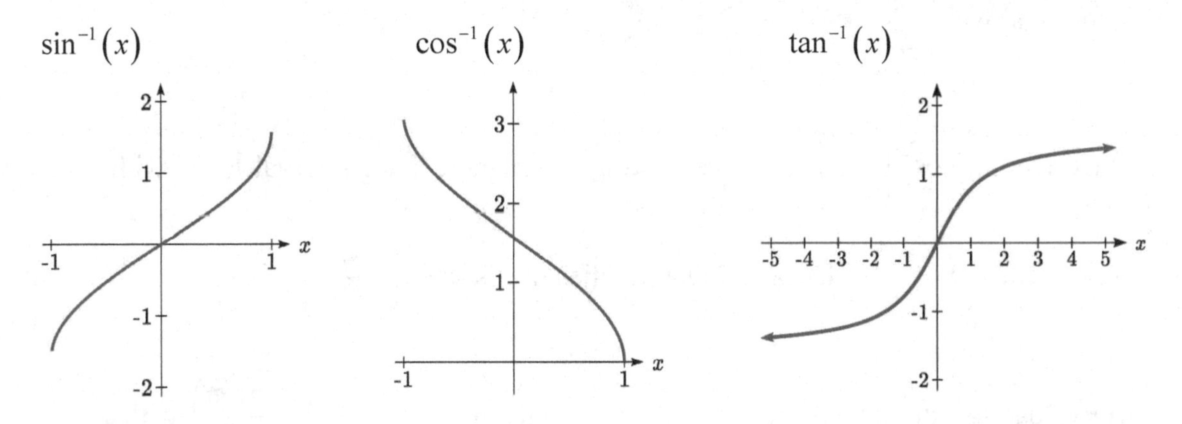

Notice that the output of each of these inverse functions is an *angle*.

Example 1

Evaluate

a) $\sin^{-1}\left(\frac{1}{2}\right)$ b) $\sin^{-1}\left(-\frac{\sqrt{2}}{2}\right)$ c) $\cos^{-1}\left(-\frac{\sqrt{3}}{2}\right)$ d) $\tan^{-1}(1)$

a) Evaluating $\sin^{-1}\left(\frac{1}{2}\right)$ is the same as asking what angle would have a sine value of $\frac{1}{2}$.

In other words, what angle θ would satisfy $\sin(\theta)=\frac{1}{2}$?

There are multiple angles that would satisfy this relationship, such as $\frac{\pi}{6}$ and $\frac{5\pi}{6}$, but we know we need the angle in the range of $\sin^{-1}(x)$, the interval $\left[-\frac{\pi}{2},\frac{\pi}{2}\right]$, so the answer will be $\sin^{-1}\left(\frac{1}{2}\right)=\frac{\pi}{6}$.

Remember that the inverse is a *function* so for each input, we will get exactly one output.

b) Evaluating $\sin^{-1}\left(-\frac{\sqrt{2}}{2}\right)$, we know that $\frac{5\pi}{4}$ and $\frac{7\pi}{4}$ both have a sine value of $-\frac{\sqrt{2}}{2}$, but neither is in the interval $\left[-\frac{\pi}{2},\frac{\pi}{2}\right]$. For that, we need the negative angle coterminal with $\frac{7\pi}{4}$. $\sin^{-1}\left(-\frac{\sqrt{2}}{2}\right)=-\frac{\pi}{4}$.

c) Evaluating $\cos^{-1}\left(-\frac{\sqrt{3}}{2}\right)$, we are looking for an angle in the interval $[0,\pi]$ with a cosine value of $-\frac{\sqrt{3}}{2}$. The angle that satisfies this is $\cos^{-1}\left(-\frac{\sqrt{3}}{2}\right)=\frac{5\pi}{6}$.

d) Evaluating $\tan^{-1}(1)$, we are looking for an angle in the interval $\left(-\frac{\pi}{2},\frac{\pi}{2}\right)$ with a tangent value of 1. The correct angle is $\tan^{-1}(1)=\frac{\pi}{4}$.

Try It Now

1. Evaluate

a) $\sin^{-1}(-1)$ b) $\tan^{-1}(-1)$ c) $\cos^{-1}(-1)$ d) $\cos^{-1}\left(\frac{1}{2}\right)$

Example 2

Evaluate $\sin^{-1}(0.97)$ using your calculator.

Since the output of the inverse function is an angle, your calculator will give you a degree value if in degree mode, and a radian value if in radian mode.

In radian mode, $\sin^{-1}(0.97)\approx 1.3252$ In degree mode, $\sin^{-1}(0.97)\approx 75.93°$

Try it Now

2. Evaluate $\cos^{-1}(-0.4)$ using your calculator.

In Section 5.5, we worked with trigonometry on a right triangle to solve for the sides of a triangle given one side and an additional angle. Using the inverse trig functions, we can solve for the angles of a right triangle given two sides.

Example 3

Solve the triangle for the angle θ.

Since we know the hypotenuse and the side adjacent to the angle, it makes sense for us to use the cosine function.

12

θ

9

$\cos(\theta) = \frac{9}{12}$ Using the definition of the inverse,

$\theta = \cos^{-1}\left(\frac{9}{12}\right)$ Evaluating

$\theta \approx 0.7227$, or about $41.4096°$

There are times when we need to compose a trigonometric function with an inverse trigonometric function. In these cases, we can find exact values for the resulting expressions

Example 4

Evaluate $\sin^{-1}\left(\cos\left(\frac{13\pi}{6}\right)\right)$.

a) Here, we can directly evaluate the inside of the composition.

$$\cos\left(\frac{13\pi}{6}\right) = \frac{\sqrt{3}}{2}$$

Now, we can evaluate the inverse function as we did earlier.

$$\sin^{-1}\left(\frac{\sqrt{3}}{2}\right) = \frac{\pi}{3}$$

Try it Now

3. Evaluate $\cos^{-1}\left(\sin\left(-\frac{11\pi}{4}\right)\right)$.

Example 5

Find an exact value for $\sin\left(\cos^{-1}\left(\frac{4}{5}\right)\right)$.

Beginning with the inside, we can say there is some angle so $\theta = \cos^{-1}\left(\frac{4}{5}\right)$, which means $\cos(\theta) = \frac{4}{5}$, and we are looking for $\sin(\theta)$. We can use the Pythagorean identity to do this.

$\sin^2(\theta) + \cos^2(\theta) = 1$ Using our known value for cosine

$\sin^2(\theta) + \left(\frac{4}{5}\right)^2 = 1$ Solving for sine

$\sin^2(\theta) = 1 - \frac{16}{25}$

$\sin(\theta) = \pm\sqrt{\frac{9}{25}} = \pm\frac{3}{5}$

Since we know that the inverse cosine always gives an angle on the interval $[0, \pi]$, we know that the sine of that angle must be positive, so $\sin\left(\cos^{-1}\left(\frac{4}{5}\right)\right) = \sin(\theta) = \frac{3}{5}$

Example 6

Find an exact value for $\sin\left(\tan^{-1}\left(\frac{7}{4}\right)\right)$.

While we could use a similar technique as in the last example, we will demonstrate a different technique here. From the inside, we know there is an angle so $\tan(\theta) = \frac{7}{4}$. We can envision this as the opposite and adjacent sides on a right triangle.

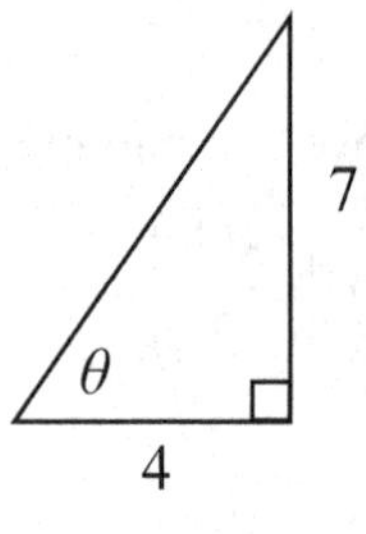

Using the Pythagorean Theorem, we can find the hypotenuse of this triangle:

$4^2 + 7^2 = hypotenuse^2$

$hypotenuse = \sqrt{65}$

Now, we can represent the sine of the angle as opposite side divided by hypotenuse.

$$\sin(\theta) = \frac{7}{\sqrt{65}}$$

This gives us our desired composition

$$\sin\left(\tan^{-1}\left(\frac{7}{4}\right)\right) = \sin(\theta) = \frac{7}{\sqrt{65}}.$$

Try it Now

4. Evaluate $\cos\left(\sin^{-1}\left(\frac{7}{9}\right)\right)$.

We can also find compositions involving algebraic expressions

Example 7

Find a simplified expression for $\cos\left(\sin^{-1}\left(\frac{x}{3}\right)\right)$, for $-3 \le x \le 3$.

We know there is an angle θ so that $\sin(\theta) = \frac{x}{3}$. Using the Pythagorean Theorem,

$\sin^2(\theta) + \cos^2(\theta) = 1$ Using our known expression for sine

$\left(\frac{x}{3}\right)^2 + \cos^2(\theta) = 1$ Solving for cosine

$$\cos^2(\theta) = 1 - \frac{x^2}{9}$$

$$\cos(\theta) = \pm\sqrt{\frac{9 - x^2}{9}} = \pm\frac{\sqrt{9 - x^2}}{3}$$

Since we know that the inverse sine must give an angle on the interval $\left[-\frac{\pi}{2}, \frac{\pi}{2}\right]$, we can deduce that the cosine of that angle must be positive. This gives us

$$\cos\left(\sin^{-1}\left(\frac{x}{3}\right)\right) = \frac{\sqrt{9 - x^2}}{3}$$

Try it Now

5. Find a simplified expression for $\sin\left(\tan^{-1}(4x)\right)$, for $-\frac{1}{4} \le x \le \frac{1}{4}$.

Important Topics of This Section
Inverse trig functions: arcsine, arccosine and arctangent
Domain restrictions
Evaluating inverses using unit circle values and the calculator
Simplifying numerical expressions involving the inverse trig functions
Simplifying algebraic expressions involving the inverse trig functions

---

Try it Now Answers

1. a) $-\frac{\pi}{2}$ b) $-\frac{\pi}{4}$ c) π d) $\frac{\pi}{3}$

2. 1.9823 or 113.578°

3. $\sin\left(-\frac{11\pi}{4}\right) = -\frac{\sqrt{2}}{2}$. $\cos^{-1}\left(-\frac{\sqrt{2}}{2}\right) = \frac{3\pi}{4}$

4. Let $\theta = \sin^{-1}\left(\frac{7}{9}\right)$ so $\sin(\theta) = \frac{7}{9}$. .

   Using Pythagorean Identity, $\sin^2\theta + \cos^2\theta = 1$, so $\left(\frac{7}{9}\right)^2 + \cos^2\theta = 1$.

   Solving, $\cos\left(\sin^{-1}\left(\frac{7}{9}\right)\right) = \cos(\theta) = \frac{4\sqrt{2}}{9}$.

5. Let $\theta = \tan^{-1}(4x)$, so $\tan(\theta) = 4x$. We can represent this on a triangle as $\tan(\theta) = \frac{4x}{1}$.

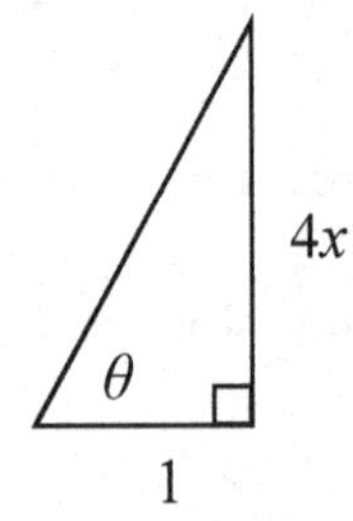

   The hypotenuse of the triangle would be $\sqrt{(4x)^2 + 1}$.

   $\sin\left(\tan^{-1}(4x)\right) = \sin(\theta) = \frac{4x}{\sqrt{16x^2 + 1}}$

Section 6.3 Exercises

Evaluate the following expressions, giving the answer in radians.

1. $\sin^{-1}\left(\frac{\sqrt{2}}{2}\right)$
2. $\sin^{-1}\left(\frac{\sqrt{3}}{2}\right)$
3. $\sin^{-1}\left(-\frac{1}{2}\right)$
4. $\sin^{-1}\left(-\frac{\sqrt{2}}{2}\right)$
5. $\cos^{-1}\left(\frac{1}{2}\right)$
6. $\cos^{-1}\left(\frac{\sqrt{2}}{2}\right)$
7. $\cos^{-1}\left(-\frac{\sqrt{2}}{2}\right)$
8. $\cos^{-1}\left(-\frac{\sqrt{3}}{2}\right)$
9. $\tan^{-1}(1)$
10. $\tan^{-1}\left(\sqrt{3}\right)$
11. $\tan^{-1}\left(-\sqrt{3}\right)$
12. $\tan^{-1}(-1)$

Use your calculator to evaluate each expression, giving the answer in radians.

13. $\cos^{-1}(-0.4)$
14. $\cos^{-1}(0.8)$
15. $\sin^{-1}(-0.8)$
16. $\tan^{-1}(6)$

Find the angle θ in degrees.

17.

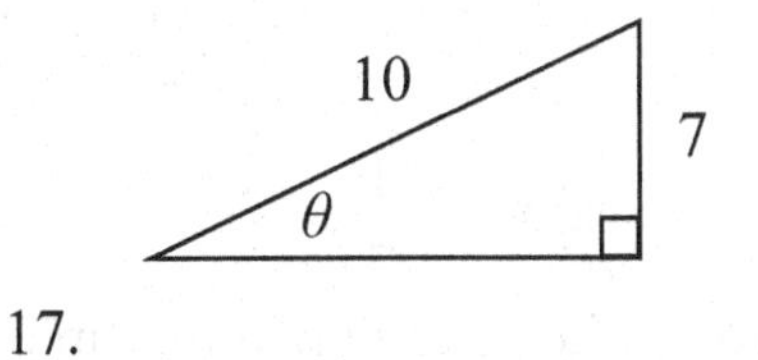

18.

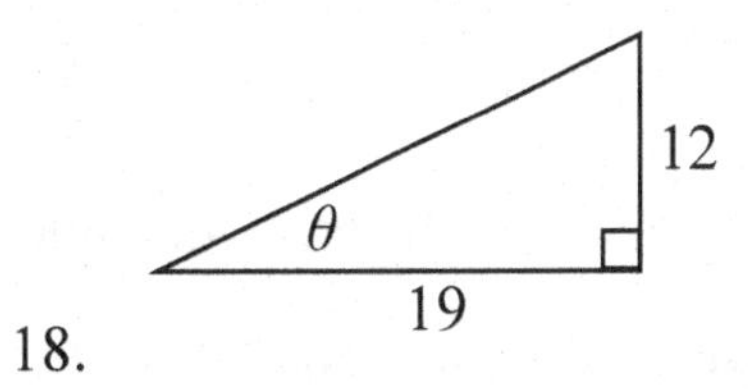

Evaluate the following expressions.

19. $\sin^{-1}\left(\cos\left(\frac{\pi}{4}\right)\right)$
20. $\cos^{-1}\left(\sin\left(\frac{\pi}{6}\right)\right)$
21. $\sin^{-1}\left(\cos\left(\frac{4\pi}{3}\right)\right)$
22. $\cos^{-1}\left(\sin\left(\frac{5\pi}{4}\right)\right)$
23. $\cos\left(\sin^{-1}\left(\frac{3}{7}\right)\right)$
24. $\sin\left(\cos^{-1}\left(\frac{4}{9}\right)\right)$
25. $\cos\left(\tan^{-1}(4)\right)$
26. $\tan\left(\sin^{-1}\left(\frac{1}{3}\right)\right)$

Find a simplified expression for each of the following.

27. $\sin\left(\cos^{-1}\left(\frac{x}{5}\right)\right)$, for $-5 \le x \le 5$
28. $\tan\left(\cos^{-1}\left(\frac{x}{2}\right)\right)$, for $-2 \le x \le 2$
29. $\sin\left(\tan^{-1}(3x)\right)$
30. $\cos\left(\tan^{-1}(4x)\right)$

Section 6.4 Solving Trig Equations

In Section 6.1, we determined the height of a rider on the London Eye Ferris wheel could be determined by the equation $h(t) = -65\cos\left(\frac{\pi}{15}t\right) + 70$.

If we wanted to know length of time during which the rider is more than 100 meters above ground, we would need to solve equations involving trig functions.

Solving using known values

In the last chapter, we learned sine and cosine values at commonly encountered angles. We can use these to solve sine and cosine equations involving these common angles.

Example 1

Solve $\sin(t) = \frac{1}{2}$ for all possible values of *t*.

Notice this is asking us to identify all angles, *t*, that have a sine value of $\frac{1}{2}$. While evaluating a function always produces one result, solving for an input can yield multiple solutions. Two solutions should immediately jump to mind from the last chapter: $t = \frac{\pi}{6}$ and $t = \frac{5\pi}{6}$ because they are the common angles on the unit circle with a sin of $\frac{1}{2}$.

Looking at a graph confirms that there are more than these two solutions. While eight are seen on this graph, there are an infinite number of solutions!

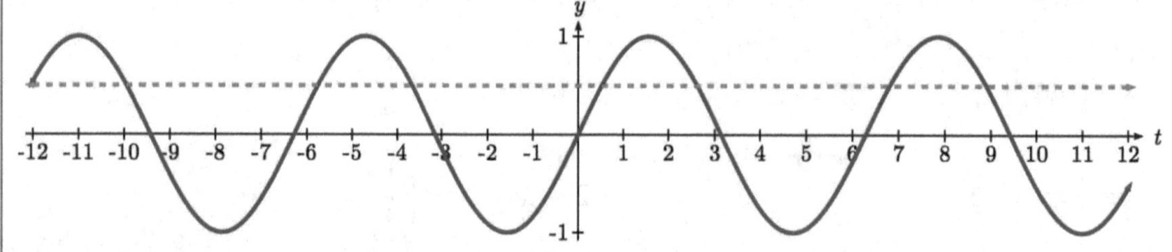

Remember that any coterminal angle will also have the same sine value, so any angle coterminal with these our first two solutions is also a solution. Coterminal angles can be found by adding full rotations of 2π, so we can write the full set of solutions:

$t = \frac{\pi}{6} + 2\pi k$ where *k* is an integer, and $t = \frac{5\pi}{6} + 2\pi k$ where *k* is an integer.

Example 2

A circle of radius $5\sqrt{2}$ intersects the line $x = -5$ at two points. Find the angles θ on the interval $0 \le \theta < 2\pi$, where the circle and line intersect.

The x coordinate of a point on a circle can be found as $x = r\cos(\theta)$, so the x coordinate of points on this circle would be $x = 5\sqrt{2}\cos(\theta)$. To find where the line $x = -5$ intersects the circle, we can solve for where the x value on the circle would be -5.

$-5 = 5\sqrt{2}\cos(\theta)$ Isolating the cosine

$\dfrac{-1}{\sqrt{2}} = \cos(\theta)$ Recall that $\dfrac{-1}{\sqrt{2}} = \dfrac{-\sqrt{2}}{2}$, so we are solving

$\cos(\theta) = \dfrac{-\sqrt{2}}{2}$

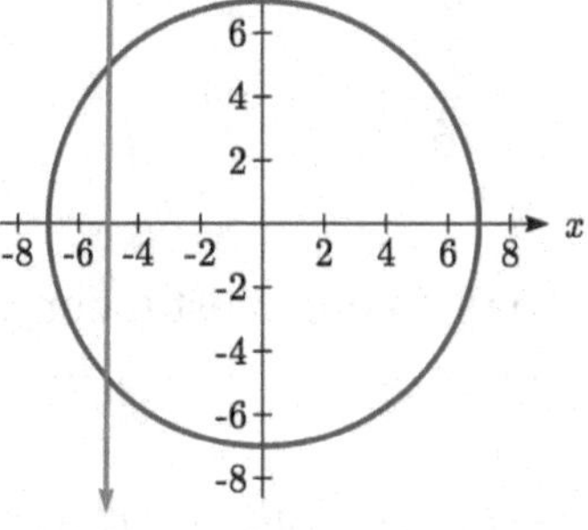

We can recognize this as one of our special cosine values from our unit circle, and it corresponds with angles $\theta = \dfrac{3\pi}{4}$ and $\theta = \dfrac{5\pi}{4}$.

Try it Now

1. Solve $\tan(t) = 1$ for all possible values of t.

Example 3

The depth of water at a dock rises and falls with the tide, following the equation $f(t) = 4\sin\left(\dfrac{\pi}{12}t\right) + 7$, where t is measured in hours after midnight. A boat requires a depth of 9 feet to tie up at the dock. Between what times will the depth be 9 feet?

To find when the depth is 9 feet, we need to solve $f(t) = 9$.

$4\sin\left(\dfrac{\pi}{12}t\right) + 7 = 9$ Isolating the sine

$4\sin\left(\dfrac{\pi}{12}t\right) = 2$ Dividing by 4

$\sin\left(\dfrac{\pi}{12}t\right) = \dfrac{1}{2}$ We know $\sin(\theta) = \dfrac{1}{2}$ when $\theta = \dfrac{\pi}{6}$ *or* $\theta = \dfrac{5\pi}{6}$

While we know what angles have a sine value of $\frac{1}{2}$, because of the horizontal stretch/compression it is less clear how to proceed.

To deal with this, we can make a substitution, defining a new temporary variable u to be $u = \frac{\pi}{12}t$, so our equation $\sin\left(\frac{\pi}{12}t\right) = \frac{1}{2}$ becomes

$\sin(u) = \frac{1}{2}$

From earlier, we saw the solutions to this equation were

$u = \frac{\pi}{6} + 2\pi k$ where k is an integer, and

$u = \frac{5\pi}{6} + 2\pi k$ where k is an integer

To undo our substitution, we replace the u in the solutions with $u = \frac{\pi}{12}t$ and solve for t.

$\frac{\pi}{12}t = \frac{\pi}{6} + 2\pi k$ where k is an integer, and $\frac{\pi}{12}t = \frac{5\pi}{6} + 2\pi k$ where k is an integer.

Dividing by π/12, we obtain solutions

$t = 2 + 24k$ where k is an integer, and
$t = 10 + 24k$ where k is an integer.

The depth will be 9 feet and the boat will be able to approach the dock between 2am and 10am.

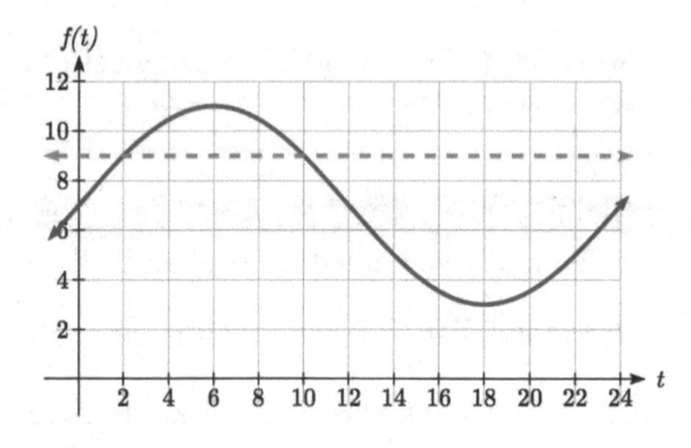

Notice how in both scenarios, the $24k$ shows how every 24 hours the cycle will be repeated.

In the previous example, looking back at the original simplified equation $\sin\left(\frac{\pi}{12}t\right) = \frac{1}{2}$, we can use the ratio of the "normal period" to the stretch factor to find the period: $\frac{2\pi}{\left(\frac{\pi}{12}\right)} = 2\pi\left(\frac{12}{\pi}\right) = 24$. Notice that the sine function has a period of 24, which is reflected in the solutions: there were two unique solutions on one full cycle of the sine function, and additional solutions were found by adding multiples of a full period.

Try it Now

2. Solve $4\sin(5t)-1=1$ for all possible values of t.

Solving using the inverse trig functions

Not all equations involve the "special" values of the trig functions to we have learned. To find the solutions to these equations, we need to use the inverse trig functions.

Example 4

Use the inverse sine function to find one solution to $\sin(\theta)=0.8$.

Since this is not a known unit circle value, calculating the inverse, $\theta=\sin^{-1}(0.8)$. This requires a calculator and we must approximate a value for this angle. If your calculator is in degree mode, your calculator will give you an angle in degrees as the output. If your calculator is in radian mode, your calculator will give you an angle in radians. In radians, $\theta=\sin^{-1}(0.8)\approx 0.927$, or in degrees, $\theta=\sin^{-1}(0.8)\approx 53.130°$.

If you are working with a composed trig function and you are not solving for an angle, you will want to ensure that you are working in radians. In calculus, we will almost always want to work with radians since they are unit-less.

Notice that the inverse trig functions do exactly what you would expect of any function – for each input they give exactly one output. While this is necessary for these to be a function, it means that to find *all* the solutions to an equation like $\sin(\theta)=0.8$, we need to do more than just evaluate the inverse function.

To find additional solutions, it is good to remember four things:

- The sine is the y-value of a point on the unit circle
- The cosine is the x-value of a point on the unit circle
- The tangent is the slope of a line at a given angle
- Other angles with the same sin/cos/tan will have the same reference angle

Example 5

Find all solutions to $\sin(\theta)=0.8$.

We would expect two unique angles on one cycle to have this sine value. In the previous example, we found one solution to be $\theta=\sin^{-1}(0.8)\approx 0.927$. To find the other, we need to answer the question "what other angle has the same sine value as an angle of 0.927?"

We can think of this as finding all the angles where the *y*-value on the unit circle is 0.8. Drawing a picture of the circle helps how the symmetry.
On a unit circle, we would recognize that the second angle would have the same reference angle and reside in the second quadrant. This second angle would be located at $\theta = \pi - \sin^{-1}(0.8)$, or approximately $\theta \approx \pi - 0.927 = 2.214$.

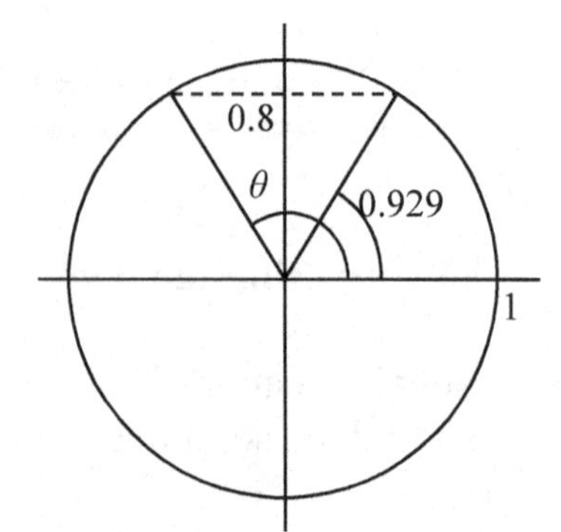

To find more solutions we recall that angles coterminal with these two would have the same sine value, so we can add full cycles of 2π.

$\theta = \sin^{-1}(0.8) + 2\pi k$ and $\theta = \pi - \sin^{-1}(0.8) + 2\pi k$ where *k* is an integer,
or approximately, $\theta = 0.927 + 2\pi k$ and $\theta = 2.214 + 2\pi k$ where *k* is an integer.

Example 6

Find all solutions to $\sin(x) = -\frac{8}{9}$ on the interval $0° \le x < 360°$.

We are looking for the angles with a *y*-value of -8/9 on the unit circle. Immediately we can see the solutions will be in the third and fourth quadrants.

First, we will turn our calculator to degree mode. Using the inverse, we can find one solution $x = \sin^{-1}\left(-\frac{8}{9}\right) \approx -62.734°$.
While this angle satisfies the equation, it does not lie in the domain we are looking for. To find the angles in the desired domain, we start looking for additional solutions.

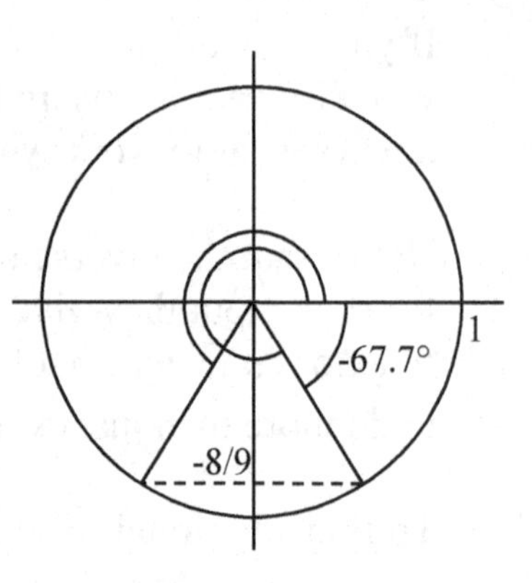

First, an angle coterminal with $-62.734°$ will have the same sine. By adding a full rotation, we can find an angle in the desired domain with the same sine.
$x = -62.734° + 360° = 297.266°$

There is a second angle in the desired domain that lies in the third quadrant. Notice that $62.734°$ is the reference angle for all solutions, so this second solution would be $62.734°$ past $180°$
$x = 62.734° + 180° = 242.734°$

The two solutions on $0° \le x < 360°$ are $x =$ 297.266° and $x =$ 242.734°

Example 7

Find all solutions to $\tan(x) = 3$ on $0 \le x < 2\pi$.

Using the inverse tangent function, we can find one solution $x = \tan^{-1}(3) \approx 1.249$. Unlike the sine and cosine, the tangent function only attains any output value once per cycle, so there is no second solution in any one cycle.

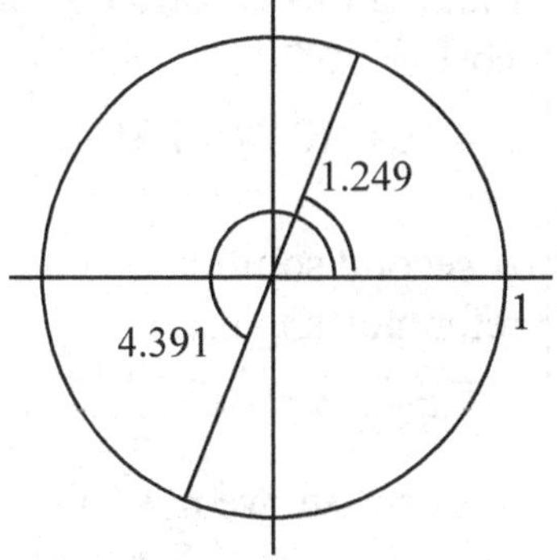

By adding π, a full period of tangent function, we can find a second angle with the same tangent value. Notice this gives another angle where the line has the same slope.

If additional solutions were desired, we could continue to add multiples of π, so all solutions would take on the form $x = 1.249 + k\pi$, however we are only interested in $0 \le x < 2\pi$.

$x = 1.249 + \pi = 4.391$

The two solutions on $0 \le x < 2\pi$ are $x = 1.249$ and $x = 4.391$.

Try it Now

3. Find all solutions to $\tan(x) = 0.7$ on $0° \le x < 360°$.

Example 8

Solve $3\cos(t) + 4 = 2$ for all solutions on one cycle, $0 \le t < 2\pi$

$3\cos(t) + 4 = 2$ Isolating the cosine

$3\cos(t) = -2$

$\cos(t) = -\dfrac{2}{3}$ Using the inverse, we can find one solution

$t = \cos^{-1}\left(-\dfrac{2}{3}\right) \approx 2.301$

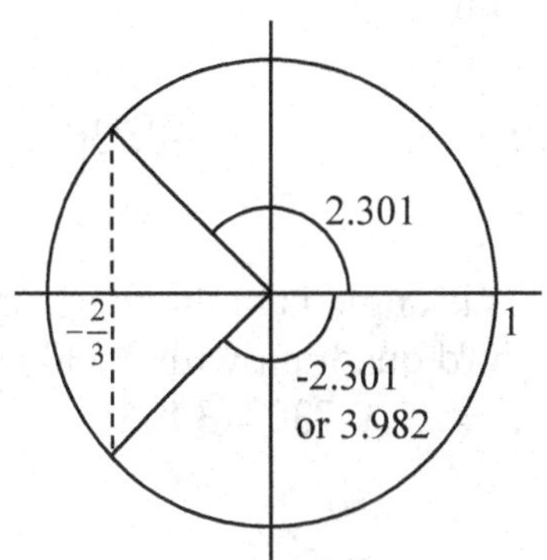

We're looking for two angles where the x-coordinate on a unit circle is -2/3. A second angle with the same cosine would be located in the third quadrant. Notice that the location of this angle could be represented as $t = -2.301$. To represent this as a positive angle we could find a coterminal angle by adding a full cycle.

$t = -2.301 + 2\pi = 3.982$

The equation has two solutions between 0 and 2π, at $t = 2.301$ and $t = 3.982$.

Example 9

Solve $\cos(3t) = 0.2$ for all solutions on two cycles, $0 \le t < \frac{4\pi}{3}$.

As before, with a horizontal compression it can be helpful to make a substitution, $u = 3t$
Making this substitution simplifies the equation to a form we have already solved.

$\cos(u) = 0.2$

$u = \cos^{-1}(0.2) \approx 1.369$

A second solution on one cycle would be located in the fourth quadrant with the same reference angle.

$u = 2\pi - 1.369 = 4.914$

In this case, we need all solutions on two cycles, so we need to find the solutions on the second cycle. We can do this by adding a full rotation to the previous two solutions.

$u = 1.369 + 2\pi = 7.653$

$u = 4.914 + 2\pi = 11.197$

Undoing the substitution, we obtain our four solutions:

$3t = 1.369$, so $t = 0.456$
$3t = 4.914$ so $t = 1.638$
$3t = 7.653$, so $t = 2.551$
$3t = 11.197$, so $t = 3.732$

Example 10

Solve $3\sin(\pi t) = -2$ for all solutions.

$3\sin(\pi t) = -2$ — Isolating the sine

$\sin(\pi t) = -\frac{2}{3}$ — We make the substitution $u = \pi t$

$\sin(u) = -\frac{2}{3}$ — Using the inverse, we find one solution

$u = \sin^{-1}\left(-\frac{2}{3}\right) \approx -0.730$

This angle is in the fourth quadrant. A second angle with the same sine would be in the third quadrant with 0.730 as a reference angle:

$u = \pi + 0.730 = 3.871$

We can write all solutions to the equation $\sin(u) = -\frac{2}{3}$ as

$u = -0.730 + 2\pi k$ or $u = 3.871 + 2\pi k$, where k is an integer.

Undoing our substitution, we can replace u in our solutions with $u = \pi t$ and solve for t

$\pi t = -0.730 + 2\pi k$ or $\pi t = 3.871 + 2\pi k$ Divide by π
$t = -0.232 + 2k$ or $t = 1.232 + 2k$

Try it Now

4. Solve $5\sin\left(\frac{\pi}{2}t\right) + 3 = 0$ for all solutions on one cycle, $0 \le t < 4$.

Solving Trig Equations
1) Isolate the trig function on one side of the equation
2) Make a substitution for the inside of the sine, cosine, or tangent (or other trig function)
3) Use inverse trig functions to find one solution
4) Use symmetries to find a second solution on one cycle (when a second exists)
5) Find additional solutions if needed by adding full periods
6) Undo the substitution

We now can return to the question we began the section with.

Example 11

The height of a rider on the London Eye Ferris wheel can be determined by the equation $h(t) = -65\cos\left(\frac{\pi}{15}t\right) + 70$. How long is the rider more than 100 meters above ground?

To find how long the rider is above 100 meters, we first find the times at which the rider is at a height of 100 meters by solving *h(t)* = 100.

$100 = -65\cos\left(\frac{\pi}{15}t\right) + 70$ Isolating the cosine

$30 = -65\cos\left(\frac{\pi}{15}t\right)$

$\frac{30}{-65} = \cos\left(\frac{\pi}{15}t\right)$ We make the substitution $u = \frac{\pi}{15}t$

$\frac{30}{-65} = \cos(u)$ Using the inverse, we find one solution

$u = \cos^{-1}\left(\frac{30}{-65}\right) \approx 2.051$

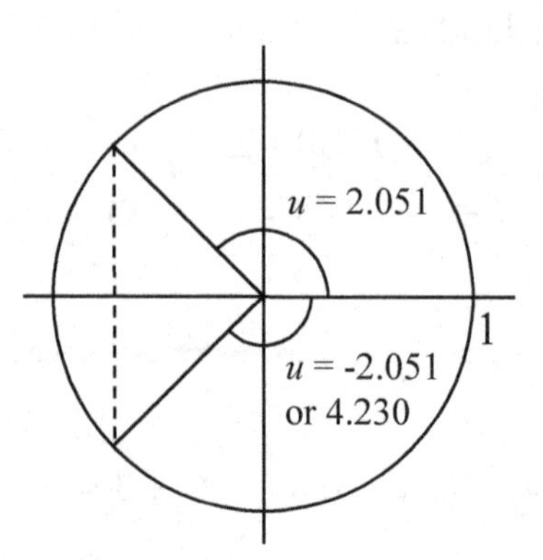

This angle is in the second quadrant. A second angle with the same cosine would be symmetric in the third quadrant. This angle could be represented as $u = -2.051$, but we need a coterminal positive angle, so we add 2π:
$u = 2\pi - 2.051 \approx 4.230$

Now we can undo the substitution to solve for t

$\frac{\pi}{15}t = 2.051$ so $t = 9.793$ minutes after the start of the ride

$\frac{\pi}{15}t = 4.230$ so $t = 20.197$ minutes after the start of the ride

A rider will be at 100 meters after 9.793 minutes, and again after 20.197 minutes. From the behavior of the height graph, we know the rider will be above 100 meters between these times. A rider will be above 100 meters for 20.197 - 9.793 = 10.404 minutes of the ride.

Important Topics of This Section
Solving trig equations using known values
Using substitution to solve equations
Finding answers in one cycle or period vs. finding all possible solutions
Method for solving trig equations

Try it Now Answers

1. From our special angles, we know one answer is $t = \frac{\pi}{4}$. Tangent equations only have one unique solution per cycle or period, so additional solutions can be found by adding multiples of a full period, π. $t = \frac{\pi}{4} + \pi k$.

2. $4\sin(5t) - 1 = 1$

   $\sin(5t) = \frac{1}{2}$. Let $u = 5t$ so this becomes $\sin(u) = \frac{1}{2}$, which has solutions

   $u = \frac{\pi}{6} + 2\pi k, \frac{5\pi}{6} + 2\pi k$. Solving $5t = u = \frac{\pi}{6} + 2\pi k, \frac{5\pi}{6} + 2\pi k$ gives the solutions

   $t = \frac{\pi}{30} + \frac{2\pi}{5}k \qquad t = \frac{\pi}{6} + \frac{2\pi}{5}k$

3. The first solution is $x = \tan^{-1}(0.7) \approx 34.992°$.
For a standard tangent, the second solution can be found by adding a full period, 180°, giving $x = 180° + 34.99° = 214.992°$.

4. $\sin\left(\frac{\pi}{2}t\right) = -\frac{3}{5}$. Let $u = \frac{\pi}{2}t$, so this becomes $\sin(u) = -\frac{3}{5}$.

Using the inverse, $u = \sin^{-1}\left(-\frac{3}{5}\right) \approx -0.6435$. Since we want positive solutions, we can find the coterminal solution by adding a full cycle: $u = -0.6435 + 2\pi = 5.6397$.

Another angle with the same sin would be in the third quadrant with the reference angle 0.6435. $u = \pi + 0.6435 = 3.7851$.

Solving for t, $u = \frac{\pi}{2}t = 5.6397$, so $t = 5.6397\left(\frac{2}{\pi}\right) = 3.5903$

and $u = \frac{\pi}{2}t = 3.7851$, so $t = 3.7851\left(\frac{2}{\pi}\right) = 2.4097$.

$t = 2.4097$ or $t = 3.5903$.

Section 6.4 Exercises

Give all answers in radians unless otherwise indicated.

Find all solutions on the interval $0 \le \theta < 2\pi$.

1. $2\sin(\theta) = -\sqrt{2}$
2. $2\sin(\theta) = \sqrt{3}$
3. $2\cos(\theta) = 1$
4. $2\cos(\theta) = -\sqrt{2}$
5. $\sin(\theta) = 1$
6. $\sin(\theta) = 0$
7. $\cos(\theta) = 0$
8. $\cos(\theta) = -1$

Find all solutions.

9. $2\cos(\theta) = \sqrt{2}$
10. $2\cos(\theta) = -1$
11. $2\sin(\theta) = -1$
12. $2\sin(\theta) = -\sqrt{3}$

Find all solutions.

13. $2\sin(3\theta) = 1$
14. $2\sin(2\theta) = \sqrt{3}$
15. $2\sin(3\theta) = -\sqrt{2}$
16. $2\sin(3\theta) = -1$
17. $2\cos(2\theta) = 1$
18. $2\cos(2\theta) = \sqrt{3}$
19. $2\cos(3\theta) = -\sqrt{2}$
20. $2\cos(2\theta) = -1$
21. $\cos\left(\frac{\pi}{4}\theta\right) = -1$
22. $\sin\left(\frac{\pi}{3}\theta\right) = -1$
23. $2\sin(\pi\theta) = 1$.
24. $2\cos\left(\frac{\pi}{5}\theta\right) = \sqrt{3}$

Find all solutions on the interval $0 \le x < 2\pi$.

25. $\sin(x) = 0.27$
26. $\sin(x) = 0.48$
27. $\sin(x) = -0.58$
28. $\sin(x) = -0.34$
29. $\cos(x) = -0.55$
30. $\sin(x) = 0.28$
31. $\cos(x) = 0.71$
32. $\cos(x) = -0.07$

Find the first two positive solutions.

33. $7\sin(6x) = 2$
34. $7\sin(5x) = 6$
35. $5\cos(3x) = -3$
36. $3\cos(4x) = 2$
37. $3\sin\left(\frac{\pi}{4}x\right) = 2$
38. $7\sin\left(\frac{\pi}{5}x\right) = 6$
39. $5\cos\left(\frac{\pi}{3}x\right) = 1$
40. $3\cos\left(\frac{\pi}{2}x\right) = -2$

Section 6.5 Modeling with Trigonometric Functions

Solving right triangles for angles

In Section 5.5, we used trigonometry on a right triangle to solve for the sides of a triangle given one side and an additional angle. Using the inverse trig functions, we can solve for the angles of a right triangle given two sides.

Example 1

An airplane needs to fly to an airfield located 300 miles east and 200 miles north of its current location. At what heading should the airplane fly? In other words, if we ignore air resistance or wind speed, how many degrees north of east should the airplane fly?

We might begin by drawing a picture and labeling all of the known information. Drawing a triangle, we see we are looking for the angle α. In this triangle, the side opposite the angle α is 200 miles and the side adjacent is 300 miles. Since we know the values for the opposite and adjacent sides, it makes sense to use the tangent function.

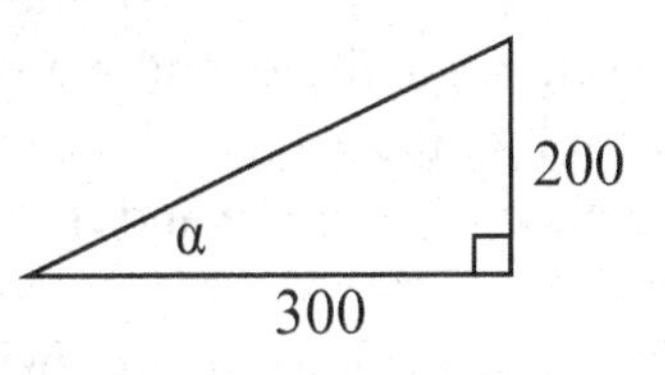

$\tan(\alpha) = \dfrac{200}{300}$ Using the inverse,

$\alpha = \tan^{-1}\left(\dfrac{200}{300}\right) \approx 0.588$, or equivalently about 33.7 degrees.

The airplane needs to fly at a heading of 33.7 degrees north of east.

Example 2

OSHA safety regulations require that the base of a ladder be placed 1 foot from the wall for every 4 feet of ladder length[3]. Find the angle such a ladder forms with the ground.

For any length of ladder, the base needs to be one quarter of the distance the foot of the ladder is away from the wall. Equivalently, if the base is a feet from the wall, the ladder can be $4a$ feet long. Since a is the side adjacent to the angle and $4a$ is the hypotenuse, we use the cosine function.

$\cos(\theta) = \dfrac{a}{4a} = \dfrac{1}{4}$ Using the inverse

$\theta = \cos^{-1}\left(\dfrac{1}{4}\right) \approx 75.52$ degrees

The ladder forms a 75.52 degree angle with the ground.

[3] http://www.osha.gov/SLTC/etools/construction/falls/4ladders.html

Try it Now

1. A cable that anchors the center of the London Eye Ferris wheel to the ground must be replaced. The center of the Ferris wheel is 70 meters above the ground and the second anchor on the ground is 23 meters from the base of the wheel. What is the angle from the ground up to the center of the Ferris wheel and how long is the cable?

Example 3

In a video game design, a map shows the location of other characters relative to the player, who is situated at the origin, and the direction they are facing. A character currently shows on the map at coordinates (-3, 5). If the player rotates counterclockwise by 20 degrees, then the objects in the map will correspondingly rotate 20 degrees clockwise. Find the new coordinates of the character.

To rotate the position of the character, we can imagine it as a point on a circle, and we will change the angle of the point by 20 degrees. To do so, we first need to find the radius of this circle and the original angle.

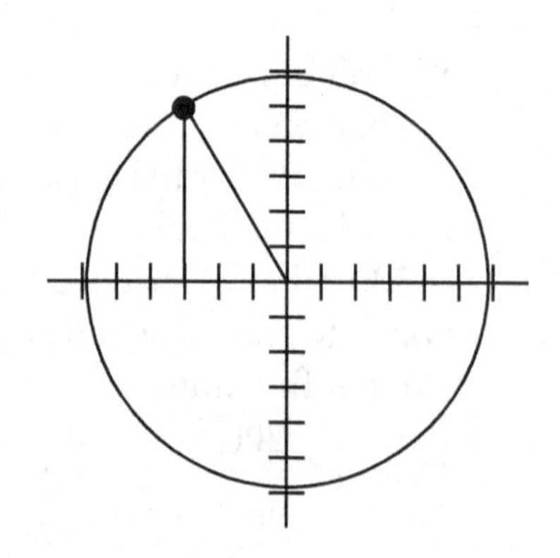

Drawing a right triangle inside the circle, we can find the radius using the Pythagorean Theorem:

$$(-3)^2 + 5^2 = r^2$$

$$r = \sqrt{9+25} = \sqrt{34}$$

To find the angle, we need to decide first if we are going to find the acute angle of the triangle, the reference angle, or if we are going to find the angle measured in standard position. While either approach will work, in this case we will do the latter. Since for any point on a circle we know $x = r\cos(\theta)$, using our given information we get

$$-3 = \sqrt{34}\cos(\theta)$$

$$\frac{-3}{\sqrt{34}} = \cos(\theta)$$

$$\theta = \cos^{-1}\left(\frac{-3}{\sqrt{34}}\right) \approx 120.964°$$

While there are two angles that have this cosine value, the angle of 120.964 degrees is in the second quadrant as desired, so it is the angle we were looking for.

Rotating the point clockwise by 20 degrees, the angle of the point will decrease to 100.964 degrees. We can then evaluate the coordinates of the rotated point

$$x = \sqrt{34}\cos(100.964°) \approx -1.109$$

$$y = \sqrt{34}\sin(100.964°) \approx 5.725$$

The coordinates of the character on the rotated map will be (-1.109, 5.725).

Modeling with sinusoidal functions

Many modeling situations involve functions that are periodic. Previously we learned that sinusoidal functions are a special type of periodic function. Problems that involve quantities that oscillate can often be modeled by a sine or cosine function and once we create a suitable model for the problem we can use that model to answer various questions.

Example 4

The hours of daylight in Seattle oscillate from a low of 8.5 hours in January to a high of 16 hours in July[4]. When should you plant a garden if you want to do it during a month where there are 14 hours of daylight?

To model this, we first note that the hours of daylight oscillate with a period of 12 months. $B = \frac{2\pi}{12} = \frac{\pi}{6}$ corresponds to the horizontal stretch, found by using the ratio of the original period to the new period.

With a low of 8.5 and a high of 16, the midline will be halfway between these values, at $\frac{16+8.5}{2} = 12.25$.

The amplitude will be half the difference between the highest and lowest values: $\frac{16-8.5}{2} = 3.75$, or equivalently the distance from the midline to the high or low value, 16-12.25=3.75.

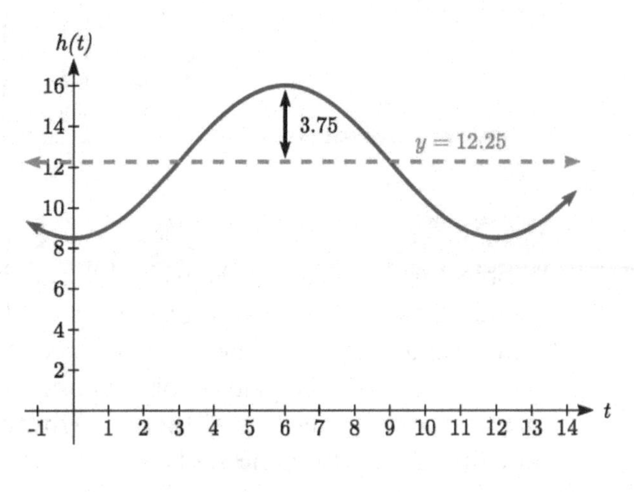

Letting January be $t = 0$, the graph starts at the lowest value, so it can be modeled as a flipped cosine graph. Putting this together, we get a model:

$$h(t) = -3.75\cos\left(\frac{\pi}{6}t\right) + 12.25$$

h(t) is our model for hours of day light *t* months after January.

To find when there will be 14 hours of daylight, we solve $h(t) = 14$.

$14 = -3.75\cos\left(\frac{\pi}{6}t\right) + 12.25$ Isolating the cosine

[4] http://www.mountaineers.org/seattle/climbing/Reference/DaylightHrs.html

$1.75 = -3.75\cos\left(\frac{\pi}{6}t\right)$ Subtracting 12.25 and dividing by -3.75

$-\frac{1.75}{3.75} = \cos\left(\frac{\pi}{6}t\right)$ Using the inverse

$\frac{\pi}{6}t = \cos^{-1}\left(-\frac{1.75}{3.75}\right) \approx 2.0563$ multiplying by the reciprocal

$t = 2.0563 \cdot \frac{6}{\pi} = 3.927$ t=3.927 months past January

There will be 14 hours of daylight 3.927 months into the year, or near the end of April.

While there would be a second time in the year when there are 14 hours of daylight, since we are planting a garden, we would want to know the first solution, in spring, so we do not need to find the second solution in this case.

Try it Now

2. The author's monthly gas usage (in therms) is shown here. Find a function to model the data.

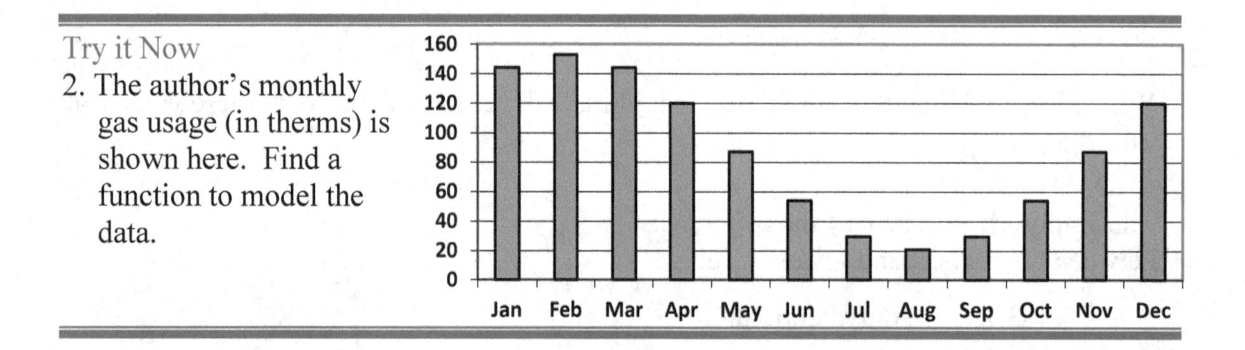

Example 6

An object is connected to the wall with a spring that has a natural length of 20 cm. The object is pulled back 8 cm past the natural length and released. The object oscillates 3 times per second. Find an equation for the horizontal position of the object ignoring the effects of friction. How much time during each cycle is the object more than 27 cm from the wall?

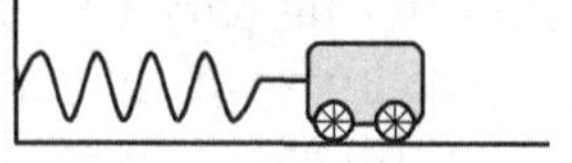

If we use the distance from the wall, x, as the desired output, then the object will oscillate equally on either side of the spring's natural length of 20, putting the midline of the function at 20 cm.

If we release the object 8 cm past the natural length, the amplitude of the oscillation will be 8 cm.

We are beginning at the largest value and so this function can most easily be modeled using a cosine function.

Since the object oscillates 3 times per second, it has a frequency of 3 and the period of one oscillation is 1/3 of second. Using this we find the horizontal compression using the ratios of the periods: $\dfrac{2\pi}{1/3} = 6\pi$.

Using all this, we can build our model:
$x(t) = 8\cos(6\pi t) + 20$

To find when the object is 27 cm from the wall, we can solve $x(t) = 27$

$27 = 8\cos(6\pi t) + 20$ Isolating the cosine

$7 = 8\cos(6\pi t)$

$\dfrac{7}{8} = \cos(6\pi t)$ Using the inverse

$6\pi t = \cos^{-1}\left(\dfrac{7}{8}\right) \approx 0.505$

$t = \dfrac{0.505}{6\pi} = 0.0268$

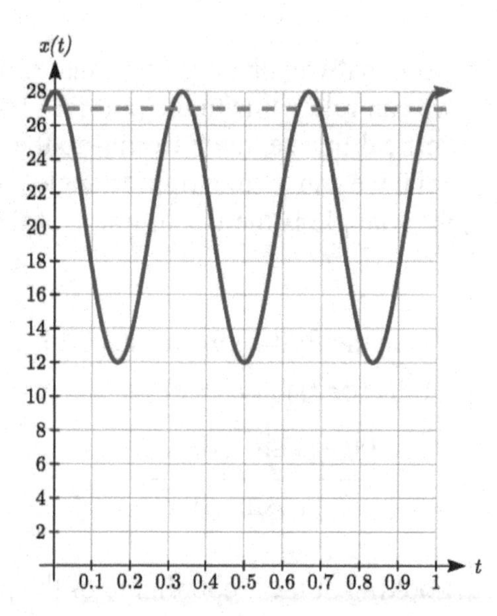

Based on the shape of the graph, we can conclude that the object will spend the first 0.0268 seconds more than 27 cm from the wall. Based on the symmetry of the function, the object will spend another 0.0268 seconds more than 27 cm from the wall at the end of the cycle. Altogether, the object spends 0.0536 seconds each cycle at a distance greater than 27 cm from the wall.

In some problems, we can use trigonometric functions to model behaviors more complicated than the basic sinusoidal function.

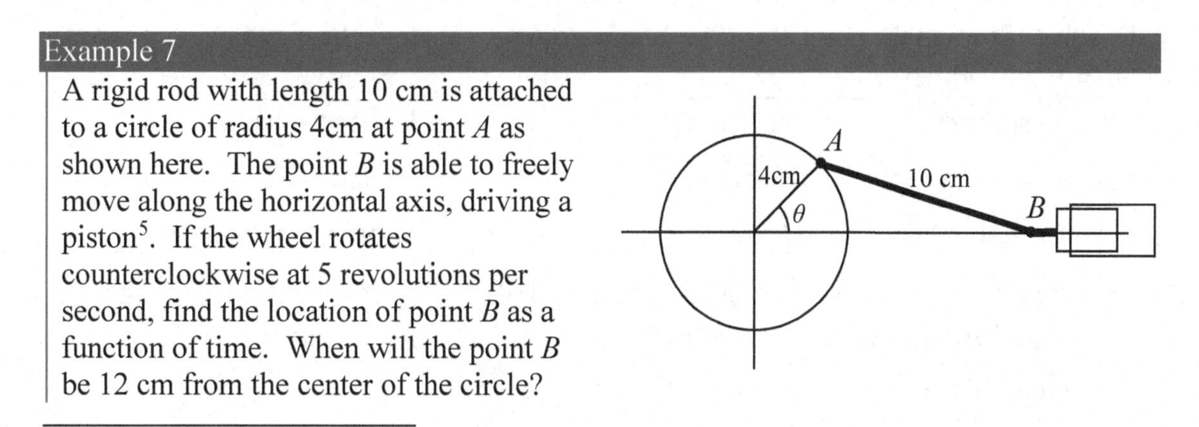

Example 7

A rigid rod with length 10 cm is attached to a circle of radius 4cm at point A as shown here. The point B is able to freely move along the horizontal axis, driving a piston[5]. If the wheel rotates counterclockwise at 5 revolutions per second, find the location of point B as a function of time. When will the point B be 12 cm from the center of the circle?

[5] For an animation of this situation, see http://www.mathdemos.org/mathdemos/sinusoidapp/engine1.gif

To find the position of point B, we can begin by finding the coordinates of point A. Since it is a point on a circle with radius 4, we can express its coordinates as $(4\cos(\theta), 4\sin(\theta))$, where θ is the angle shown.

The angular velocity is 5 revolutions per second, or equivalently 10π radians per second. After t seconds, the wheel will rotate by $\theta = 10\pi t$ radians. Substituting this, we can find the coordinates of A in terms of t.
$(4\cos(10\pi t), 4\sin(10\pi t))$

Notice that this is the same value we would have obtained by observing that the period of the rotation is 1/5 of a second and calculating the stretch/compression factor:
$\frac{\text{"original"}}{\text{"new"}} \frac{2\pi}{1/5} = 10\pi$.

Now that we have the coordinates of the point A, we can relate this to the point B. By drawing a vertical line segment from A to the horizontal axis, we can form a right triangle. The height of the triangle is the y coordinate of the point A: $4\sin(10\pi t)$.

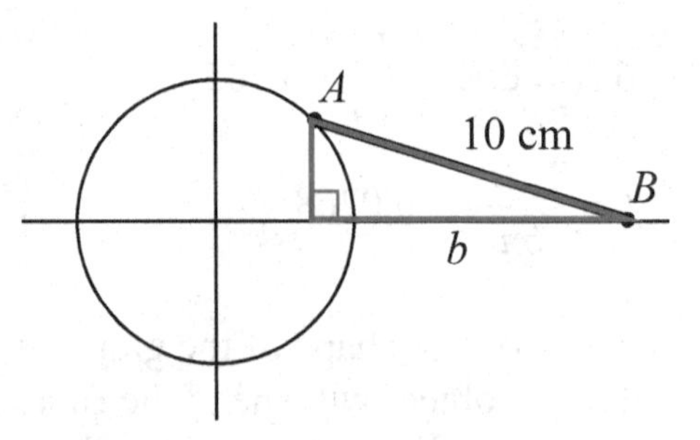

Using the Pythagorean Theorem, we can find the base length of the triangle:
$$(4\sin(10\pi t))^2 + b^2 = 10^2$$
$$b^2 = 100 - 16\sin^2(10\pi t)$$
$$b = \sqrt{100 - 16\sin^2(10\pi t)}$$

Looking at the x coordinate of the point A, we can see that the triangle we drew is shifted to the right of the y axis by $4\cos(10\pi t)$. Combining this offset with the length of the base of the triangle gives the x coordinate of the point B:
$$x(t) = 4\cos(10\pi t) + \sqrt{100 - 16\sin^2(10\pi t)}$$

To solve for when the point B will be 12 cm from the center of the circle, we need to solve $x(t) = 12$.

$12 = 4\cos(10\pi t) + \sqrt{100 - 16\sin^2(10\pi t)}$	Isolate the square root
$12 - 4\cos(10\pi t) = \sqrt{100 - 16\sin^2(10\pi t)}$	Square both sides
$(12 - 4\cos(10\pi t))^2 = 100 - 16\sin^2(10\pi t)$	Expand the left side
$144 - 96\cos(10\pi t) + 16\cos^2(10\pi t) = 100 - 16\sin^2(10\pi t)$	Move all terms to the left
$44 - 96\cos(10\pi t) + 16\cos^2(10\pi t) + 16\sin^2(10\pi t) = 0$	Factor out 16
$44 - 96\cos(10\pi t) + 16\left(\cos^2(10\pi t) + \sin^2(10\pi t)\right) = 0$	

At this point, we can utilize the Pythagorean Identity, which tells us that
$\cos^2(10\pi t)+\sin^2(10\pi t)=1$.

Using this identity, our equation simplifies to

$44-96\cos(10\pi t)+16=0$	Combine the constants and move to the right side
$-96\cos(10\pi t)=-60$	Divide
$\cos(10\pi t)=\dfrac{60}{96}$	Make a substitution
$\cos(u)=\dfrac{60}{96}$	
$u=\cos^{-1}\left(\dfrac{60}{96}\right)\approx 0.896$	By symmetry we can find a second solution
$u=2\pi-0.896=5.388$	Undoing the substitution

$10\pi t=0.896$, so $t=0.0285$
$10\pi t=5.388$, so $t=0.1715$

The point B will be 12 cm from the center of the circle 0.0285 seconds after the process begins, 0.1715 seconds after the process begins, and every 1/5 of a second after each of those values.

Important Topics of This Section
Modeling with trig equations
Modeling with sinusoidal functions
Solving right triangles for angles in degrees and radians

Try it Now Answers
1. Angle of elevation for the cable is 71.81 degrees and the cable is 73.68 m long
2. Approximately $G(t)=66\cos\left(\dfrac{\pi}{6}(t-1)\right)+87$

Section 6.5 Exercises

In each of the following triangles, solve for the unknown side and angles.

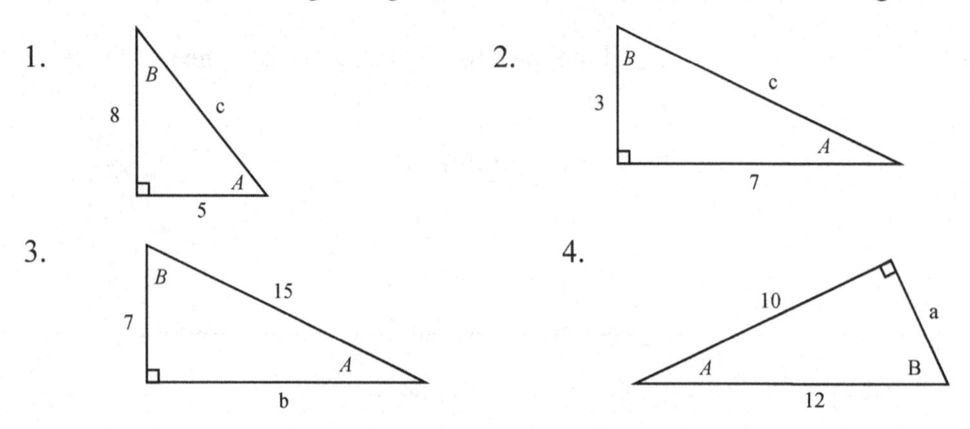

Find a possible formula for the trigonometric function whose values are in the following tables.

5.

x	0	1	2	3	4	5	6
y	-2	4	10	4	-2	4	10

6.

x	0	1	2	3	4	5	6
y	1	-3	-7	-3	1	-3	-7

7. Outside temperature over the course of a day can be modeled as a sinusoidal function. Suppose you know the high temperature for the day is 63 degrees and the low temperature of 37 degrees occurs at 5 AM. Assuming t is the number of hours since midnight, find an equation for the temperature, D, in terms of t.

8. Outside temperature over the course of a day can be modeled as a sinusoidal function. Suppose you know the high temperature for the day is 92 degrees and the low temperature of 78 degrees occurs at 4 AM. Assuming t is the number of hours since midnight, find an equation for the temperature, D, in terms of t.

9. A population of rabbits oscillates 25 above and below an average of 129 during the year, hitting the lowest value in January ($t = 0$).
    a. Find an equation for the population, P, in terms of the months since January, t.
    b. What if the lowest value of the rabbit population occurred in April instead?

10. A population of elk oscillates 150 above and below an average of 720 during the year, hitting the lowest value in January ($t = 0$).
    a. Find an equation for the population, P, in terms of the months since January, t.
    b. What if the lowest value of the rabbit population occurred in March instead?

11. Outside temperature over the course of a day can be modeled as a sinusoidal function. Suppose you know the high temperature of 105 degrees occurs at 5 PM and the average temperature for the day is 85 degrees. Find the temperature, to the nearest degree, at 9 AM.

12. Outside temperature over the course of a day can be modeled as a sinusoidal function. Suppose you know the high temperature of 84 degrees occurs at 6 PM and the average temperature for the day is 70 degrees. Find the temperature, to the nearest degree, at 7 AM.

13. Outside temperature over the course of a day can be modeled as a sinusoidal function. Suppose you know the temperature varies between 47 and 63 degrees during the day and the average daily temperature first occurs at 10 AM. How many hours after midnight does the temperature first reach 51 degrees?

14. Outside temperature over the course of a day can be modeled as a sinusoidal function. Suppose you know the temperature varies between 64 and 86 degrees during the day and the average daily temperature first occurs at 12 AM. How many hours after midnight does the temperature first reach 70 degrees?

15. A Ferris wheel is 20 meters in diameter and boarded from a platform that is 2 meters above the ground. The six o'clock position on the Ferris wheel is level with the loading platform. The wheel completes 1 full revolution in 6 minutes. How many minutes of the ride are spent higher than 13 meters above the ground?

16. A Ferris wheel is 45 meters in diameter and boarded from a platform that is 1 meter above the ground. The six o'clock position on the Ferris wheel is level with the loading platform. The wheel completes 1 full revolution in 10 minutes. How many minutes of the ride are spent higher than 27 meters above the ground?

17. The sea ice area around the North Pole fluctuates between about 6 million square kilometers in September to 14 million square kilometers in March. Assuming sinusoidal fluctuation, during how many months are there less than 9 million square kilometers of sea ice?

18. The sea ice area around the South Pole fluctuates between about 18 million square kilometers in September to 3 million square kilometers in March. Assuming sinusoidal fluctuation, during how many months are there more than 15 million square kilometers of sea ice?

19. A respiratory ailment called "Cheyne-Stokes Respiration" causes the volume per breath to increase and decrease in a sinusoidal manner, as a function of time. For one particular patient with this condition, a machine begins recording a plot of volume per breath versus time (in seconds). Let $b(t)$ be a function of time t that tells us the volume (in liters) of a breath that starts at time t. During the test, the smallest volume per breath is 0.6 liters and this first occurs for a breath that starts 5 seconds into the test. The largest volume per breath is 1.8 liters and this first occurs for a breath beginning 55 seconds into the test. [UW]
    a. Find a formula for the function $b(t)$ whose graph will model the test data for this patient.
    b. If the patient begins a breath every 5 seconds, what are the breath volumes during the first minute of the test?

20. Suppose the high tide in Seattle occurs at 1:00 a.m. and 1:00 p.m, at which time the water is 10 feet above the height of low tide. Low tides occur 6 hours after high tides. Suppose there are two high tides and two low tides every day and the height of the tide varies sinusoidally. [UW]
    a. Find a formula for the function $y = h(t)$ that computes the height of the tide above low tide at time t. (In other words, $y = 0$ corresponds to low tide.)
    b. What is the tide height at 11:00 a.m.?

21. A communications satellite orbits the earth t miles above the surface. Assume the radius of the earth is 3,960 miles. The satellite can only "see" a portion of the earth's surface, bounded by what is called a horizon circle. This leads to a two-dimensional cross-sectional picture we can use to study the size of the horizon slice: [UW]

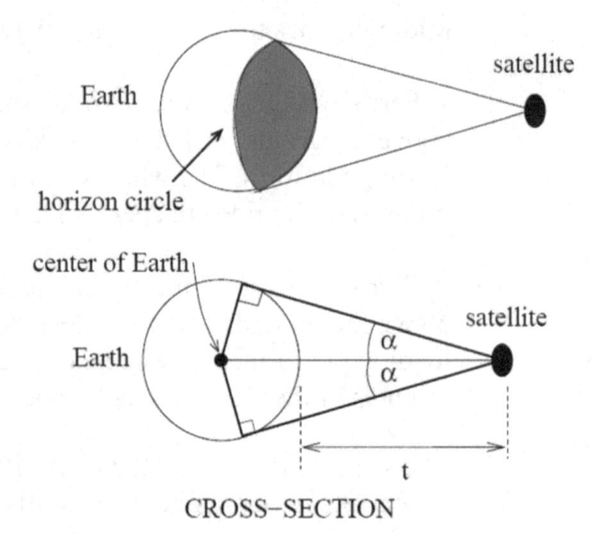

    a. Find a formula for α in terms of t.
    b. If $t = 30{,}000$ miles, what is α? What percentage of the circumference of the earth is covered by the satellite? What would be the minimum number of such satellites required to cover the circumference?
    c. If $t = 1{,}000$ miles, what is α? What percentage of the circumference of the earth is covered by the satellite? What would be the minimum number of such satellites required to cover the circumference?
    d. Suppose you wish to place a satellite into orbit so that 20% of the circumference is covered by the satellite. What is the required distance t?

22. Tiffany is a model rocket enthusiast. She has been working on a pressurized rocket filled with nitrous oxide. According to her design, if the atmospheric pressure exerted on the rocket is less than 10 pounds/sq.in., the nitrous oxide chamber inside the rocket will explode. Tiff worked from a formula $p = 14.7e^{-h/10}$ pounds/sq.in. for the atmospheric pressure h miles above sea level. Assume that the rocket is launched at an angle of α above level ground at sea level with an initial speed of 1400 feet/sec. Also, assume the height (in feet) of the rocket at time t seconds is given by the equation $y(t) = -16t^2 + 1400\sin(\alpha)t$. [UW]
    a. At what altitude will the rocket explode?
    b. If the angle of launch is α = 12°, determine the minimum atmospheric pressure exerted on the rocket during its flight. Will the rocket explode in midair?
    c. If the angle of launch is α = 82°, determine the minimum atmospheric pressure exerted on the rocket during its flight. Will the rocket explode in midair?
    d. Find the largest launch angle α so that the rocket will not explode.

Chapter 7: Trigonometric Equations and Identities

In the last two chapters we have used basic definitions and relationships to simplify trigonometric expressions and solve trigonometric equations. In this chapter we will look at more complex relationships. By conducting a deeper study of trigonometric identities we can learn to simplify complicated expressions, allowing us to solve more interesting applications.

Section 7.1 Solving Trigonometric Equations with Identities

In the last chapter, we solved basic trigonometric equations. In this section, we explore the techniques needed to solve more complicated trig equations. Building from what we already know makes this a much easier task.

Consider the function $f(x) = 2x^2 + x$. If you were asked to solve $f(x) = 0$, it requires simple algebra:

$2x^2 + x = 0$ Factor
$x(2x+1) = 0$ Giving solutions
$x = 0$ or $x = -\frac{1}{2}$

Similarly, for $g(t) = \sin(t)$, if we asked you to solve $g(t) = 0$, you can solve this using unit circle values:
$\sin(t) = 0$ for $t = 0, \pi, 2\pi$ and so on.

Using these same concepts, we consider the composition of these two functions:
$f(g(t)) = 2(\sin(t))^2 + (\sin(t)) = 2\sin^2(t) + \sin(t)$

This creates an equation that is a polynomial trig function. With these types of functions, we use algebraic techniques like factoring and the quadratic formula, along with trigonometric identities and techniques, to solve equations.

As a reminder, here are some of the essential trigonometric identities that we have learned so far:

Identities

Pythagorean Identities

$\cos^2(t) + \sin^2(t) = 1$ $\quad$ $1 + \cot^2(t) = \csc^2(t)$ $\quad$ $1 + \tan^2(t) = \sec^2(t)$

Negative Angle Identities

$\sin(-t) = -\sin(t)$ $\quad$ $\cos(-t) = \cos(t)$ $\quad$ $\tan(-t) = -\tan(t)$

$\csc(-t) = -\csc(t)$ $\quad$ $\sec(-t) = \sec(t)$ $\quad$ $\cot(-t) = -\cot(t)$

Reciprocal Identities

$\sec(t) = \dfrac{1}{\cos(t)}$ $\quad$ $\csc(t) = \dfrac{1}{\sin(t)}$ $\quad$ $\tan(t) = \dfrac{\sin(t)}{\cos(t)}$ $\quad$ $\cot(t) = \dfrac{1}{\tan(t)}$

Example 1

Solve $2\sin^2(t) + \sin(t) = 0$ for all solutions with $0 \le t < 2\pi$.

This equation kind of looks like a quadratic equation, but with sin(t) in place of an algebraic variable (we often call such an equation "quadratic in sine"). As with all quadratic equations, we can use factoring techniques or the quadratic formula. This expression factors nicely, so we proceed by factoring out the common factor of sin(t):

$\sin(t)(2\sin(t) + 1) = 0$

Using the zero product theorem, we know that the product on the left will equal zero if either factor is zero, allowing us to break this equation into two cases:

$\sin(t) = 0$ $\quad$ or $\quad$ $2\sin(t) + 1 = 0$

We can solve each of these equations independently, using our knowledge of special angles.

$\sin(t) = 0$ $\qquad$ $2\sin(t) + 1 = 0$

$t = 0$ or $t = \pi$ $\qquad$ $\sin(t) = -\dfrac{1}{2}$

$t = \dfrac{7\pi}{6}$ or $t = \dfrac{11\pi}{6}$

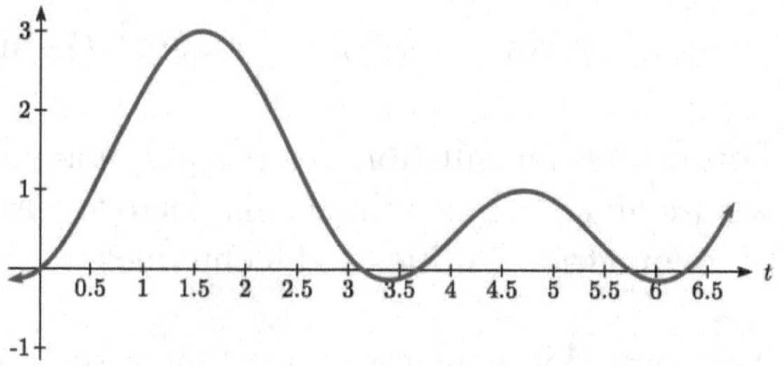

Together, this gives us four solutions to the equation on $0 \le t < 2\pi$:

$t = 0, \pi, \dfrac{7\pi}{6}, \dfrac{11\pi}{6}$

We could check these answers are reasonable by graphing the function and comparing the zeros.

Example 2

Solve $3\sec^2(t) - 5\sec(t) - 2 = 0$ for all solutions with $0 \le t < 2\pi$.

Since the left side of this equation is quadratic in secant, we can try to factor it, and hope it factors nicely.

If it is easier to for you to consider factoring without the trig function present, consider using a substitution $u = \sec(t)$, resulting in $3u^2 - 5u - 2 = 0$, and then try to factor:

$3u^2 - 5u - 2 = (3u + 1)(u - 2)$

Undoing the substitution,

$(3\sec(t) + 1)(\sec(t) - 2) = 0$

Since we have a product equal to zero, we break it into the two cases and solve each separately.

$3\sec(t) + 1 = 0$	Isolate the secant
$\sec(t) = -\frac{1}{3}$	Rewrite as a cosine
$\frac{1}{\cos(t)} = -\frac{1}{3}$	Invert both sides
$\cos(t) = -3$	

Since the cosine has a range of [-1, 1], the cosine will never take on an output of -3. There are no solutions to this case.

Continuing with the second case,

$\sec(t) - 2 = 0$	Isolate the secant
$\sec(t) = 2$	Rewrite as a cosine
$\frac{1}{\cos(t)} = 2$	Invert both sides
$\cos(t) = \frac{1}{2}$	This gives two solutions

$t = \frac{\pi}{3}$ or $t = \frac{5\pi}{3}$

These are the only two solutions on the interval. By utilizing technology to graph $f(t) = 3\sec^2(t) - 5\sec(t) - 2$, a look at a graph confirms there are only two zeros for this function on the interval [0, 2π), which assures us that we didn't miss anything.

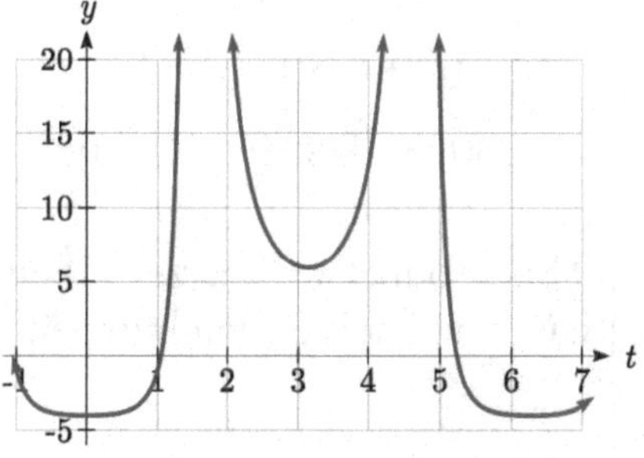

Try it Now

1. Solve $2\sin^2(t)+3\sin(t)+1=0$ for all solutions with $0 \le t < 2\pi$.

When solving some trigonometric equations, it becomes necessary to first rewrite the equation using trigonometric identities. One of the most common is the Pythagorean Identity, $\sin^2(\theta)+\cos^2(\theta)=1$ which allows you to rewrite $\sin^2(\theta)$ in terms of $\cos^2(\theta)$ or vice versa,

Identities

Alternate Forms of the Pythagorean Identity

$\sin^2(\theta)=1-\cos^2(\theta)$

$\cos^2(\theta)=1-\sin^2(\theta)$

These identities become very useful whenever an equation involves a combination of sine and cosine functions.

Example 3

Solve $2\sin^2(t)-\cos(t)=1$ for all solutions with $0 \le t < 2\pi$.

Since this equation has a mix of sine and cosine functions, it becomes more complicated to solve. It is usually easier to work with an equation involving only one trig function. This is where we can use the Pythagorean Identity.

$2\sin^2(t)-\cos(t)=1$ Using $\sin^2(\theta)=1-\cos^2(\theta)$

$2\left(1-\cos^2(t)\right)-\cos(t)=1$ Distributing the 2

$2-2\cos^2(t)-\cos(t)=1$

Since this is now quadratic in cosine, we rearrange the equation so one side is zero and factor.

$-2\cos^2(t)-\cos(t)+1=0$ Multiply by -1 to simplify the factoring

$2\cos^2(t)+\cos(t)-1=0$ Factor

$(2\cos(t)-1)(\cos(t)+1)=0$

This product will be zero if either factor is zero, so we can break this into two separate cases and solve each independently.

$2\cos(t)-1=0$ or $\cos(t)+1=0$

$\cos(t)=\frac{1}{2}$ or $\cos(t)=-1$

$t=\frac{\pi}{3}$ or $t=\frac{5\pi}{3}$ or $t=\pi$

Try it Now

2. Solve $2\sin^2(t)=3\cos(t)$ for all solutions with $0\le t<2\pi$.

In addition to the Pythagorean Identity, it is often necessary to rewrite the tangent, secant, cosecant, and cotangent as part of solving an equation.

Example 4

Solve $\tan(x)=3\sin(x)$ for all solutions with $0\le x<2\pi$.

With a combination of tangent and sine, we might try rewriting tangent

$\tan(x)=3\sin(x)$

$\frac{\sin(x)}{\cos(x)}=3\sin(x)$ Multiplying both sides by cosine

$\sin(x)=3\sin(x)\cos(x)$

At this point, you may be tempted to divide both sides of the equation by sin(x). **Resist the urge**. When we divide both sides of an equation by a quantity, we are assuming the quantity is never zero. In this case, when sin(x) = 0 the equation is satisfied, so we'd lose those solutions if we divided by the sine.

To avoid this problem, we can rearrange the equation so that one side is zero[1].

$\sin(x)-3\sin(x)\cos(x)=0$ Factoring out sin(x) from both parts

$\sin(x)(1-3\cos(x))=0$

From here, we can see we get solutions when $\sin(x)=0$ or $1-3\cos(x)=0$.

Using our knowledge of the special angles of the unit circle,

$\sin(x)=0$ when $x=0$ or $x=\pi$.

[1] You technically *can* divide by sin(x), as long as you separately consider the case where sin(x) = 0. Since it is easy to forget this step, the factoring approach used in the example is recommended.

For the second equation, we will need the inverse cosine.

$1-3\cos(x)=0$

$\cos(x)=\frac{1}{3}$ — Using our calculator or technology

$x=\cos^{-1}\left(\frac{1}{3}\right)\approx 1.231$ — Using symmetry to find a second solution

$x=2\pi-1.231=5.052$

We have four solutions on $0\le x<2\pi$:

$x = 0, 1.231, \pi, 5.052$

Try it Now

3. Solve $\sec(\theta)=2\cos(\theta)$ to find the first four positive solutions.

Example 5

Solve $\frac{4}{\sec^2(\theta)}+3\cos(\theta)=2\cot(\theta)\tan(\theta)$ for all solutions with $0\le\theta<2\pi$.

$\frac{4}{\sec^2(\theta)}+3\cos(\theta)=2\cot(\theta)\tan(\theta)$ — Using the reciprocal identities

$4\cos^2(\theta)+3\cos(\theta)=2\frac{1}{\tan(\theta)}\tan(\theta)$ — Simplifying

$4\cos^2(\theta)+3\cos(\theta)=2$ — Subtracting 2 from each side

$4\cos^2(\theta)+3\cos(\theta)-2=0$

This does not appear to factor nicely so we use the quadratic formula, remembering that we are solving for $\cos(\theta)$.

$$\cos(\theta)=\frac{-3\pm\sqrt{3^2-4(4)(-2)}}{2(4)}=\frac{-3\pm\sqrt{41}}{8}$$

Using the negative square root first,

$$\cos(\theta)=\frac{-3-\sqrt{41}}{8}=-1.175$$

This has no solutions, since the cosine can't be less than -1.

Using the positive square root,

$$\cos(\theta) = \frac{-3+\sqrt{41}}{8} = 0.425$$

$$\theta = \cos^{-1}(0.425) = 1.131$$

By symmetry, a second solution can be found

$$\theta = 2\pi - 1.131 = 5.152$$

Important Topics of This Section
Review of Trig Identities
Solving Trig Equations
By Factoring
Using the Quadratic Formula
Utilizing Trig Identities to simplify

Try it Now Answers

1. Factor as $(2\sin(t)+1)(\sin(t)+1) = 0$

   $2\sin(t)+1=0$ at $t = \frac{7\pi}{6}, \frac{11\pi}{6}$

   $\sin(t)+1=0$ at $t = \frac{3\pi}{2}$

   $t = \frac{7\pi}{6}, \frac{3\pi}{2}, \frac{11\pi}{6}$

2. $2(1-\cos^2(t)) = 3\cos(t)$

   $2\cos^2(t)+3\cos(t)-2=0$

   $(2\cos(t)-1)(\cos(t)+2)=0$

   $\cos(t)+2=0$ has no solutions

   $2\cos(t)-1=0$ at $t = \frac{\pi}{3}, \frac{5\pi}{3}$

3. $\frac{1}{\cos(\theta)} = 2\cos(\theta)$

   $\frac{1}{2} = \cos^2(\theta)$

   $\cos(\theta) = \pm\sqrt{\frac{1}{2}} = \pm\frac{\sqrt{2}}{2}$

   $\theta = \frac{\pi}{4}, \frac{3\pi}{4}, \frac{5\pi}{4}, \frac{7\pi}{4}$

Section 7.1 Exercises

Find all solutions on the interval $0 \le \theta < 2\pi$.

1. $2\sin(\theta) = -1$
2. $2\sin(\theta) = \sqrt{3}$
3. $2\cos(\theta) = 1$
4. $2\cos(\theta) = -\sqrt{2}$

Find all solutions.

5. $2\sin\left(\frac{\pi}{4}x\right) = 1$
6. $2\sin\left(\frac{\pi}{3}x\right) = \sqrt{2}$
7. $2\cos(2t) = -\sqrt{3}$
8. $2\cos(3t) = -1$
9. $3\cos\left(\frac{\pi}{5}x\right) = 2$
10. $8\cos\left(\frac{\pi}{2}x\right) = 6$
11. $7\sin(3t) = -2$
12. $4\sin(4t) = 1$

Find all solutions on the interval $[0, 2\pi)$.

13. $10\sin(x)\cos(x) = 6\cos(x)$
14. $-3\sin(t) = 15\cos(t)\sin(t)$
15. $\csc(2x) - 9 = 0$
16. $\sec(2\theta) = 3$
17. $\sec(x)\sin(x) - 2\sin(x) = 0$
18. $\tan(x)\sin(x) - \sin(x) = 0$
19. $\sin^2 x = \frac{1}{4}$
20. $\cos^2\theta = \frac{1}{2}$
21. $\sec^2 x = 7$
22. $\csc^2 t = 3$
23. $2\sin^2 w + 3\sin w + 1 = 0$
24. $8\sin^2 x + 6\sin(x) + 1 = 0$
25. $2\cos^2 t + \cos(t) = 1$
26. $8\cos^2(\theta) = 3 - 2\cos(\theta)$
27. $4\cos^2(x) - 4 = 15\cos(x)$
28. $9\sin(w) - 2 = 4\sin^2(w)$
29. $12\sin^2(t) + \cos(t) - 6 = 0$
30. $6\cos^2(x) + 7\sin(x) - 8 = 0$
31. $\cos^2\phi = -6\sin\phi$
32. $\sin^2 t = \cos t$
33. $\tan^3(x) = 3\tan(x)$
34. $\cos^3(t) = -\cos(t)$
35. $\tan^5(x) = \tan(x)$
36. $\tan^5(x) - 9\tan(x) = 0$
37. $4\sin(x)\cos(x) + 2\sin(x) - 2\cos(x) - 1 = 0$
38. $2\sin(x)\cos(x) - \sin(x) + 2\cos(x) - 1 = 0$
39. $\tan(x) - 3\sin(x) = 0$
40. $3\cos(x) = \cot(x)$
41. $2\tan^2(t) = 3\sec(t)$
42. $1 - 2\tan(w) = \tan^2(w)$

Section 7.2 Addition and Subtraction Identities

In this section, we begin expanding our repertoire of trigonometric identities.

Identities
The sum and difference identities
$\cos(\alpha-\beta)=\cos(\alpha)\cos(\beta)+\sin(\alpha)\sin(\beta)$
$\cos(\alpha+\beta)=\cos(\alpha)\cos(\beta)-\sin(\alpha)\sin(\beta)$
$\sin(\alpha+\beta)=\sin(\alpha)\cos(\beta)+\cos(\alpha)\sin(\beta)$
$\sin(\alpha-\beta)=\sin(\alpha)\cos(\beta)-\cos(\alpha)\sin(\beta)$

We will prove the difference of angles identity for cosine. The rest of the identities can be derived from this one.

Proof of the difference of angles identity for cosine
Consider two points on a unit circle:
P at an angle of α from the positive x axis with coordinates $(\cos(\alpha),\sin(\alpha))$, and Q at an angle of β with coordinates $(\cos(\beta),\sin(\beta))$.

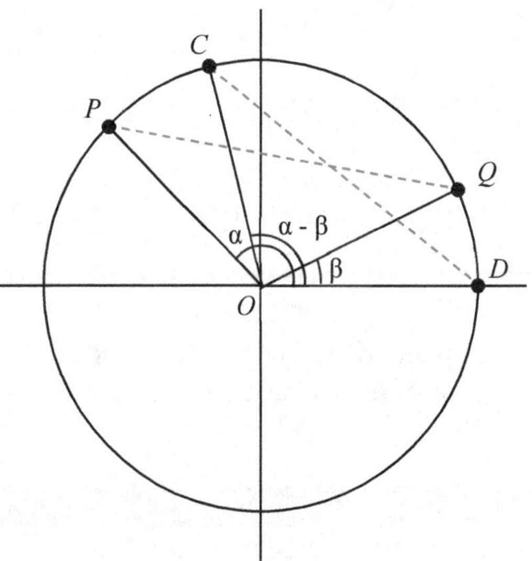

Notice the measure of angle POQ is α – β.
Label two more points:
C at an angle of α – β, with coordinates $(\cos(\alpha-\beta),\sin(\alpha-\beta))$,
D at the point (1, 0).

Notice that the distance from C to D is the same as the distance from P to Q because triangle COD is a rotation of triangle POQ.

Using the distance formula to find the distance from P to Q yields

$$\sqrt{(\cos(\alpha)-\cos(\beta))^2+(\sin(\alpha)-\sin(\beta))^2}$$

Expanding this

$$\sqrt{\cos^2(\alpha)-2\cos(\alpha)\cos(\beta)+\cos^2(\beta)+\sin^2(\alpha)-2\sin(\alpha)\sin(\beta)+\sin^2(\beta)}$$

Applying the Pythagorean Identity and simplifying

$$\sqrt{2-2\cos(\alpha)\cos(\beta)-2\sin(\alpha)\sin(\beta)}$$

Similarly, using the distance formula to find the distance from C to D

$$\sqrt{(\cos(\alpha-\beta)-1)^2+(\sin(\alpha-\beta)-0)^2}$$

Expanding this

$$\sqrt{\cos^2(\alpha-\beta)-2\cos(\alpha-\beta)+1+\sin^2(\alpha-\beta)}$$

Applying the Pythagorean Identity and simplifying

$$\sqrt{-2\cos(\alpha-\beta)+2}$$

Since the two distances are the same we set these two formulas equal to each other and simplify

$$\sqrt{2-2\cos(\alpha)\cos(\beta)-2\sin(\alpha)\sin(\beta)}=\sqrt{-2\cos(\alpha-\beta)+2}$$

$$2-2\cos(\alpha)\cos(\beta)-2\sin(\alpha)\sin(\beta)=-2\cos(\alpha-\beta)+2$$

$$\cos(\alpha)\cos(\beta)+\sin(\alpha)\sin(\beta)=\cos(\alpha-\beta)$$

This establishes the identity.

Try it Now

1. By writing $\cos(\alpha+\beta)$ as $\cos(\alpha-(-\beta))$, show the sum of angles identity for cosine follows from the difference of angles identity proven above.

The sum and difference of angles identities are often used to rewrite expressions in other forms, or to rewrite an angle in terms of simpler angles.

Example 1

Find the exact value of $\cos(75°)$.

Since $75° = 30° + 45°$, we can evaluate $\cos(75°)$ as

$\cos(75°) = \cos(30° + 45°)$ — Apply the cosine sum of angles identity

$= \cos(30°)\cos(45°) - \sin(30°)\sin(45°)$ — Evaluate

$= \frac{\sqrt{3}}{2}\cdot\frac{\sqrt{2}}{2} - \frac{1}{2}\cdot\frac{\sqrt{2}}{2}$ — Simply

$= \frac{\sqrt{6}-\sqrt{2}}{4}$

Try it Now

2. Find the exact value of $\sin\left(\frac{\pi}{12}\right)$.

Example 2

Rewrite $\sin\left(x-\frac{\pi}{4}\right)$ in terms of $\sin(x)$ and $\cos(x)$.

$\sin\left(x-\frac{\pi}{4}\right)$ Use the difference of angles identity for sine

$=\sin(x)\cos\left(\frac{\pi}{4}\right)-\cos(x)\sin\left(\frac{\pi}{4}\right)$ Evaluate the cosine and sine and rearrange

$=\frac{\sqrt{2}}{2}\sin(x)-\frac{\sqrt{2}}{2}\cos(x)$

Additionally, these identities can be used to simplify expressions or prove new identities

Example 3

Prove $\frac{\sin(a+b)}{\sin(a-b)}=\frac{\tan(a)+\tan(b)}{\tan(a)-\tan(b)}$.

As with any identity, we need to first decide which side to begin with. Since the left side involves sum and difference of angles, we might start there

$\frac{\sin(a+b)}{\sin(a-b)}$ Apply the sum and difference of angle identities

$=\frac{\sin(a)\cos(b)+\cos(a)\sin(b)}{\sin(a)\cos(b)-\cos(a)\sin(b)}$

Since it is not immediately obvious how to proceed, we might start on the other side, and see if the path is more apparent.

$\frac{\tan(a)+\tan(b)}{\tan(a)-\tan(b)}$ Rewriting the tangents using the tangent identity

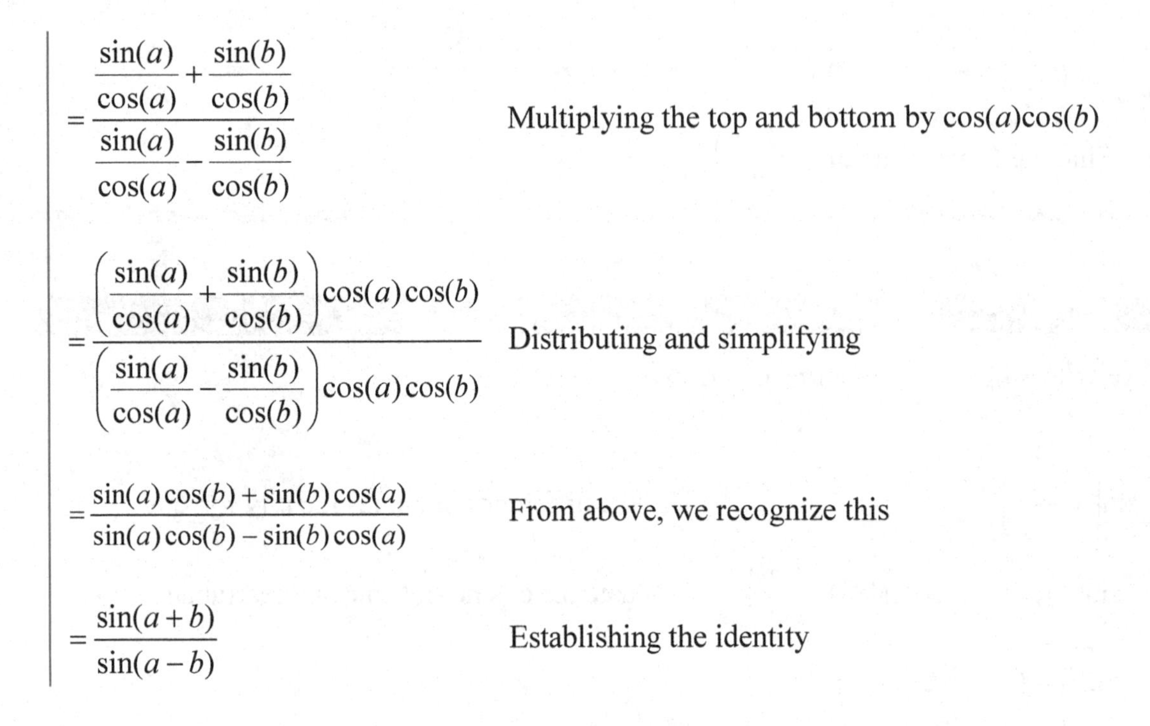

$$= \frac{\dfrac{\sin(a)}{\cos(a)} + \dfrac{\sin(b)}{\cos(b)}}{\dfrac{\sin(a)}{\cos(a)} - \dfrac{\sin(b)}{\cos(b)}}$$ Multiplying the top and bottom by $\cos(a)\cos(b)$

$$= \frac{\left(\dfrac{\sin(a)}{\cos(a)} + \dfrac{\sin(b)}{\cos(b)}\right)\cos(a)\cos(b)}{\left(\dfrac{\sin(a)}{\cos(a)} - \dfrac{\sin(b)}{\cos(b)}\right)\cos(a)\cos(b)}$$ Distributing and simplifying

$$= \frac{\sin(a)\cos(b) + \sin(b)\cos(a)}{\sin(a)\cos(b) - \sin(b)\cos(a)}$$ From above, we recognize this

$$= \frac{\sin(a+b)}{\sin(a-b)}$$ Establishing the identity

These identities can also be used to solve equations.

Example 4

Solve $\sin(x)\sin(2x) + \cos(x)\cos(2x) = \dfrac{\sqrt{3}}{2}$.

By recognizing the left side of the equation as the result of the difference of angles identity for cosine, we can simplify the equation

$$\sin(x)\sin(2x) + \cos(x)\cos(2x) = \frac{\sqrt{3}}{2}$$ Apply the difference of angles identity

$$\cos(x - 2x) = \frac{\sqrt{3}}{2}$$

$$\cos(-x) = \frac{\sqrt{3}}{2}$$ Use the negative angle identity

$$\cos(x) = \frac{\sqrt{3}}{2}$$

Since this is a special cosine value we recognize from the unit circle, we can quickly write the answers:

$$x = \frac{\pi}{6} + 2\pi k$$
$$x = \frac{11\pi}{6} + 2\pi k$$
, where k is an integer

Combining Waves of Equal Period

A sinusoidal function of the form $f(x) = A\sin(Bx + C)$ can be rewritten using the sum of angles identity.

Example 5

Rewrite $f(x) = 4\sin\left(3x + \frac{\pi}{3}\right)$ as a sum of sine and cosine.

$4\sin\left(3x + \frac{\pi}{3}\right)$ Using the sum of angles identity

$= 4\left(\sin(3x)\cos\left(\frac{\pi}{3}\right) + \cos(3x)\sin\left(\frac{\pi}{3}\right)\right)$ Evaluate the sine and cosine

$= 4\left(\sin(3x)\cdot\frac{1}{2} + \cos(3x)\cdot\frac{\sqrt{3}}{2}\right)$ Distribute and simplify

$= 2\sin(3x) + 2\sqrt{3}\cos(3x)$

Notice that the result is a stretch of the sine added to a different stretch of the cosine, but both have the same horizontal compression, which results in the same period.

We might ask now whether this process can be reversed – can a combination of a sine and cosine of the same period be written as a single sinusoidal function? To explore this, we will look in general at the procedure used in the example above.

$f(x) = A\sin(Bx + C)$ Use the sum of angles identity

$= A(\sin(Bx)\cos(C) + \cos(Bx)\sin(C))$ Distribute the A

$= A\sin(Bx)\cos(C) + A\cos(Bx)\sin(C)$ Rearrange the terms a bit

$= A\cos(C)\sin(Bx) + A\sin(C)\cos(Bx)$

Based on this result, if we have an expression of the form $m\sin(Bx) + n\cos(Bx)$, we could rewrite it as a single sinusoidal function if we can find values A and C so that $m\sin(Bx) + n\cos(Bx) = A\cos(C)\sin(Bx) + A\sin(C)\cos(Bx)$, which will require that:

$$m = A\cos(C) \qquad \frac{m}{A} = \cos(C)$$
$$n = A\sin(C) \quad \text{which can be rewritten as} \quad \frac{n}{A} = \sin(C)$$

To find A,

$$m^2 + n^2 = (A\cos(C))^2 + (A\sin(C))^2$$
$$= A^2\cos^2(C) + A^2\sin^2(C)$$
$$= A^2(\cos^2(C) + \sin^2(C))$$ Apply the Pythagorean Identity and simplify
$$= A^2$$

Rewriting a Sum of Sine and Cosine as a Single Sine

To rewrite $m\sin(Bx) + n\cos(Bx)$ as $A\sin(Bx + C)$

$A^2 = m^2 + n^2$, $\cos(C) = \dfrac{m}{A}$, and $\sin(C) = \dfrac{n}{A}$

You can use either of the last two equations to solve for possible values of C. Since there will usually be two possible solutions, we will need to look at both to determine which quadrant C is in and determine which solution for C satisfies both equations.

Example 6

Rewrite $4\sqrt{3}\sin(2x) - 4\cos(2x)$ as a single sinusoidal function.

Using the formulas above, $A^2 = (4\sqrt{3})^2 + (-4)^2 = 16\cdot 3 + 16 = 64$, so $A = 8$.

Solving for C,

$\cos(C) = \dfrac{4\sqrt{3}}{8} = \dfrac{\sqrt{3}}{2}$, so $C = \dfrac{\pi}{6}$ or $C = \dfrac{11\pi}{6}$.

However, notice $\sin(C) = \dfrac{-4}{8} = -\dfrac{1}{2}$. Sine is negative in the third and fourth quadrant, so the angle that works for both is $C = \dfrac{11\pi}{6}$.

Combining these results gives us the expression

$$8\sin\left(2x + \frac{11\pi}{6}\right)$$

Try it Now

3. Rewrite $-3\sqrt{2}\sin(5x) + 3\sqrt{2}\cos(5x)$ as a single sinusoidal function.

Rewriting a combination of sine and cosine of equal periods as a single sinusoidal function provides an approach for solving some equations.

Example 7

Solve $3\sin(2x)+4\cos(2x)=1$ to find two positive solutions.

Since the sine and cosine have the same period, we can rewrite them as a single sinusoidal function.

$A^2=(3)^2+(4)^2=25$, so $A=5$

$\cos(C)=\frac{3}{5}$, so $C=\cos^{-1}\left(\frac{3}{5}\right)\approx 0.927$ or $C=2\pi-0.927=5.356$

Since $\sin(C)=\frac{4}{5}$, a positive value, we need the angle in the first quadrant, $C=0.927$.

Using this, our equation becomes

$5\sin(2x+0.927)=1$	Divide by 5
$\sin(2x+0.927)=\frac{1}{5}$	Make the substitution $u=2x+0.927$
$\sin(u)=\frac{1}{5}$	The inverse gives a first solution
$u=\sin^{-1}\left(\frac{1}{5}\right)\approx 0.201$	By symmetry, the second solution is
$u=\pi-0.201=2.940$	A third solution would be
$u=2\pi+0.201=6.485$	

Undoing the substitution, we can find two positive solutions for x.

$2x+0.927=0.201$	or	$2x+0.927=2.940$	or	$2x+0.927=6.485$
$2x=-0.726$		$2x=2.013$		$2x=5.558$
$x=-0.363$		$x=1.007$		$x=2.779$

Since the first of these is negative, we eliminate it and keep the two positive solutions, $x=1.007$ and $x=2.779$.

The Product-to-Sum and Sum-to-Product Identities

Identities
The Product-to-Sum Identities
$\sin(\alpha)\cos(\beta) = \frac{1}{2}(\sin(\alpha+\beta) + \sin(\alpha-\beta))$
$\sin(\alpha)\sin(\beta) = \frac{1}{2}(\cos(\alpha-\beta) - \cos(\alpha+\beta))$
$\cos(\alpha)\cos(\beta) = \frac{1}{2}(\cos(\alpha+\beta) + \cos(\alpha-\beta))$

We will prove the first of these, using the sum and difference of angles identities from the beginning of the section. The proofs of the other two identities are similar and are left as an exercise.

Proof of the product-to-sum identity for $\sin(\alpha)\cos(\beta)$

Recall the sum and difference of angles identities from earlier

$$\sin(\alpha+\beta) = \sin(\alpha)\cos(\beta) + \cos(\alpha)\sin(\beta)$$
$$\sin(\alpha-\beta) = \sin(\alpha)\cos(\beta) - \cos(\alpha)\sin(\beta)$$

Adding these two equations, we obtain

$$\sin(\alpha+\beta) + \sin(\alpha-\beta) = 2\sin(\alpha)\cos(\beta)$$

Dividing by 2, we establish the identity

$$\sin(\alpha)\cos(\beta) = \frac{1}{2}(\sin(\alpha+\beta) + \sin(\alpha-\beta))$$

Example 8

Write $\sin(2t)\sin(4t)$ as a sum or difference.

Using the product-to-sum identity for a product of sines

$$\sin(2t)\sin(4t) = \frac{1}{2}(\cos(2t-4t) - \cos(2t+4t))$$

$$= \frac{1}{2}(\cos(-2t) - \cos(6t))$$ If desired, apply the negative angle identity

$$= \frac{1}{2}(\cos(2t) - \cos(6t))$$ Distribute

$$= \frac{1}{2}\cos(2t) - \frac{1}{2}\cos(6t)$$

Try it Now

4. Evaluate $\cos\left(\frac{11\pi}{12}\right)\cos\left(\frac{\pi}{12}\right)$.

Identities

The Sum-to-Product Identities

$$\sin(u)+\sin(v)=2\sin\left(\frac{u+v}{2}\right)\cos\left(\frac{u-v}{2}\right)$$

$$\sin(u)-\sin(v)=2\sin\left(\frac{u-v}{2}\right)\cos\left(\frac{u+v}{2}\right)$$

$$\cos(u)+\cos(v)=2\cos\left(\frac{u+v}{2}\right)\cos\left(\frac{u-v}{2}\right)$$

$$\cos(u)-\cos(v)=-2\sin\left(\frac{u+v}{2}\right)\sin\left(\frac{u-v}{2}\right)$$

We will again prove one of these and leave the rest as an exercise.

Proof of the sum-to-product identity for sine functions

We define two new variables:

$u=\alpha+\beta$

$v=\alpha-\beta$

Adding these equations yields $u+v=2\alpha$, giving $\alpha=\frac{u+v}{2}$

Subtracting the equations yields $u-v=2\beta$, or $\beta=\frac{u-v}{2}$

Substituting these expressions into the product-to-sum identity

$\sin(\alpha)\cos(\beta)=\frac{1}{2}(\sin(\alpha+\beta)+\sin(\alpha-\beta))$ gives

$\sin\left(\frac{u+v}{2}\right)\cos\left(\frac{u-v}{2}\right)=\frac{1}{2}(\sin(u)+\sin(v))$ Multiply by 2 on both sides

$2\sin\left(\frac{u+v}{2}\right)\cos\left(\frac{u-v}{2}\right)=\sin(u)+\sin(v)$ Establishing the identity

Try it Now

5. Notice that, using the negative angle identity, $\sin(u)-\sin(v)=\sin(u)+\sin(-v)$. Use this along with the sum of sines identity to prove the sum-to-product identity for $\sin(u)-\sin(v)$.

Example 9

Evaluate $\cos(15°)-\cos(75°)$.

Using the sum-to-product identity for the difference of cosines,

$\cos(15°)-\cos(75°)$

$=-2\sin\left(\frac{15°+75°}{2}\right)\sin\left(\frac{15°-75°}{2}\right)$ Simplify

$=-2\sin(45°)\sin(-30°)$ Evaluate

$=-2\cdot\frac{\sqrt{2}}{2}\cdot\frac{-1}{2}=\frac{\sqrt{2}}{2}$

Example 10

Prove the identity $\frac{\cos(4t)-\cos(2t)}{\sin(4t)+\sin(2t)}=-\tan(t)$.

Since the left side seems more complicated, we can start there and simplify.

$\frac{\cos(4t)-\cos(2t)}{\sin(4t)+\sin(2t)}$ Use the sum-to-product identities

$=\frac{-2\sin\left(\frac{4t+2t}{2}\right)\sin\left(\frac{4t-2t}{2}\right)}{2\sin\left(\frac{4t+2t}{2}\right)\cos\left(\frac{4t-2t}{2}\right)}$ Simplify

$=\frac{-2\sin(3t)\sin(t)}{2\sin(3t)\cos(t)}$ Simplify further

$=\frac{-\sin(t)}{\cos(t)}$ Rewrite as a tangent

$=-\tan(t)$ Establishing the identity

Example 11

Solve $\sin(\pi t)+\sin(3\pi t)=\cos(\pi t)$ for all solutions with $0 \le t < 2$.

In an equation like th is, it is not immediately obvious how to proceed. One option would be to combine the two sine functions on the left side of the equation. Another would be to move the cosine to the left side of the equation, and combine it with one of the sines. For no particularly good reason, we'll begin by combining the sines on the left side of the equation and see how things work out.

$\sin(\pi t)+\sin(3\pi t)=\cos(\pi t)$ Apply the sum to product identity on the left

$2\sin\left(\frac{\pi t+3\pi t}{2}\right)\cos\left(\frac{\pi t-3\pi t}{2}\right)=\cos(\pi t)$ Simplify

$2\sin(2\pi t)\cos(-\pi t)=\cos(\pi t)$ Apply the negative angle identity

$2\sin(2\pi t)\cos(\pi t)=\cos(\pi t)$ Rearrange the equation to be 0 on one side

$2\sin(2\pi t)\cos(\pi t)-\cos(\pi t)=0$ Factor out the cosine

$\cos(\pi t)(2\sin(2\pi t)-1)=0$

Using the Zero Product Theorem we know that at least one of the two factors must be zero. The first factor, $\cos(\pi t)$, has period $P=\frac{2\pi}{\pi}=2$, so the solution interval of $0 \le t < 2$ represents one full cycle of this function.

$\cos(\pi t)=0$ Substitute $u=\pi t$

$\cos(u)=0$ On one cycle, this has solutions

$u=\frac{\pi}{2}$ or $u=\frac{3\pi}{2}$ Undo the substitution

$\pi t=\frac{\pi}{2}$, so $t=\frac{1}{2}$

$\pi t=\frac{3\pi}{2}$, so $t=\frac{3}{2}$

The second factor, $2\sin(2\pi t)-1$, has period of $P=\frac{2\pi}{2\pi}=1$, so the solution interval $0 \le t < 2$ contains two complete cycles of this function.

$2\sin(2\pi t)-1=0$ Isolate the sine

$\sin(2\pi t)=\frac{1}{2}$ Substitute $u=2\pi t$

$\sin(u) = \frac{1}{2}$ On one cycle, this has solutions

$u = \frac{\pi}{6}$ or $u = \frac{5\pi}{6}$ On the second cycle, the solutions are

$u = 2\pi + \frac{\pi}{6} = \frac{13\pi}{6}$ or $u = 2\pi + \frac{5\pi}{6} = \frac{17\pi}{6}$ Undo the substitution

$2\pi t = \frac{\pi}{6}$, so $t = \frac{1}{12}$

$2\pi t = \frac{5\pi}{6}$, so $t = \frac{5}{12}$

$2\pi t = \frac{13\pi}{6}$, so $t = \frac{13}{12}$

$2\pi t = \frac{17\pi}{6}$, so $t = \frac{17}{12}$

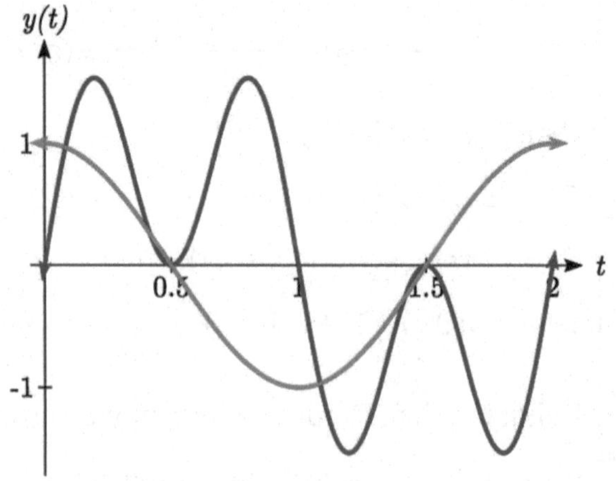

Altogether, we found six solutions on $0 \le t < 2$, which we can confirm by looking at the graph.

$t = \frac{1}{12}, \frac{5}{12}, \frac{1}{2}, \frac{13}{12}, \frac{3}{2}, \frac{17}{12}$

y(t)
1
-1
0.5
1
1.5
2
t

Important Topics of This Section
The sum and difference identities
Combining waves of equal periods
Product-to-sum identities
Sum-to-product identities
Completing proofs

Try it Now Answers

1.
$$\cos(\alpha + \beta) = \cos(\alpha - (-\beta))$$
$$\cos(\alpha)\cos(-\beta) + \sin(\alpha)\sin(-\beta)$$
$$\cos(\alpha)\cos(\beta) + \sin(\alpha)(-\sin(\beta))$$
$$\cos(\alpha)\cos(\beta) - \sin(\alpha)\sin(\beta)$$

2. $\sin\left(\frac{\pi}{12}\right)=\sin\left(\frac{\pi}{3}-\frac{\pi}{4}\right)=\sin\left(\frac{\pi}{3}\right)\cos\left(\frac{\pi}{4}\right)-\cos\left(\frac{\pi}{3}\right)\sin\left(\frac{\pi}{4}\right)$
$=\frac{\sqrt{3}}{2}\frac{\sqrt{2}}{2}-\frac{1}{2}\frac{\sqrt{2}}{2}\ \frac{\sqrt{6}-\sqrt{2}}{4}$

3. $A^2=\left(-3\sqrt{2}\right)^2+\left(3\sqrt{2}\right)^2=36$. $A=6$
$\cos(C)=\frac{-3\sqrt{2}}{6}=\frac{-\sqrt{2}}{2}$, $\sin(C)=\frac{3\sqrt{2}}{6}=\frac{\sqrt{2}}{2}$. $C=\frac{3\pi}{4}$
$6\sin\left(5x+\frac{3\pi}{4}\right)$

4. $\cos\left(\frac{11\pi}{12}\right)\cos\left(\frac{\pi}{12}\right)=\frac{1}{2}\left(\cos\left(\frac{11\pi}{12}+\frac{\pi}{12}\right)+\cos\left(\frac{11\pi}{12}-\frac{\pi}{12}\right)\right)$
$=\frac{1}{2}\left(\cos(\pi)+\cos\left(\frac{5\pi}{6}\right)\right)=\frac{1}{2}\left(-1-\frac{\sqrt{3}}{2}\right)$
$=\frac{-2-\sqrt{3}}{4}$

5. $\sin(u)-\sin(v)$ — Use negative angle identity for sine
$\sin(u)+\sin(-v)$ — Use sum-to-product identity for sine
$2\sin\left(\frac{u+(-v)}{2}\right)\cos\left(\frac{u-(-v)}{2}\right)$ — Eliminate the parenthesis
$2\sin\left(\frac{u-v}{2}\right)\cos\left(\frac{u+v}{2}\right)$ — Establishing the identity

Section 7.2 Exercises

Find an exact value for each of the following.

1. $\sin(75°)$
2. $\sin(195°)$
3. $\cos(165°)$
4. $\cos(345°)$
5. $\cos\left(\frac{7\pi}{12}\right)$
6. $\cos\left(\frac{\pi}{12}\right)$
7. $\sin\left(\frac{5\pi}{12}\right)$
8. $\sin\left(\frac{11\pi}{12}\right)$

Rewrite in terms of $\sin(x)$ and $\cos(x)$.

9. $\sin\left(x+\frac{11\pi}{6}\right)$
10. $\sin\left(x-\frac{3\pi}{4}\right)$
11. $\cos\left(x-\frac{5\pi}{6}\right)$
12. $\cos\left(x+\frac{2\pi}{3}\right)$

Simplify each expression.

13. $\csc\left(\frac{\pi}{2}-t\right)$
14. $\sec\left(\frac{\pi}{2}-w\right)$
15. $\cot\left(\frac{\pi}{2}-x\right)$
16. $\tan\left(\frac{\pi}{2}-x\right)$

Rewrite the product as a sum.

17. $16\sin(16x)\sin(11x)$
18. $20\cos(36t)\cos(6t)$
19. $2\sin(5x)\cos(3x)$
20. $10\cos(5x)\sin(10x)$

Rewrite the sum as a product.

21. $\cos(6t)+\cos(4t)$
22. $\cos(6u)+\cos(4u)$
23. $\sin(3x)+\sin(7x)$
24. $\sin(h)+\sin(3h)$

25. Given $\sin(a)=\frac{2}{3}$ and $\cos(b)=-\frac{1}{4}$, with a and b both in the interval $\left[\frac{\pi}{2},\pi\right)$:
    a. Find $\sin(a+b)$
    b. Find $\cos(a-b)$

26. Given $\sin(a)=\frac{4}{5}$ and $\cos(b)=\frac{1}{3}$, with a and b both in the interval $\left[0,\frac{\pi}{2}\right)$:
    a. Find $\sin(a-b)$
    b. Find $\cos(a+b)$

Solve each equation for all solutions.

27. $\sin(3x)\cos(6x)-\cos(3x)\sin(6x)=-0.9$
28. $\sin(6x)\cos(11x)-\cos(6x)\sin(11x)=-0.1$
29. $\cos(2x)\cos(x)+\sin(2x)\sin(x)=1$
30. $\cos(5x)\cos(3x)-\sin(5x)\sin(3x)=\frac{\sqrt{3}}{2}$

Solve each equation for all solutions.

31. $\cos(5x) = -\cos(2x)$

32. $\sin(5x) = \sin(3x)$

33. $\cos(6\theta) - \cos(2\theta) = \sin(4\theta)$

34. $\cos(8\theta) - \cos(2\theta) = \sin(5\theta)$

Rewrite as a single function of the form $A\sin(Bx + C)$.

35. $4\sin(x) - 6\cos(x)$

36. $-\sin(x) - 5\cos(x)$

37. $5\sin(3x) + 2\cos(3x)$

38. $-3\sin(5x) + 4\cos(5x)$

Solve for the first two positive solutions.

39. $-5\sin(x) + 3\cos(x) = 1$

40. $3\sin(x) + \cos(x) = 2$

41. $3\sin(2x) - 5\cos(2x) = 3$

42. $-3\sin(4x) - 2\cos(4x) = 1$

Simplify.

43. $\dfrac{\sin(7t) + \sin(5t)}{\cos(7t) + \cos(5t)}$

44. $\dfrac{\sin(9t) - \sin(3t)}{\cos(9t) + \cos(3t)}$

Prove the identity.

44. $\tan\left(x + \dfrac{\pi}{4}\right) = \dfrac{\tan(x) + 1}{1 - \tan(x)}$

45. $\tan\left(\dfrac{\pi}{4} - t\right) = \dfrac{1 - \tan(t)}{1 + \tan(t)}$

46. $\cos(a + b) + \cos(a - b) = 2\cos(a)\cos(b)$

47. $\dfrac{\cos(a + b)}{\cos(a - b)} = \dfrac{1 - \tan(a)\tan(b)}{1 + \tan(a)\tan(b)}$

48. $\dfrac{\tan(a + b)}{\tan(a - b)} = \dfrac{\sin(a)\cos(a) + \sin(b)\cos(b)}{\sin(a)\cos(a) - \sin(b)\cos(b)}$

49. $2\sin(a + b)\sin(a - b) = \cos(2b) - \cos(2a)$

50. $\dfrac{\sin(x) + \sin(y)}{\cos(x) + \cos(y)} = \tan\left(\dfrac{1}{2}(x + y)\right)$

Prove the identity.

51. $\dfrac{\cos(a+b)}{\cos(a)\cos(b)} = 1 - \tan(a)\tan(b)$

52. $\cos(x+y)\cos(x-y) = \cos^2 x - \sin^2 y$

53. Use the sum and difference identities to establish the product-to-sum identity
$\sin(\alpha)\sin(\beta) = \frac{1}{2}(\cos(\alpha - \beta) - \cos(\alpha + \beta))$

54. Use the sum and difference identities to establish the product-to-sum identity
$\cos(\alpha)\cos(\beta) = \frac{1}{2}(\cos(\alpha + \beta) + \cos(\alpha - \beta))$

55. Use the product-to-sum identities to establish the sum-to-product identity
$\cos(u) + \cos(v) = 2\cos\left(\frac{u+v}{2}\right)\cos\left(\frac{u-v}{2}\right)$

56. Use the product-to-sum identities to establish the sum-to-product identity
$\cos(u) - \cos(v) = -2\sin\left(\frac{u+v}{2}\right)\sin\left(\frac{u-v}{2}\right)$

Section 7.3 Double Angle Identities

Two special cases of the sum of angles identities arise often enough that we choose to state these identities separately.

Identities
The double angle identities
$\sin(2\alpha) = 2\sin(\alpha)\cos(\alpha)$
$\cos(2\alpha) = \cos^2(\alpha) - \sin^2(\alpha)$
$= 1 - 2\sin^2(\alpha)$
$= 2\cos^2(\alpha) - 1$

These identities follow from the sum of angles identities.

Proof of the sine double angle identity

$\sin(2\alpha)$

$= \sin(\alpha + \alpha)$	Apply the sum of angles identity
$= \sin(\alpha)\cos(\alpha) + \cos(\alpha)\sin(\alpha)$	Simplify
$= 2\sin(\alpha)\cos(\alpha)$	Establishing the identity

Try it Now

1. Show $\cos(2\alpha) = \cos^2(\alpha) - \sin^2(\alpha)$ by using the sum of angles identity for cosine.

For the cosine double angle identity, there are three forms of the identity stated because the basic form, $\cos(2\alpha) = \cos^2(\alpha) - \sin^2(\alpha)$, can be rewritten using the Pythagorean Identity. Rearranging the Pythagorean Identity results in the equality $\cos^2(\alpha) = 1 - \sin^2(\alpha)$, and by substituting this into the basic double angle identity, we obtain the second form of the double angle identity.

$\cos(2\alpha) = \cos^2(\alpha) - \sin^2(\alpha)$	Substituting using the Pythagorean identity
$\cos(2\alpha) = 1 - \sin^2(\alpha) - \sin^2(\alpha)$	Simplifying
$\cos(2\alpha) = 1 - 2\sin^2(\alpha)$	

Example 1

If $\sin(\theta) = \frac{3}{5}$ and θ is in the second quadrant, find exact values for $\sin(2\theta)$ and $\cos(2\theta)$.

To evaluate $\cos(2\theta)$, since we know the value for $\sin(\theta)$ we can use the version of the double angle that only involves sine.

$$\cos(2\theta) = 1 - 2\sin^2(\theta) = 1 - 2\left(\frac{3}{5}\right)^2 = 1 - \frac{18}{25} = \frac{7}{25}$$

Since the double angle for sine involves both sine and cosine, we'll need to first find $\cos(\theta)$, which we can do using the Pythagorean Identity.

$$\sin^2(\theta) + \cos^2(\theta) = 1$$

$$\left(\frac{3}{5}\right)^2 + \cos^2(\theta) = 1$$

$$\cos^2(\theta) = 1 - \frac{9}{25}$$

$$\cos(\theta) = \pm\sqrt{\frac{16}{25}} = \pm\frac{4}{5}$$

Since θ is in the second quadrant, we know that $\cos(\theta) < 0$, so

$$\cos(\theta) = -\frac{4}{5}$$

Now we can evaluate the sine double angle

$$\sin(2\theta) = 2\sin(\theta)\cos(\theta) = 2\left(\frac{3}{5}\right)\left(-\frac{4}{5}\right) = -\frac{24}{25}$$

Example 2

Simplify the expressions

a) $2\cos^2(12°) - 1$ b) $8\sin(3x)\cos(3x)$

a) Notice that the expression is in the same form as one version of the double angle identity for cosine: $\cos(2\theta) = 2\cos^2(\theta) - 1$. Using this,

$$2\cos^2(12°) - 1 = \cos(2 \cdot 12°) = \cos(24°)$$

b) This expression looks similar to the result of the double angle identity for sine.

$8\sin(3x)\cos(3x)$ Factoring a 4 out of the original expression

$4 \cdot 2\sin(3x)\cos(3x)$ Applying the double angle identity

$4\sin(6x)$

We can use the double angle identities to simplify expressions and prove identities.

Example 2

Simplify $\dfrac{\cos(2t)}{\cos(t)-\sin(t)}$.

With three choices for how to rewrite the double angle, we need to consider which will be the most useful. To simplify this expression, it would be great if the denominator would cancel with something in the numerator, which would require a factor of $\cos(t)-\sin(t)$ in the numerator, which is most likely to occur if we rewrite the numerator with a mix of sine and cosine.

$\dfrac{\cos(2t)}{\cos(t)-\sin(t)}$ Apply the double angle identity

$=\dfrac{\cos^2(t)-\sin^2(t)}{\cos(t)-\sin(t)}$ Factor the numerator

$=\dfrac{(\cos(t)-\sin(t))(\cos(t)+\sin(t))}{\cos(t)-\sin(t)}$ Cancelling the common factor

$=\cos(t)+\sin(t)$ Resulting in the most simplified form

Example 3

Prove $\sec(2\alpha)=\dfrac{\sec^2(\alpha)}{2-\sec^2(\alpha)}$.

Since the right side seems a bit more complicated than the left side, we begin there.

$\dfrac{\sec^2(\alpha)}{2-\sec^2(\alpha)}$ Rewrite the secants in terms of cosine

$$=\frac{\dfrac{1}{\cos^2(\alpha)}}{2-\dfrac{1}{\cos^2(\alpha)}}$$

At this point, we could rewrite the bottom with common denominators, subtract the terms, invert and multiply, then simplify. Alternatively, we can multiple both the top and bottom by $\cos^2(\alpha)$, the common denominator:

$$=\frac{\dfrac{1}{\cos^2(\alpha)}\cdot\cos^2(\alpha)}{\left(2-\dfrac{1}{\cos^2(\alpha)}\right)\cdot\cos^2(\alpha)}$$ Distribute on the bottom

$= \dfrac{\dfrac{\cos^2(\alpha)}{\cos^2(\alpha)}}{2\cos^2(\alpha) - \dfrac{\cos^2(\alpha)}{\cos^2(\alpha)}}$	Simplify
$= \dfrac{1}{2\cos^2(\alpha)-1}$	Rewrite the denominator as a double angle
$= \dfrac{1}{\cos(2\alpha)}$	Rewrite as a secant
$= \sec(2\alpha)$	Establishing the identity

Try it Now

2. Use an identity to find the exact value of $\cos^2(75°) - \sin^2(75°)$.

As with other identities, we can also use the double angle identities for solving equations.

Example 4

Solve $\cos(2t) = \cos(t)$ for all solutions with $0 \le t < 2\pi$.

In general when solving trig equations, it makes things more complicated when we have a mix of sines and cosines and when we have a mix of functions with different periods. In this case, we can use a double angle identity to rewrite the $\cos(2t)$. When choosing which form of the double angle identity to use, we notice that we have a cosine on the right side of the equation. We try to limit our equation to one trig function, which we can do by choosing the version of the double angle formula for cosine that only involves cosine.

$\cos(2t) = \cos(t)$	Apply the double angle identity
$2\cos^2(t) - 1 = \cos(t)$	This is quadratic in cosine, so make one side 0
$2\cos^2(t) - \cos(t) - 1 = 0$	Factor
$(2\cos(t)+1)(\cos(t)-1) = 0$	Break this apart to solve each part separately

$2\cos(t) + 1 = 0$ or $\cos(t) - 1 = 0$

$\cos(t) = -\dfrac{1}{2}$ or $\cos(t) = 1$

$t = \dfrac{2\pi}{3}$ or $t = \dfrac{4\pi}{3}$ or $t = 0$

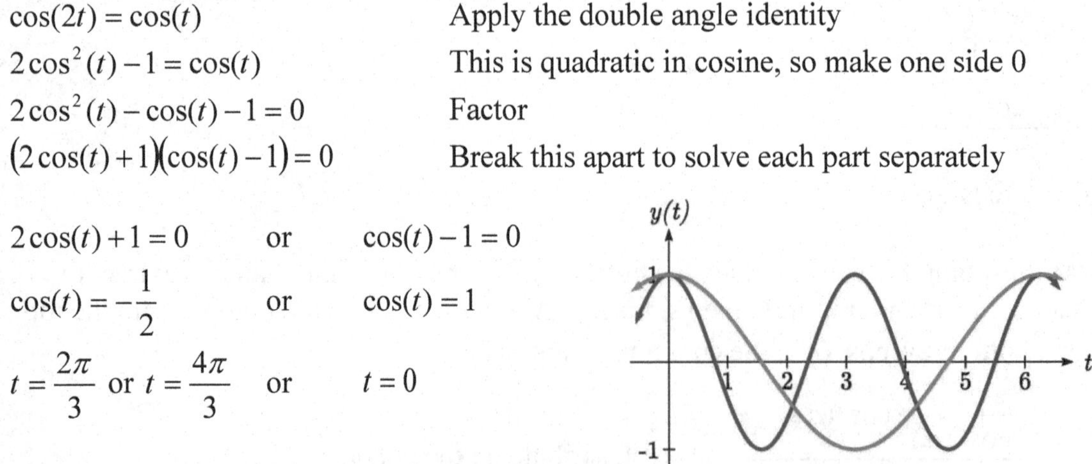

Looking at a graph of $\cos(2t)$ and $\cos(t)$ shown together, we can verify that these three solutions on $[0, 2\pi)$ seem reasonable.

Example 5

A cannonball is fired with velocity of 100 meters per second. If it is launched at an angle of θ, the vertical component of the velocity will be $100\sin(\theta)$ and the horizontal component will be $100\cos(\theta)$. Ignoring wind resistance, the height of the cannonball will follow the equation $h(t) = -4.9t^2 + 100\sin(\theta)t$ and horizontal position will follow the equation $x(t) = 100\cos(\theta)t$. If you want to hit a target 900 meters away, at what angle should you aim the cannon?

To hit the target 900 meters away, we want $x(t) = 900$ at the time when the cannonball hits the ground, when $h(t) = 0$. To solve this problem, we will first solve for the time, t, when the cannonball hits the ground. Our answer will depend upon the angle θ.

$h(t) = 0$

$-4.9t^2 + 100\sin(\theta)t = 0$ Factor

$t(-4.9t + 100\sin(\theta)) = 0$ Break this apart to find two solutions

$t = 0$ or $-4.9t + 100\sin(\theta) = 0$ Solve for t

$-4.9t = -100\sin(\theta)$

$$t = \frac{100\sin(\theta)}{4.9}$$

This shows that the height is 0 twice, once at $t = 0$ when the cannonball is fired, and again when the cannonball hits the ground after flying through the air. This second value of t gives the time when the ball hits the ground in terms of the angle θ. We want the horizontal distance $x(t)$ to be 900 when the ball hits the ground, in other words when $t = \frac{100\sin(\theta)}{4.9}$.

Since the target is 900 m away we start with

$x(t) = 900$ Use the formula for $x(t)$

$100\cos(\theta)t = 900$ Substitute the desired time, t from above

$100\cos(\theta)\frac{100\sin(\theta)}{4.9} = 900$ Simplify

$\frac{100^2}{4.9}\cos(\theta)\sin(\theta) = 900$ Isolate the cosine and sine product

$$\cos(\theta)\sin(\theta) = \frac{900(4.9)}{100^2}$$

The left side of this equation almost looks like the result of the double angle identity for sine: $\sin(2\theta) = 2\sin(\theta)\cos(\theta)$.
Multiplying both sides of our equation by 2,

$2\cos(\theta)\sin(\theta) = \dfrac{2(900)(4.9)}{100^2}$ Using the double angle identity on the left

$\sin(2\theta) = \dfrac{2(900)(4.9)}{100^2}$ Use the inverse sine

$2\theta = \sin^{-1}\left(\dfrac{2(900)(4.9)}{100^2}\right) \approx 1.080$ Divide by 2

$\theta = \dfrac{1.080}{2} = 0.540$, or about 30.94 degrees

Power Reduction and Half Angle Identities

Another use of the cosine double angle identities is to use them in reverse to rewrite a squared sine or cosine in terms of the double angle. Starting with one form of the cosine double angle identity:

$\cos(2\alpha) = 2\cos^2(\alpha) - 1$ Isolate the cosine squared term

$\cos(2\alpha) + 1 = 2\cos^2(\alpha)$ Add 1

$\cos^2(\alpha) = \dfrac{\cos(2\alpha)+1}{2}$ Divide by 2

$\cos^2(\alpha) = \dfrac{\cos(2\alpha)+1}{2}$ This is called a **power reduction identity**

Try it Now

3. Use another form of the cosine double angle identity to prove the identity
$\sin^2(\alpha) = \dfrac{1-\cos(2\alpha)}{2}$.

The cosine double angle identities can also be used in reverse for evaluating angles that are half of a common angle. Building from our formula $\cos^2(\alpha) = \dfrac{\cos(2\alpha)+1}{2}$, if we let $\theta = 2\alpha$, then $\alpha = \dfrac{\theta}{2}$ this identity becomes $\cos^2\left(\dfrac{\theta}{2}\right) = \dfrac{\cos(\theta)+1}{2}$. Taking the square root, we obtain

$\cos\left(\dfrac{\theta}{2}\right) = \pm\sqrt{\dfrac{\cos(\theta)+1}{2}}$, where the sign is determined by the quadrant.

This is called a **half-angle identity**.

Try it Now

4. Use your results from the last Try it Now to prove the identity
$$\sin\left(\frac{\theta}{2}\right) = \pm\sqrt{\frac{1-\cos(\theta)}{2}}.$$

Identities

Half-Angle Identities

$$\cos\left(\frac{\theta}{2}\right) = \pm\sqrt{\frac{\cos(\theta)+1}{2}} \qquad \sin\left(\frac{\theta}{2}\right) = \pm\sqrt{\frac{1-\cos(\theta)}{2}}$$

Power Reduction Identities

$$\cos^2(\alpha) = \frac{\cos(2\alpha)+1}{2} \qquad \sin^2(\alpha) = \frac{1-\cos(2\alpha)}{2}$$

Since these identities are easy to derive from the double-angle identities, the power reduction and half-angle identities are not ones you should need to memorize separately.

Example 6

Rewrite $\cos^4(x)$ without any powers.

$$\cos^4(x) = \left(\cos^2(x)\right)^2$$ Using the power reduction formula

$$= \left(\frac{\cos(2x)+1}{2}\right)^2$$ Square the numerator and denominator

$$= \frac{\left(\cos(2x)+1\right)^2}{4}$$ Expand the numerator

$$= \frac{\cos^2(2x)+2\cos(2x)+1}{4}$$ Split apart the fraction

$$= \frac{\cos^2(2x)}{4} + \frac{2\cos(2x)}{4} + \frac{1}{4}$$ Apply the formula above to $\cos^2(2x)$

$$\cos^2(2x) = \frac{\cos(2\cdot 2x)+1}{2}$$

$$= \frac{\left(\frac{\cos(4x)+1}{2}\right)}{4} + \frac{2\cos(2x)}{4} + \frac{1}{4}$$ Simplify

$$= \frac{\cos(4x)}{8} + \frac{1}{8} + \frac{1}{2}\cos(2x) + \frac{1}{4}$$ Combine the constants

$$= \frac{\cos(4x)}{8} + \frac{1}{2}\cos(2x) + \frac{3}{8}$$

Example 7

Find an exact value for $\cos(15°)$.

Since 15 degrees is half of 30 degrees, we can use our result from above:

$$\cos(15°) = \cos\left(\frac{30°}{2}\right) = \pm\sqrt{\frac{\cos(30°)+1}{2}}$$

We can evaluate the cosine. Since 15 degrees is in the first quadrant, we need the positive result.

$$\sqrt{\frac{\cos(30°)+1}{2}} = \sqrt{\frac{\frac{\sqrt{3}}{2}+1}{2}}$$

$$= \sqrt{\frac{\sqrt{3}}{4} + \frac{1}{2}}$$

Important Topics of This Section
Double angle identity
Power reduction identity
Half angle identity
Using identities
Simplify equations
Prove identities
Solve equations

Try it Now Answers

1.
$$\cos(2\alpha) = \cos(\alpha + \alpha)$$
$$\cos(\alpha)\cos(\alpha) - \sin(\alpha)\sin(\alpha)$$
$$\cos^2(\alpha) - \sin^2(\alpha)$$

2. $\cos^2(75°) - \sin^2(75°) = \cos(2 \cdot 75°) = \cos(150°) = \dfrac{-\sqrt{3}}{2}$

3.
$$\frac{1-\cos(2\alpha)}{2}$$
$$\frac{1-\left(\cos^2(\alpha)-\sin^2(\alpha)\right)}{2}$$
$$\frac{1-\cos^2(\alpha)+\sin^2(\alpha)}{2}$$
$$\frac{\sin^2(\alpha)+\sin^2(\alpha)}{2}$$
$$\frac{2\sin^2(\alpha)}{2} = \sin^2(\alpha)$$

4.
$$\sin^2(\alpha) = \frac{1-\cos(2\alpha)}{2}$$
$$\sin(\alpha) = \pm\sqrt{\frac{1-\cos(2\alpha)}{2}}$$
$$\alpha = \frac{\theta}{2}$$
$$\sin\left(\frac{\theta}{2}\right) = \pm\sqrt{\frac{1-\cos\left(2\left(\frac{\theta}{2}\right)\right)}{2}}$$
$$\sin\left(\frac{\theta}{2}\right) = \pm\sqrt{\frac{1-\cos(\theta)}{2}}$$

Section 7.3 Exercises

1. If $\sin(x) = \frac{1}{8}$ and x is in quadrant I, then find exact values for (without solving for x):

   a. $\sin(2x)$ b. $\cos(2x)$ c. $\tan(2x)$

2. If $\cos(x) = \frac{2}{3}$ and x is in quadrant I, then find exact values for (without solving for x):

   a. $\sin(2x)$ b. $\cos(2x)$ c. $\tan(2x)$

Simplify each expression.

3. $\cos^2(28°) - \sin^2(28°)$
4. $2\cos^2(37°) - 1$
5. $1 - 2\sin^2(17°)$
6. $\cos^2(37°) - \sin^2(37°)$
7. $\cos^2(9x) - \sin^2(9x)$
8. $\cos^2(6x) - \sin^2(6x)$
9. $4\sin(8x)\cos(8x)$
10. $6\sin(5x)\cos(5x)$

Solve for all solutions on the interval $[0, 2\pi)$.

11. $6\sin(2t) + 9\sin(t) = 0$
12. $2\sin(2t) + 3\cos(t) = 0$
13. $9\cos(2\theta) = 9\cos^2(\theta) - 4$
14. $8\cos(2\alpha) = 8\cos^2(\alpha) - 1$
15. $\sin(2t) = \cos(t)$
16. $\cos(2t) = \sin(t)$
17. $\cos(6x) - \cos(3x) = 0$
18. $\sin(4x) - \sin(2x) = 0$

Use a double angle, half angle, or power reduction formula to rewrite without exponents.

19. $\cos^2(5x)$
20. $\cos^2(6x)$
21. $\sin^4(8x)$
22. $\sin^4(3x)$
23. $\cos^2 x \sin^4 x$
24. $\cos^4 x \sin^2 x$

25. If $\csc(x) = 7$ and $90° < x < 180°$, then find exact values for (without solving for x):

   a. $\sin\left(\frac{x}{2}\right)$ b. $\cos\left(\frac{x}{2}\right)$ c. $\tan\left(\frac{x}{2}\right)$

26. If $\sec(x) = 4$ and $270° < x < 360°$, then find exact values for (without solving for x):

   a. $\sin\left(\frac{x}{2}\right)$ b. $\cos\left(\frac{x}{2}\right)$ c. $\tan\left(\frac{x}{2}\right)$

Prove the identity.

27. $(\sin t - \cos t)^2 = 1 - \sin(2t)$

28. $(\sin^2 x - 1)^2 = \cos(2x) + \sin^4 x$

29. $\sin(2x) = \dfrac{2\tan(x)}{1 + \tan^2(x)}$

30. $\tan(2x) = \dfrac{2\sin(x)\cos(x)}{2\cos^2(x) - 1}$

31. $\cot(x) - \tan(x) = 2\cot(2x)$

32. $\dfrac{\sin(2\theta)}{1 + \cos(2\theta)} = \tan(\theta)$

33. $\cos(2\alpha) = \dfrac{1 - \tan^2(\alpha)}{1 + \tan^2(\alpha)}$

34. $\dfrac{1 + \cos(2t)}{\sin(2t) - \cos(t)} = \dfrac{2\cos(t)}{2\sin(t) - 1}$

35. $\sin(3x) = 3\sin(x)\cos^2(x) - \sin^3(x)$

36. $\cos(3x) = \cos^3(x) - 3\sin^2(x)\cos(x)$

Section 7.4 Modeling Changing Amplitude and Midline

While sinusoidal functions can model a variety of behaviors, it is often necessary to combine sinusoidal functions with linear and exponential curves to model real applications and behaviors. We begin this section by looking at changes to the midline of a sinusoidal function. Recall that the midline describes the middle, or average value, of the sinusoidal function.

Changing Midlines

Example 1

A population of elk currently averages 2000 elk, and that average has been growing by 4% each year. Due to seasonal fluctuation, the population oscillates from 50 below average in the winter up to 50 above average in the summer. Find a function that models the number of elk after *t* years, starting in the winter.

There are two components to the behavior of the elk population: the changing average, and the oscillation. The average is an exponential growth, starting at 2000 and growing by 4% each year. Writing a formula for this:

$average = initial(1+r)^t = 2000(1+0.04)^t$

For the oscillation, since the population oscillates 50 above and below average, the amplitude will be 50. Since it takes one year for the population to cycle, the period is 1. We find the value of the horizontal stretch coefficient $B = \dfrac{\text{original period}}{\text{new period}} = \dfrac{2\pi}{1} = 2\pi$.

The function starts in winter, so the shape of the function will be a negative cosine, since it starts at the lowest value.

Putting it all together, the equation would be:

$P(t) = -50\cos(2\pi t) + midline$

Since the midline represents the average population, we substitute in the exponential function into the population equation to find our final equation:

$P(t) = -50\cos(2\pi t) + 2000(1+0.04)^t$

This is an example of changing midline – in this case an exponentially changing midline.

Changing Midline
A function of the form $f(t) = A\sin(Bt) + g(t)$ will oscillate above and below the average given by the function *g(t)*.

Changing midlines can be exponential, linear, or any other type of function. Here are some examples:

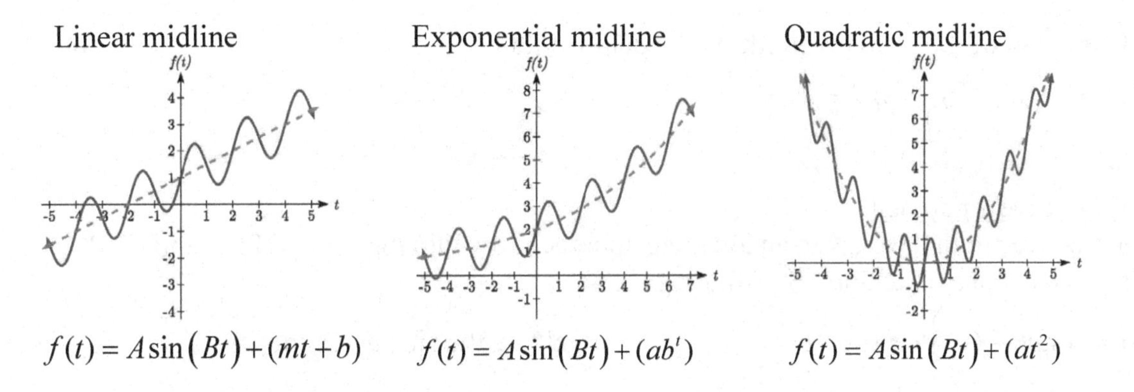

Example 2

Find a function with linear midline of the form $f(t) = A\sin\left(\frac{\pi}{2}t\right) + mt + b$ that will pass through the points given below.

t	0	1	2	3
f(t)	5	10	9	8

Since we are given the value of the horizontal compression coefficient we can calculate the period of this function: new period $= \dfrac{\text{original period}}{B} = \dfrac{2\pi}{\pi/2} = 4$.

Since the sine function is at the midline at the beginning of a cycle and halfway through a cycle, we would expect this function to be at the midline at $t = 0$ and $t = 2$, since 2 is half the full period of 4. Based on this, we expect the points (0, 5) and (2, 9) to be points on the midline. We can clearly see that this is not a constant function and so we use the two points to calculate a linear function: $midline = mt + b$. From these two points we can calculate a slope:

$$m = \frac{9-5}{2-0} = \frac{4}{2} = 2$$

Combining this with the initial value of 5, we have the midline: $midline = 2t + 5$.

The full function will have form $f(t) = A\sin\left(\frac{\pi}{2}t\right) + 2t + 5$. To find the amplitude, we can plug in a point we haven't already used, such as (1, 10).

$10 = A\sin\left(\frac{\pi}{2}(1)\right) + 2(1) + 5$ Evaluate the sine and combine like terms

$10 = A + 7$

$A = 3$

A function of the form given fitting the data would be

$f(t) = 3\sin\left(\frac{\pi}{2}t\right) + 2t + 5$

Alternative Approach

Notice we could have taken an alternate approach by plugging points (0, 5) and (2, 9) into the original equation. Substituting (0, 5),

$5 = A\sin\left(\frac{\pi}{2}(0)\right) + m(0) + b$ Evaluate the sine and simplify

$5 = b$

Substituting (2, 9)

$9 = A\sin\left(\frac{\pi}{2}(2)\right) + m(2) + 5$ Evaluate the sine and simplify

$9 = 2m + 5$

$4 = 2m$

$m = 2$, as we found above. Now we can proceed to find A the same way we did before.

Example 3

The number of tourists visiting a ski and hiking resort averages 4000 people annually and oscillates seasonally, 1000 above and below the average. Due to a marketing campaign, the average number of tourists has been increasing by 200 each year. Write an equation for the number of tourists after t years, beginning at the peak season.

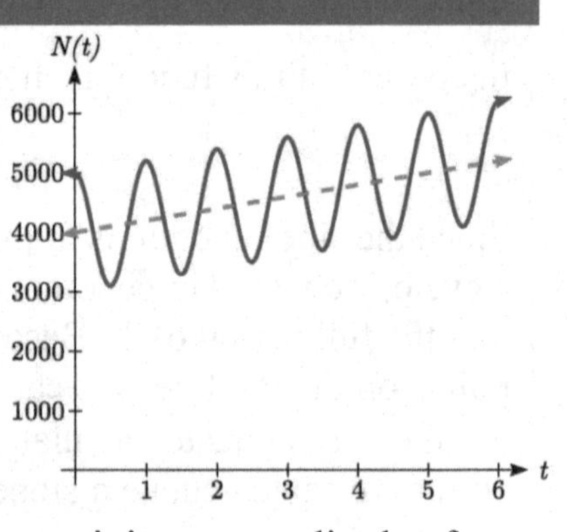

Again there are two components to this problem: the oscillation and the average. For the oscillation, the number of tourists oscillates 1000 above and below average, giving an amplitude of 1000. Since the oscillation is seasonal, it has a period of 1 year. Since we are given a starting point of "peak season", we will model this scenario with a cosine function. So far, this gives an equation in the form $N(t) = 1000\cos(2\pi t) + midline$.

The average is currently 4000, and is increasing by 200 each year. This is a constant rate of change, so this is linear growth, $average = 4000 + 200t$. This function will act as the midline.

Combining these two pieces gives a function for the number of tourists:
$N(t) = 1000\cos(2\pi t) + 4000 + 200t$

Try it Now
1. Given the function $g(x) = (x^2 - 1) + 8\cos(x)$, describe the midline and amplitude using words.

Changing Amplitude

There are also situations in which the amplitude of a sinusoidal function does not stay constant. Back in Chapter 6, we modeled the motion of a spring using a sinusoidal function, but had to ignore friction in doing so. If there were friction in the system, we would expect the amplitude of the oscillation to decrease over time. In the equation $f(t) = A\sin(Bt) + k$, A gives the amplitude of the oscillation, we can allow the amplitude to change by replacing this constant A with a function $A(t)$.

Changing Amplitude
A function of the form $f(t) = A(t)\sin(Bt) + k$ will oscillate above and below the midline with an amplitude given by $A(t)$.

Here are some examples:

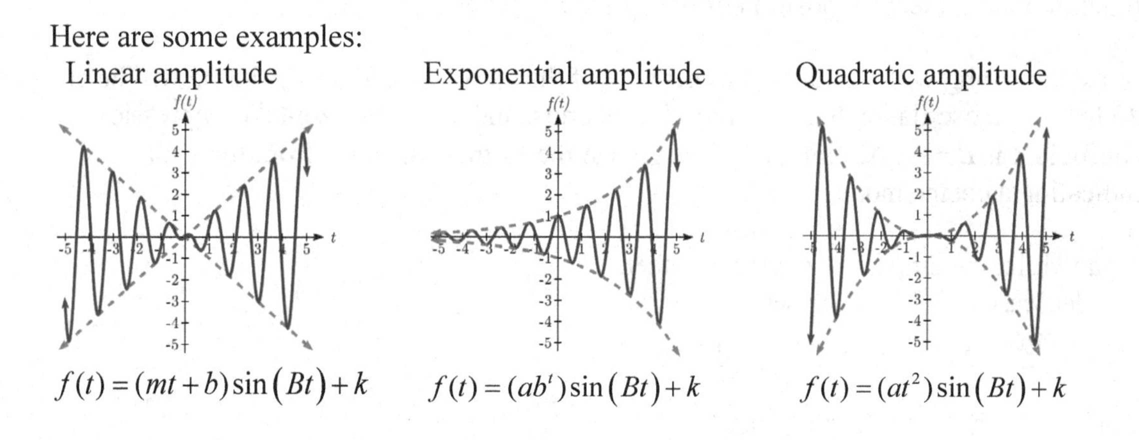

$f(t) = (mt + b)\sin(Bt) + k$ $\quad$ $f(t) = (ab^t)\sin(Bt) + k$ $\quad$ $f(t) = (at^2)\sin(Bt) + k$

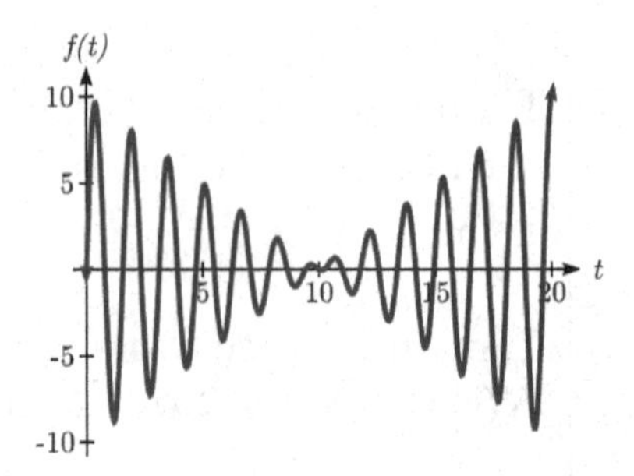

When thinking about a spring with amplitude decreasing over time, it is tempting to use the simplest tool for the job – a linear function. But if we attempt to model the amplitude with a decreasing linear function, such as $A(t) = 10 - t$, we quickly see the problem when we graph the equation $f(t) = (10 - t)\sin(4t)$.

While the amplitude decreases at first as intended, the amplitude hits zero at $t = 10$, then continues past the intercept, increasing in absolute value, which is not the expected behavior. This behavior and function may model the situation on a restricted domain and we might try to chalk the rest of it up to model breakdown, but in fact springs just don't behave like this.

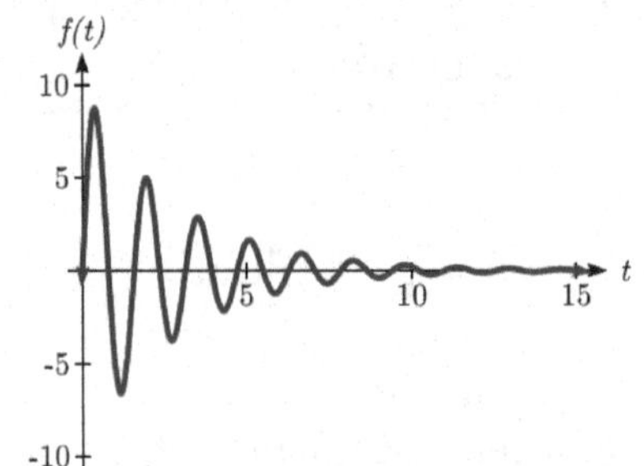

A better model, as you will learn later in physics and calculus, would show the amplitude decreasing by a fixed *percentage* each second, leading to an exponential decay model for the amplitude.

Damped Harmonic Motion
Damped harmonic motion, exhibited by springs subject to friction, follows a model of the form $f(t) = ab^t \sin(Bt) + k$ or $f(t) = ae^{rt} \sin(Bt) + k$.

Example 4

A spring with natural length of feet inches is pulled back 6 feet and released. It oscillates once every 2 seconds. Its amplitude decreases by 20% each second. Find a function that models the position of the spring t seconds after being released.

Since the spring will oscillate on either side of the natural length, the midline will be at 20 feet. The oscillation has a period of 2 seconds, and so the horizontal compression coefficient is $B = \pi$. Additionally, it begins at the furthest distance from the wall, indicating a cosine model.

Meanwhile, the amplitude begins at 6 feet, and decreases by 20% each second, giving an amplitude function of $A(t) = 6(1 - 0.20)^t$.

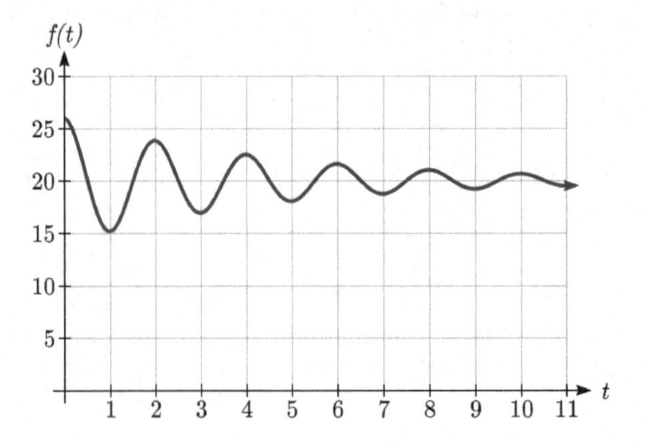

Combining this with the sinusoidal information gives a function for the position of the spring:

$f(t) = 6(0.80)^t \cos(\pi t) + 20$

Example 5

A spring with natural length of 30 cm is pulled out 10 cm and released. It oscillates 4 times per second. After 2 seconds, the amplitude has decreased to 5 cm. Find a function that models the position of the spring.

The oscillation has a period of $\frac{1}{4}$ second, so $B = \frac{2\pi}{1/4} = 8\pi$. Since the spring will oscillate on either side of the natural length, the midline will be at 30 cm. It begins at the furthest distance from the wall, suggesting a cosine model. Together, this gives $f(t) = A(t)\cos(8\pi t) + 30$.

For the amplitude function, we notice that the amplitude starts at 10 cm, and decreases to 5 cm after 2 seconds. This gives two points (0, 10) and (2, 5) that must be satisfied by an exponential function: $A(0) = 10$ and $A(2) = 5$. Since the function is exponential, we can use the form $A(t) = ab^t$. Substituting the first point, $10 = ab^0$, so $a = 10$. Substituting in the second point,

$5 = 10b^2$ Divide by 10

$\frac{1}{2} = b^2$ Take the square root

$b = \sqrt{\frac{1}{2}} \approx 0.707$

This gives an amplitude function of $A(t) = 10(0.707)^t$. Combining this with the oscillation,

$f(t) = 10(0.707)^t \cos(8\pi t) + 30$

Try it Now

2. A certain stock started at a high value of $7 per share, oscillating monthly above and below the average value, with the oscillation decreasing by 2% per year. However, the average value started at $4 per share and has grown linearly by 50 cents per year.
   a. Find a formula for the midline and the amplitude.
   b. Find a function *S(t)* that models the value of the stock after *t* years.

Example 6

In AM (Amplitude Modulated) radio, a carrier wave with a high frequency is used to transmit music or other signals by applying the to-be-transmitted signal as the amplitude of the carrier signal. A musical note with frequency 110 Hz (Hertz = cycles per second) is to be carried on a wave with frequency of 2 KHz (KiloHertz = thousands of cycles per second). If the musical wave has an amplitude of 3, write a function describing the broadcast wave.

The carrier wave, with a frequency of 2000 cycles per second, would have period $\frac{1}{2000}$ of a second, giving an equation of the form $\sin(4000\pi t)$. Our choice of a sine function here was arbitrary – it would have worked just was well to use a cosine.

The musical tone, with a frequency of 110 cycles per second, would have a period of $\frac{1}{110}$ of a second. With an amplitude of 3, this would correspond to a function of the form $3\sin(220\pi t)$. Again our choice of using a sine function is arbitrary.

The musical wave is acting as the amplitude of the carrier wave, so we will multiply the musical tone's function by the carrier wave function, resulting in the function $f(t) = 3\sin(220\pi t)\sin(4000\pi t)$

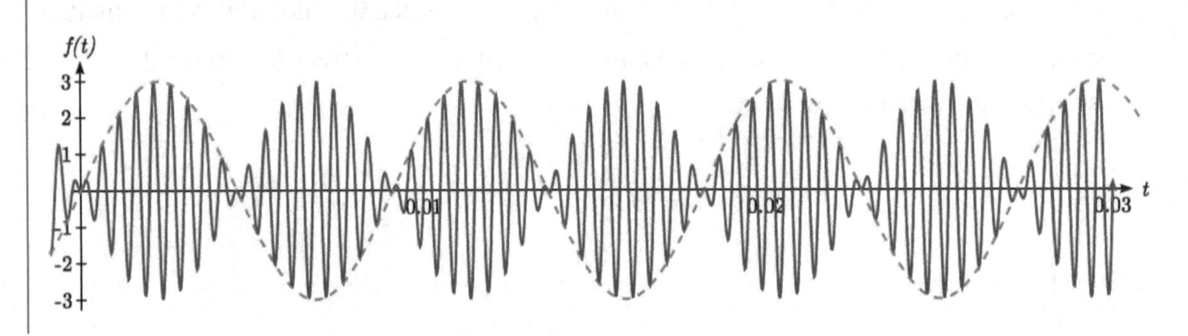

Important Topics of This Section
Changing midline
Changing amplitude
Linear Changes
Exponential Changes
Damped Harmonic Motion

Try it Now Answers

1. The midline follows the path of the quadratic $x^2 - 1$ and the amplitude is a constant value of 8.

2. $m(t) = 4 + 0.5t$
   $A(t) = 7(0.98)^t$
   $S(t) = 7(0.98)^t \cos(24\pi t) + 4 + 0.5t$

Section 7.4 Exercises

Find a possible formula for the trigonometric function whose values are given in the following tables.

1.

x	0	3	6	9	12	15	18
y	-4	-1	2	-1	-4	-1	2

2.

x	0	2	4	6	8	10	12
y	5	1	-3	1	5	1	-3

3. The displacement $h(t)$, in centimeters, of a mass suspended by a spring is modeled by the function $h(t) = 8\sin(6\pi t)$, where t is measured in seconds. Find the amplitude, period, and frequency of this displacement.

4. The displacement $h(t)$, in centimeters, of a mass suspended by a spring is modeled by the function $h(t) = 11\sin(12\pi t)$, where t is measured in seconds. Find the amplitude, period, and frequency of this displacement.

5. A population of rabbits oscillates 19 above and below average during the year, reaching the lowest value in January. The average population starts at 650 rabbits and increases by 160 each year. Find a function that models the population, P, in terms of the months since January, t.

6. A population of deer oscillates 15 above and below average during the year, reaching the lowest value in January. The average population starts at 800 deer and increases by 110 each year. Find a function that models the population, P, in terms of the months since January, t.

7. A population of muskrats oscillates 33 above and below average during the year, reaching the lowest value in January. The average population starts at 900 muskrats and increases by 7% each month. Find a function that models the population, P, in terms of the months since January, t.

8. A population of fish oscillates 40 above and below average during the year, reaching the lowest value in January. The average population starts at 800 fish and increases by 4% each month. Find a function that models the population, P, in terms of the months since January, t.

9. A spring is attached to the ceiling and pulled 10 cm down from equilibrium and released. The amplitude decreases by 15% each second. The spring oscillates 18 times each second. Find a function that models the distance, D, the end of the spring is below equilibrium in terms of seconds, t, since the spring was released.

10. A spring is attached to the ceiling and pulled 7 cm down from equilibrium and released. The amplitude decreases by 11% each second. The spring oscillates 20 times each second. Find a function that models the distance, *D,* the end of the spring is below equilibrium in terms of seconds, *t*, since the spring was released.

11. A spring is attached to the ceiling and pulled 17 cm down from equilibrium and released. After 3 seconds the amplitude has decreased to 13 cm. The spring oscillates 14 times each second. Find a function that models the distance, *D* the end of the spring is below equilibrium in terms of seconds, *t*, since the spring was released.

12. A spring is attached to the ceiling and pulled 19 cm down from equilibrium and released. After 4 seconds the amplitude has decreased to 14 cm. The spring oscillates 13 times each second. Find a function that models the distance, *D* the end of the spring is below equilibrium in terms of seconds, *t*, since the spring was released.

Match each equation form with one of the graphs.

13. a. $ab^x + \sin(5x)$ b. $\sin(5x) + mx + b$

14. a. $ab^x \sin(5x)$ b. $(mx + b)\sin(5x)$

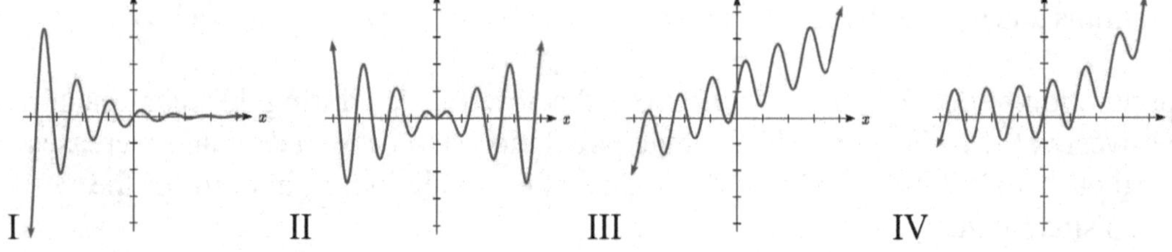

Find a function of the form $y = ab^x + c\sin\left(\frac{\pi}{2}x\right)$ that fits the data given.

15.

x	0	1	2	3
y	6	29	96	379

16.

x	0	1	2	3
y	6	34	150	746

Find a function of the form $y = a\sin\left(\frac{\pi}{2}x\right) + m + bx$ that fits the data given.

17.

x	0	1	2	3
y	7	6	11	16

18.

x	0	1	2	3
y	-2	6	4	2

Find a function of the form $y = ab^x \cos\left(\frac{\pi}{2}x\right) + c$ that fits the data given.

19.

x	0	1	2	3
y	11	3	1	3

20.

x	0	1	2	3
y	4	1	-11	1

Chapter 8: Further Applications of Trigonometry

In this chapter, we will explore additional applications of trigonometry. We will begin with an extension of the right triangle trigonometry we explored in Chapter 5 to situations involving non-right triangles. We will explore the polar coordinate system and parametric equations as new ways of describing curves in the plane. In the process, we will introduce vectors and an alternative way of writing complex numbers, two important mathematical tools we use when analyzing and modeling the world around us.

Section 8.1 Non-Right Triangles: Laws of Sines and Cosines

Although right triangles allow us to solve many applications, it is more common to find scenarios where the triangle we are interested in does not have a right angle.

Two radar stations located 20 miles apart both detect a UFO located between them. The angle of elevation measured by the first station is 35 degrees. The angle of elevation measured by the second station is 15 degrees. What is the altitude of the UFO?

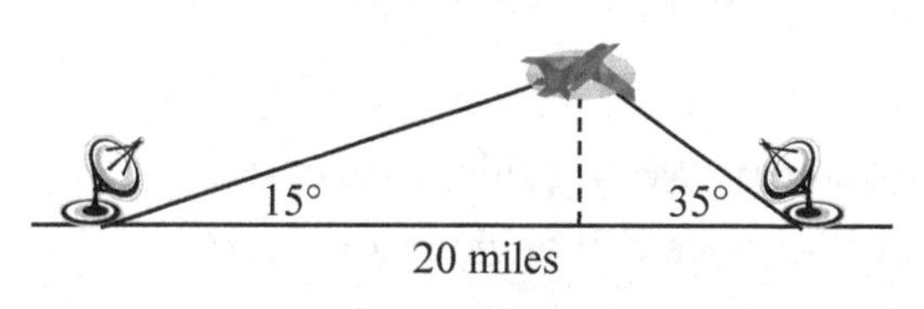

We see that the triangle formed by the UFO and the two stations is not a right triangle. Of course, in any triangle we could draw an **altitude**, a perpendicular line from one vertex to the opposite side, forming two right triangles, but it would be nice to have methods for working directly with non-right triangles. In this section, we will expand upon the right triangle trigonometry we learned in Chapter 5, and adapt it to non-right triangles.

Law of Sines

Given an arbitrary non-right triangle, we can drop an altitude, which we temporarily label h, to create two right triangles.

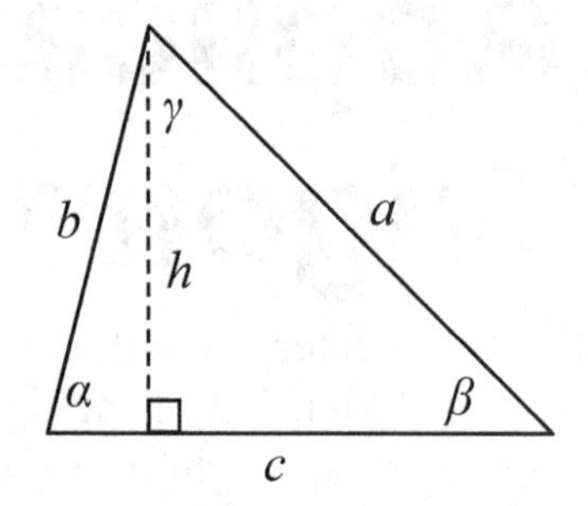

Using the right triangle relationships,
$\sin(\alpha) = \frac{h}{b}$ and $\sin(\beta) = \frac{h}{a}$.

Solving both equations for h, we get $b\sin(\alpha) = h$ and $a\sin(\beta) = h$. Since the h is the same in both equations, we establish $b\sin(\alpha) = a\sin(\beta)$. Dividing both sides by ab, we conclude that

$$\frac{\sin(\alpha)}{a} = \frac{\sin(\beta)}{b}$$

Had we drawn the altitude to be perpendicular to side b or a, we could similarly establish

$$\frac{\sin(\alpha)}{a} = \frac{\sin(\gamma)}{c} \text{ and } \frac{\sin(\beta)}{b} = \frac{\sin(\gamma)}{c}$$

Collectively, these relationships are called the Law of Sines.

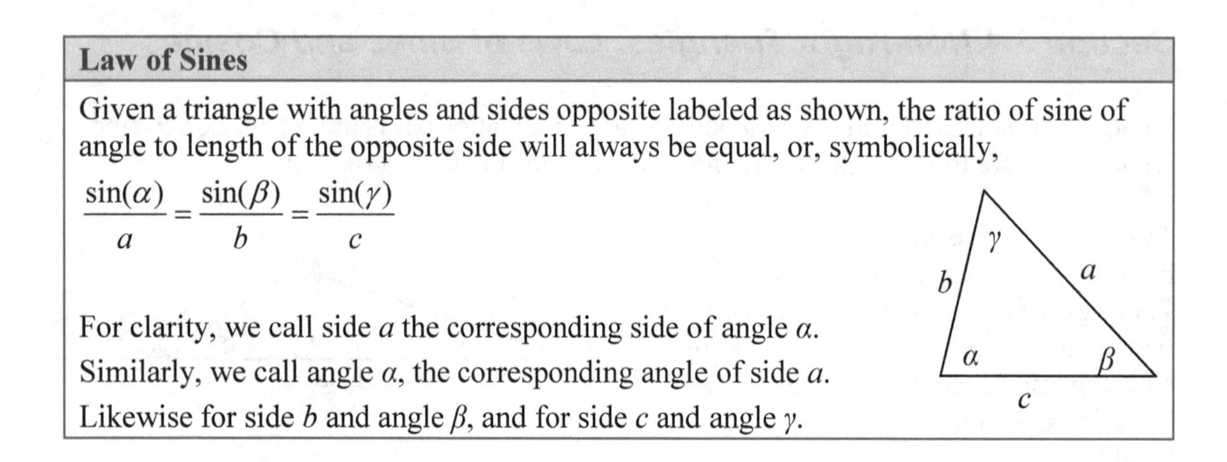

Law of Sines

Given a triangle with angles and sides opposite labeled as shown, the ratio of sine of angle to length of the opposite side will always be equal, or, symbolically,

$$\frac{\sin(\alpha)}{a} = \frac{\sin(\beta)}{b} = \frac{\sin(\gamma)}{c}$$

For clarity, we call side a the corresponding side of angle α. Similarly, we call angle α, the corresponding angle of side a. Likewise for side b and angle β, and for side c and angle γ.

When we use the law of sines, we use any pair of ratios as an equation. In the most straightforward case, we know two angles and one of the corresponding sides.

Example 1

In the triangle shown here, solve for the unknown sides and angle.

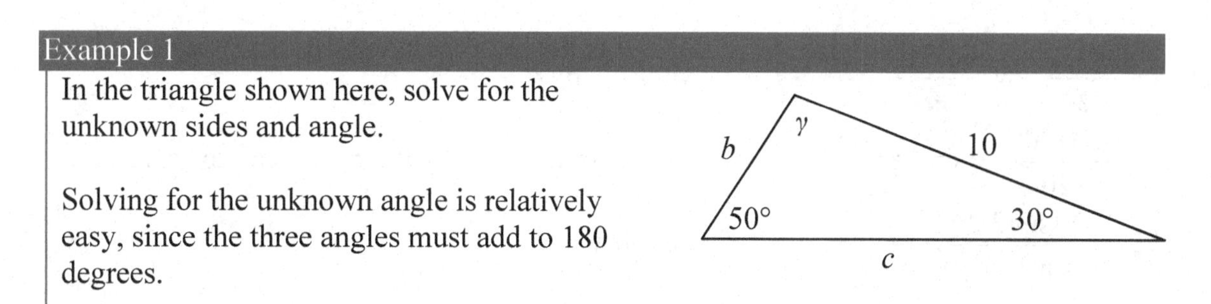

Solving for the unknown angle is relatively easy, since the three angles must add to 180 degrees.

From this, we can determine that
$\gamma = 180° - 50° - 30° = 100°$.

To find an unknown side, we need to know the corresponding angle, and we also need another known ratio.

Since we know the angle 50° and its corresponding side, we can use this for one of the two ratios. To look for side b, we would use its corresponding angle, 30°.

$$\frac{\sin(50°)}{10} = \frac{\sin(30°)}{b}$$ Multiply both sides by b

$$b\frac{\sin(50°)}{10} = \sin(30°)$$ Divide, or multiply by the reciprocal, to solve for b

$$b = \sin(30°)\frac{10}{\sin(50°)} \approx 6.527$$

Similarly, to solve for side c, we set up the equation

$$\frac{\sin(50°)}{10} = \frac{\sin(100°)}{c}$$

$$c = \sin(100°)\frac{10}{\sin(50°)} \approx 12.856$$

Example 2

Find the elevation of the UFO from the beginning of the section.

To find the elevation of the UFO, we first find the distance from one station to the UFO, such as the side a in the picture, then use right triangle relationships to find the height of the UFO, h.

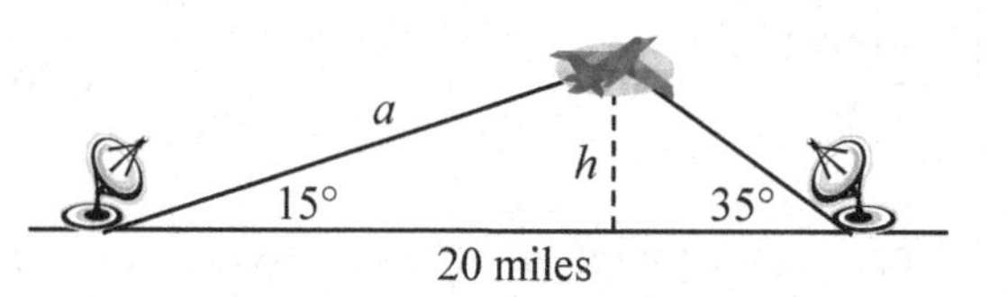

Since the angles in the triangle add to 180 degrees, the unknown angle of the triangle must be $180° - 15° - 35° = 130°$. This angle is opposite the side of length 20, allowing us to set up a Law of Sines relationship:

$$\frac{\sin(130°)}{20} = \frac{\sin(35°)}{a}$$ Multiply by a

$$a\frac{\sin(130°)}{20} = \sin(35°)$$ Divide, or multiply by the reciprocal, to solve for a

$$a = \frac{20\sin(35°)}{\sin(130°)} \approx 14.975$$ Simplify

The distance from one station to the UFO is about 15 miles. Now that we know a, we can use right triangle relationships to solve for h.

$$\sin(15°) = \frac{opposite}{hypotenuse} = \frac{h}{a} = \frac{h}{14.975}$$ Solve for h

$$h = 14.975\sin(15°) \approx 3.876$$

The UFO is at an altitude of 3.876 miles.

In addition to solving triangles in which two angles are known, the law of sines can be used to solve for an angle when two sides and one corresponding angle are known.

Example 3

In the triangle shown here, solve for the unknown sides and angles.

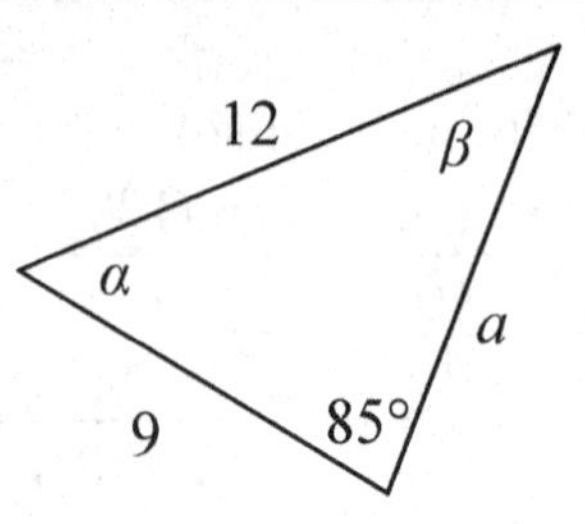

In choosing which pair of ratios from the Law of Sines to use, we always want to pick a pair where we know three of the four pieces of information in the equation. In this case, we know the angle 85° and its corresponding side, so we will use that ratio. Since our only other known information is the side with length 9, we will use that side and solve for its corresponding angle.

$$\frac{\sin(85°)}{12} = \frac{\sin(\beta)}{9}$$ Isolate the unknown

$$\frac{9\sin(85°)}{12} = \sin(\beta)$$ Use the inverse sine to find a first solution

Remember when we use the inverse function that there are two possible answers.

$$\beta = \sin^{-1}\left(\frac{9\sin(85°)}{12}\right) \approx 48.3438°$$ By symmetry we find the second possible solution

$$\beta = 180° - 48.3438° = 131.6562°$$

In this second case, if $\beta \approx 132°$, then α would be $\alpha = 180° - 85° - 132° = -37°$, which doesn't make sense, so the only possibility for this triangle is $\beta = 48.3438°$.
With a second angle, we can now easily find the third angle, since the angles must add to 180°, so $\alpha = 180° - 85° - 48.3438° = 46.6562°$.

Now that we know α, we can proceed as in earlier examples to find the unknown side *a*.

$$\frac{\sin(85°)}{12} = \frac{\sin(46.6562°)}{a}$$

$$a = \frac{12\sin(46.6562°)}{\sin(85°)} \approx 8.7603$$

Notice that in the problem above, when we use Law of Sines to solve for an unknown angle, there can be two possible solutions. This is called the **ambiguous case**, and can arise when we know two sides and a non-included angle. In the ambiguous case we may find that a particular set of given information can lead to 2, 1 or no solution at all. However, when an accurate picture of the triangle or suitable context is available, we can determine which angle is desired.

Try it Now

1. Given $\alpha = 80°, a = 120$, and $b = 121$, find the corresponding and missing side and angles. If there is more than one possible solution, show both.

Example 4

Find all possible triangles if one side has length 4 opposite an angle of 50° and a second side has length 10.

Using the given information, we can look for the angle opposite the side of length 10.

$$\frac{\sin(50°)}{4} = \frac{\sin(\alpha)}{10}$$

$$\sin(\alpha) = \frac{10\sin(50°)}{4} \approx 1.915$$

Since the range of the sine function is [-1, 1], it is impossible for the sine value to be 1.915. There are no triangles that can be drawn with the provided dimensions.

Example 5

Find all possible triangles if one side has length 6 opposite an angle of 50° and a second side has length 4.

Using the given information, we can look for the angle opposite the side of length 4.

$$\frac{\sin(50^\circ)}{6} = \frac{\sin(\alpha)}{4}$$

$$\sin(\alpha) = \frac{4\sin(50^\circ)}{6} \approx 0.511$$ Use the inverse to find one solution

$$\alpha = \sin^{-1}(0.511) \approx 30.710^\circ$$ By symmetry there is a second possible solution

$$\alpha = 180^\circ - 30.710^\circ = 149.290^\circ$$

If we use the angle 30.710°, the third angle would be $180^\circ - 50^\circ - 30.710^\circ = 99.290^\circ$. We can then use Law of Sines again to find the third side.

$$\frac{\sin(50^\circ)}{6} = \frac{\sin(99.290^\circ)}{c}$$ Solve for c

$$c = 7.730$$

If we used the angle $\alpha = 149.290^\circ$, the third angle would be $180^\circ - 50^\circ - 149.290^\circ = -19.29^\circ$, which is impossible, so the previous triangle is the only possible one.

Try it Now

2. Given $\alpha = 80^\circ, a = 100$, and $b = 10$ find the missing side and angles. If there is more than one possible solution, show both.

Law of Cosines

Suppose a boat leaves port, travels 10 miles, turns 20 degrees, and travels another 8 miles. How far from port is the boat?

8 mi
20°
10 mi

Unfortunately, while the Law of Sines lets us address many non-right triangle cases, it does not allow us to address triangles where the one known angle is included between two known sides, which means it is not a corresponding angle for a known side. For this, we need another tool.

Given an arbitrary non-right triangle, we can drop an altitude, which we temporarily label h, to create two right triangles. We will divide the base b into two pieces, one of which we will temporarily label x.

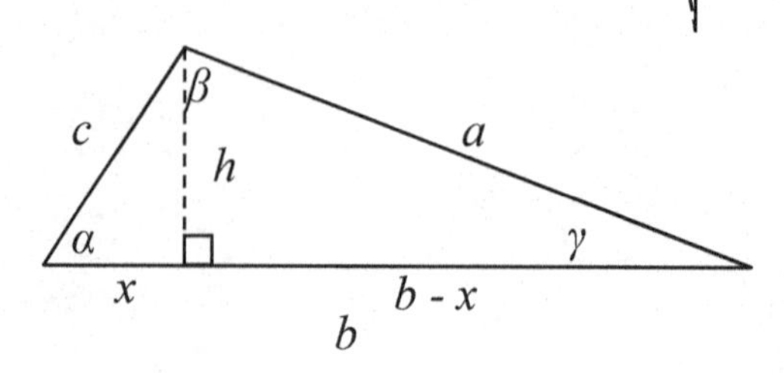

From this picture, we can establish the right triangle relationship

$\cos(\alpha) = \frac{x}{c}$, or equivalently, $x = c\cos(\alpha)$

Using the Pythagorean Theorem, we can establish

$(b-x)^2 + h^2 = a^2$ and $x^2 + h^2 = c^2$

Both of these equations can be solved for h^2

$h^2 = a^2 - (b-x)^2$ and $h^2 = c^2 - x^2$

Since the left side of each equation is h^2, the right sides must be equal

$c^2 - x^2 = a^2 - (b-x)^2$ Multiply out the right

$c^2 - x^2 = a^2 - (b^2 - 2bx + x^2)$ Simplify

$c^2 - x^2 = a^2 - b^2 + 2bx - x^2$

$c^2 = a^2 - b^2 + 2bx$ Isolate a^2

$a^2 = c^2 + b^2 - 2bx$ Substitute in $c\cos(\alpha) = x$ from above

$a^2 = c^2 + b^2 - 2bc\cos(\alpha)$

This result is called the Law of Cosines. Depending upon which side we dropped the altitude down from, we could have established this relationship using any of the angles. The important thing to note is that the right side of the equation involves an angle and the sides adjacent to that angle – the left side of the equation involves the side opposite that angle.

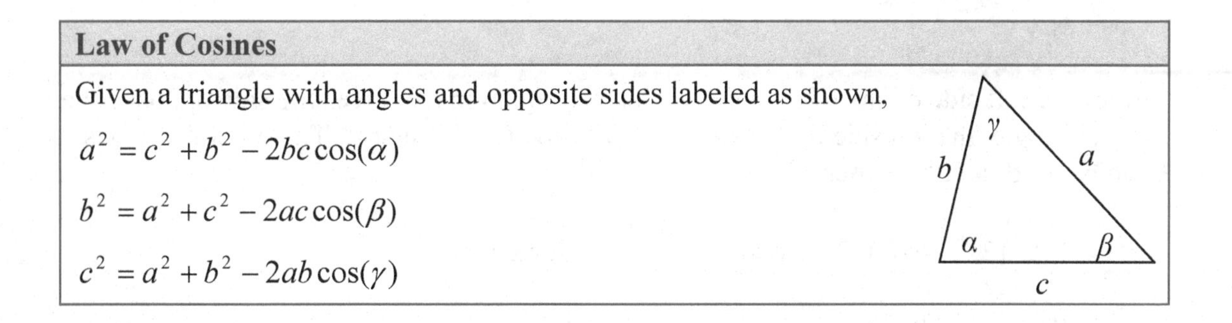

Law of Cosines

Given a triangle with angles and opposite sides labeled as shown,

$a^2 = c^2 + b^2 - 2bc\cos(\alpha)$

$b^2 = a^2 + c^2 - 2ac\cos(\beta)$

$c^2 = a^2 + b^2 - 2ab\cos(\gamma)$

Notice that if one of the angles of the triangle is 90 degrees, cos(90°) = 0, so the formula

$c^2 = a^2 + b^2 - 2ab\cos(90°)$ Simplifies to

$c^2 = a^2 + b^2$

You should recognize this as the Pythagorean Theorem. Indeed, the Law of Cosines is sometimes called the **Generalized Pythagorean Theorem**, since it extends the Pythagorean Theorem to non-right triangles.

Example 6

Returning to our question from earlier, suppose a boat leaves port, travels 10 miles, turns 20 degrees, and travels another 8 miles. How far from port is the boat?

8 mi
20°
10 mi

The boat turned 20 degrees, so the obtuse angle of the non-right triangle shown in the picture is the supplemental angle, 180° - 20° = 160°.

With this, we can utilize the Law of Cosines to find the missing side of the obtuse triangle – the distance from the boat to port.

$x^2 = 8^2 + 10^2 - 2(8)(10)\cos(160°)$ Evaluate the cosine and simplify

$x^2 = 314.3508$ Square root both sides

$x = \sqrt{314.3508} = 17.730$

The boat is 17.73 miles from port.

Example 7

Find the unknown side and angles of this triangle.

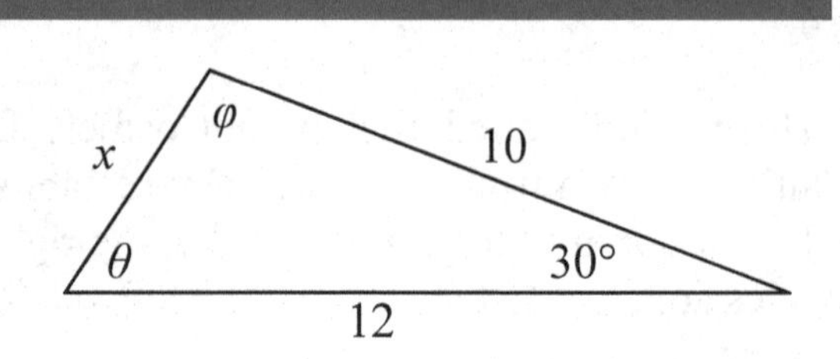

Notice that we don't have both pieces of any side/angle pair, so the Law of Sines would not work with this triangle.

Since we have the angle included between the two known sides, we can turn to Law of Cosines.

Since the left side of any of the Law of Cosines equations involves the side opposite the known angle, the left side in this situation will involve the side x. The other two sides can be used in either order.

$x^2 = 10^2 + 12^2 - 2(10)(12)\cos(30°)$ Evaluate the cosine

$x^2 = 10^2 + 12^2 - 2(10)(12)\frac{\sqrt{3}}{2}$ Simplify

$x^2 = 244 - 120\sqrt{3}$ Take the square root

$x = \sqrt{244 - 120\sqrt{3}} \approx 6.013$

Now that we know an angle and its corresponding side, we can use the Law of Sines to fill in the remaining angles of the triangle. Solving for angle θ,

$$\frac{\sin(30°)}{6.013} = \frac{\sin(\theta)}{10}$$

$$\sin(\theta) = \frac{10\sin(30°)}{6.013}$$ Use the inverse sine

$$\theta = \sin^{-1}\left(\frac{10\sin(30°)}{6.013}\right) \approx 56.256°$$

The other possibility for θ would be $\theta = 180° - 56.256° = 123.744°$. In the original picture, θ is an acute angle, so 123.744° doesn't make sense if we assume the picture is drawn to scale.

Proceeding with $\theta = 56.256°$, we can then find the third angle of the triangle: $\varphi = 180° - 30° - 56.256° = 93.744°$.

In addition to solving for the missing side opposite one known angle, the Law of Cosines allows us to find the angles of a triangle when we know all three sides.

Example 8

Solve for the angle α in the triangle shown.

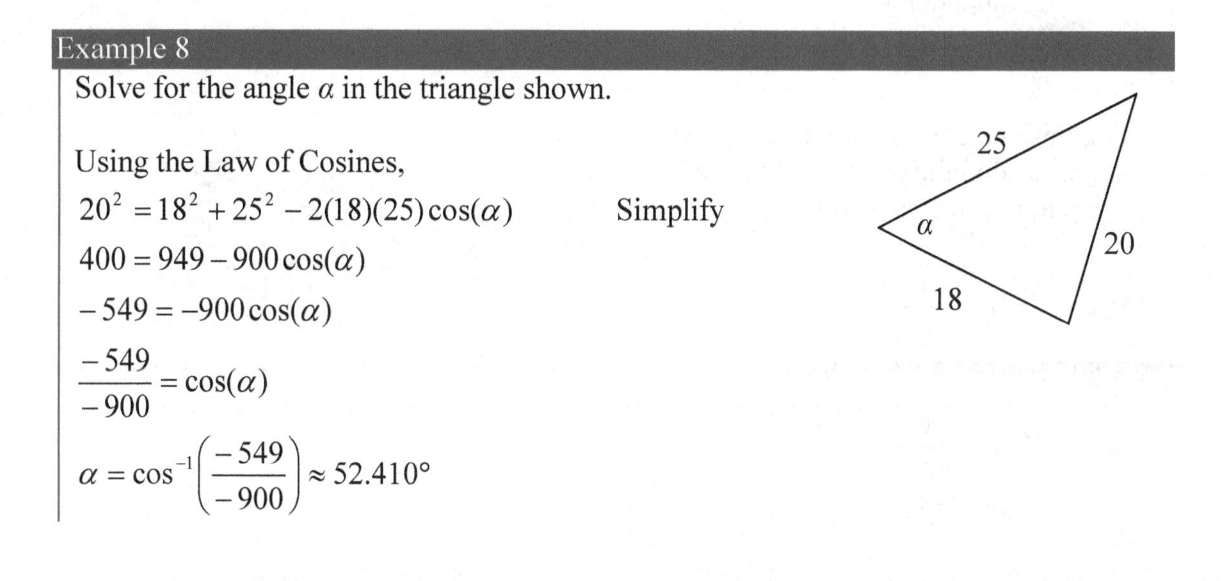

Using the Law of Cosines,

$$20^2 = 18^2 + 25^2 - 2(18)(25)\cos(\alpha)$$ Simplify

$$400 = 949 - 900\cos(\alpha)$$

$$-549 = -900\cos(\alpha)$$

$$\frac{-549}{-900} = \cos(\alpha)$$

$$\alpha = \cos^{-1}\left(\frac{-549}{-900}\right) \approx 52.410°$$

Try it Now

3. Given $\alpha = 25°, b = 10$, and $c = 20$ find the missing side and angles.

Notice that since the inverse cosine can return any angle between 0 and 180 degrees, there will not be any ambiguous cases when using Law of Cosines to find an angle.

Example 9

On many cell phones with GPS, an approximate location can be given before the GPS signal is received. This is done by a process called triangulation, which works by using the distance from two known points. Suppose there are two cell phone towers within range of you, located 6000 feet apart along a straight highway that runs east to west, and you know you are north of the highway. Based on the signal delay, it can be determined you are 5050 feet from the first tower, and 2420 feet from the second. Determine your position north and east of the first tower, and determine how far you are from the highway.

For simplicity, we start by drawing a picture and labeling our given information. Using the Law of Cosines, we can solve for the angle θ.

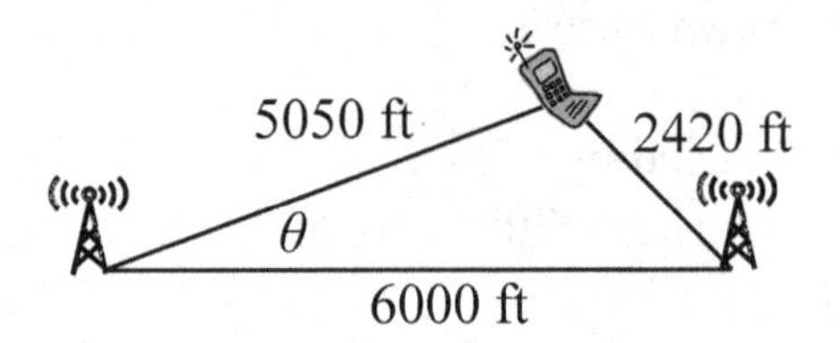

$$2420^2 = 6000^2 + 5050^2 - 2(5050)(6000)\cos(\theta)$$
$$5856400 = 61501500 - 60600000\cos(\theta)$$
$$-554646100 = -60600000\cos(\theta)$$
$$\cos(\theta) = \frac{-554646100}{-60600000} = 0.9183$$
$$\theta = \cos^{-1}(0.9183) = 23.328°$$

Using this angle, we could then use right triangles to find the position of the cell phone relative to the western tower.

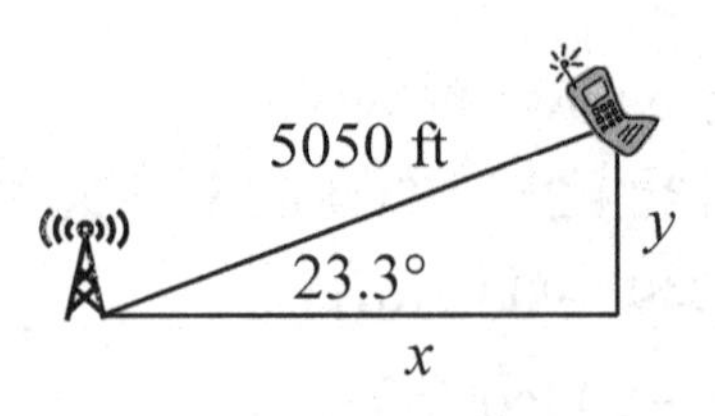

$$\cos(23.328°) = \frac{x}{5050}$$
$$x = 5050\cos(23.328°) \approx 4637.2 \text{ feet}$$

$$\sin(23.328°) = \frac{y}{5050}$$
$$y = 5050\sin(23.328°) \approx 1999.8 \text{ feet}$$

You are 5050 ft from the tower and 23.328° north of east (or, equivalently, 66.672° east of north). Specifically, you are about 4637 feet east and 2000 feet north of the first tower.

Note that if you didn't know whether you were north or south of the towers, our calculations would have given two possible locations, one north of the highway and one south. To resolve this ambiguity in real world situations, locating a position using triangulation requires a signal from a third tower.

Example 10

To measure the height of a hill, a woman measures the angle of elevation to the top of the hill to be 24 degrees. She then moves back 200 feet and measures the angle of elevation to be 22 degrees. Find the height of the hill.

As with many problems of this nature, it will be helpful to draw a picture.

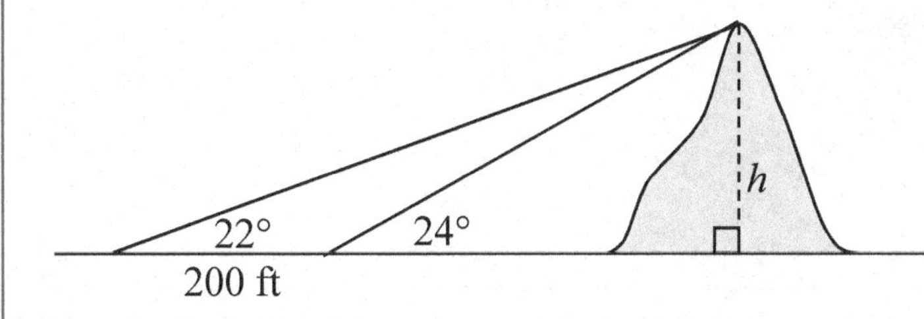

Notice there are three triangles formed here – the right triangle including the height h and the 22 degree angle, the right triangle including the height h and the 24 degree angle, and the (non-right) obtuse triangle including the 200 ft side. Since this is the triangle we have the most information for, we will begin with it. It may seem odd to work with this triangle since it does not include the desired side h, but we don't have enough information to work with either of the right triangles yet.

We can find the obtuse angle of the triangle, since it and the angle of 24 degrees complete a straight line – a 180 degree angle. The obtuse angle must be 180° - 24° = 156°. From this, we can determine that the third angle is 2°. We know one side is 200 feet, and its corresponding angle is 2°, so by introducing a temporary variable x for one of the other sides (as shown below), we can use Law of Sines to solve for this length x.

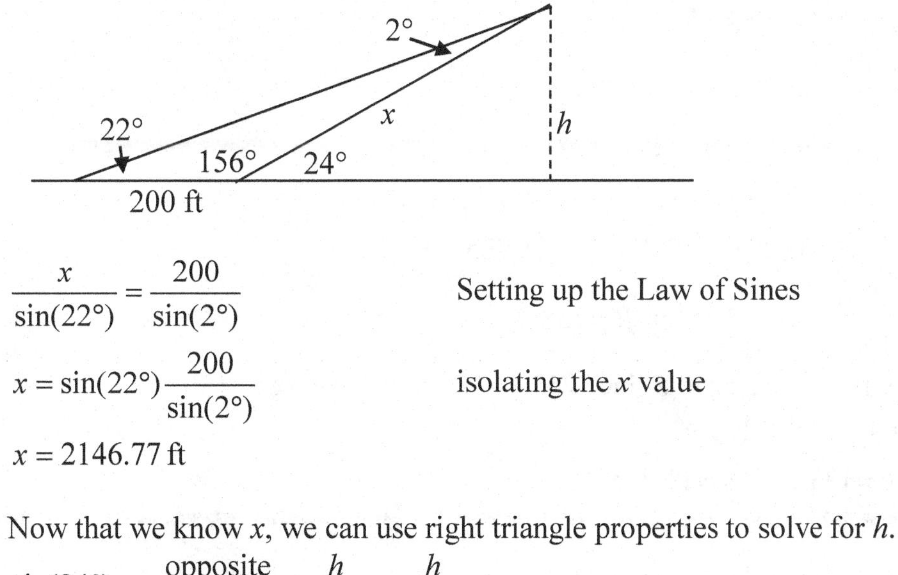

$$\frac{x}{\sin(22°)} = \frac{200}{\sin(2°)}$$ Setting up the Law of Sines

$$x = \sin(22°)\frac{200}{\sin(2°)}$$ isolating the x value

$$x = 2146.77 \text{ ft}$$

Now that we know x, we can use right triangle properties to solve for h.

$$\sin(24°) = \frac{\text{opposite}}{\text{hypotenuse}} = \frac{h}{x} = \frac{h}{2146.77}$$

$h = 2146.77\sin(24°) = 873.17$ ft. The hill is 873 feet high.

Important Topics of This Section
Law of Sines
Solving for sides
Solving for angles
Ambiguous case, 0, 1 or 2 solutions
Law of Cosines
Solving for sides
Solving for angles
Generalized Pythagorean Theorem

Try it Now Answers

1. $\frac{\sin(80°)}{120} = \frac{\sin(\beta)}{121}$

   $\beta = 83.2°$ $\qquad$ $\beta = 96.8°$

   1st possible solution $\gamma = 16.8°$ $\qquad$ 2nd solution $\gamma = 3.2°$

   $c = 35.2$ $\qquad$ $c = 6.9$

   If we were given a picture of the triangle it may be possible to eliminate one of these

2. $\frac{\sin(80°)}{120} = \frac{\sin(\beta)}{10}$. $\beta = 5.65°$ or $\beta = 174.35°$; only the first is reasonable.

   $\gamma = 180° - 5.65° - 80° = 94.35°$

   $\frac{\sin(80°)}{120} = \frac{\sin(94.35°)}{c}$

   $\beta = 5.65°, \gamma = 94.35°, c = 101.25$

3. $a^2 = 10^2 + 20^2 - 2(10)(20)\cos(25°)$. $a = 11.725$

   $\frac{\sin(25°)}{11.725} = \frac{\sin(\beta)}{10}$. $\beta = 21.1°$ or $\beta = 158.9°$;

   only the first is reasonable since 25° + 158.9° would exceed 180°.

   $\gamma = 180° - 21.1° - 25° = 133.9°$

   $\beta = 21.1°, \quad \gamma = 133.9°, \quad a = 11.725$

Section 8.1 Exercises

Solve for the unknown sides and angles of the triangles shown.

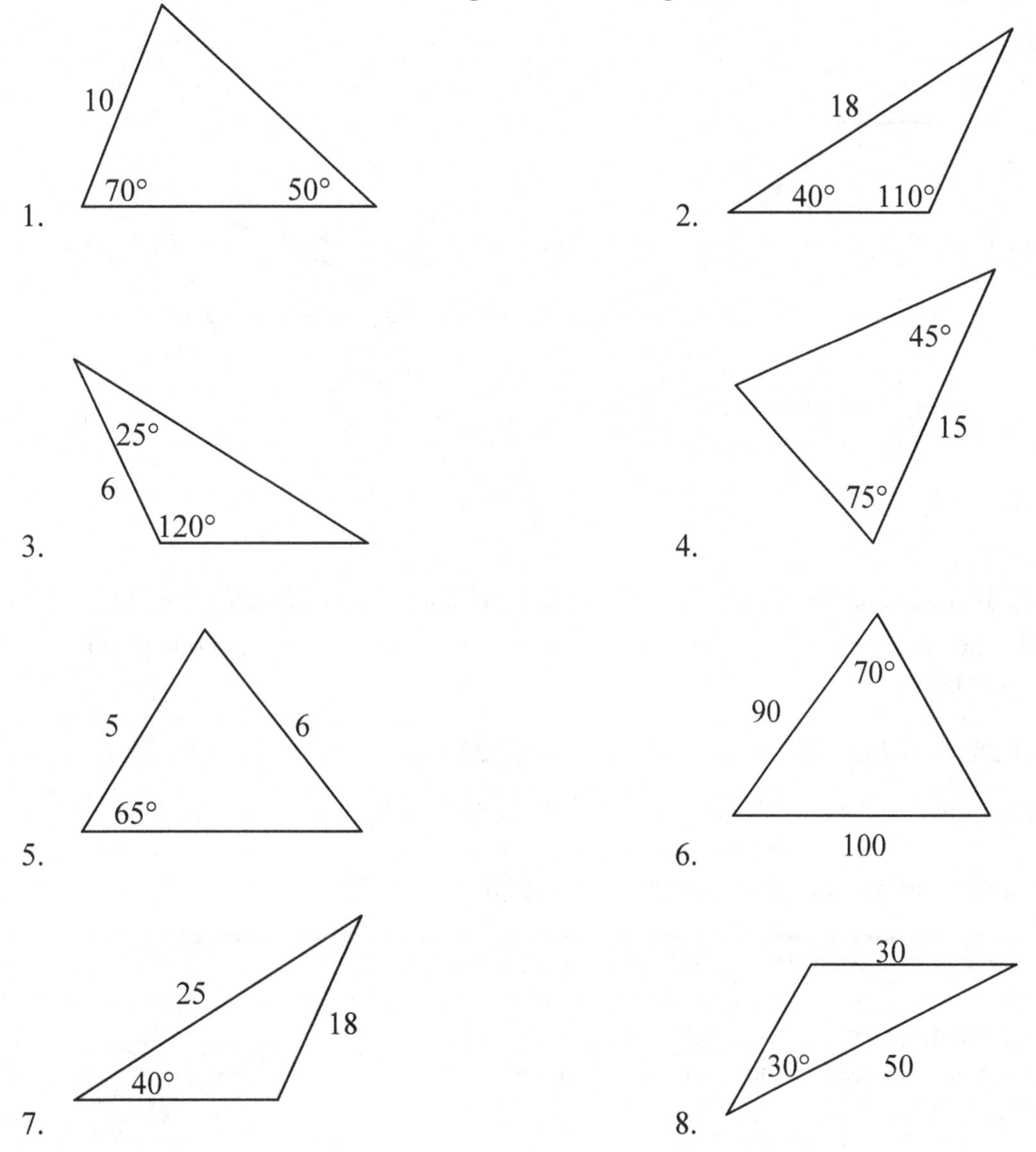

Assume α is opposite side a, β is opposite side b, and γ is opposite side c. Solve each triangle for the unknown sides and angles if possible. If there is more than one possible solution, give both.

9. $\alpha = 43°, \gamma = 69°, b = 20$

10. $\alpha = 35°, \gamma = 73°, b = 19$

11. $\alpha = 119°, a = 26, b = 14$

12. $\gamma = 113°, b = 10, c = 32$

13. $\beta = 50°, a = 105, b = 45$

14. $\beta = 67°, a = 49, b = 38$

15. $\alpha = 43.1°, a = 184.2, b = 242.8$

16. $\alpha = 36.6°, a = 186.2, b = 242.2$

Solve for the unknown sides and angles of the triangles shown.

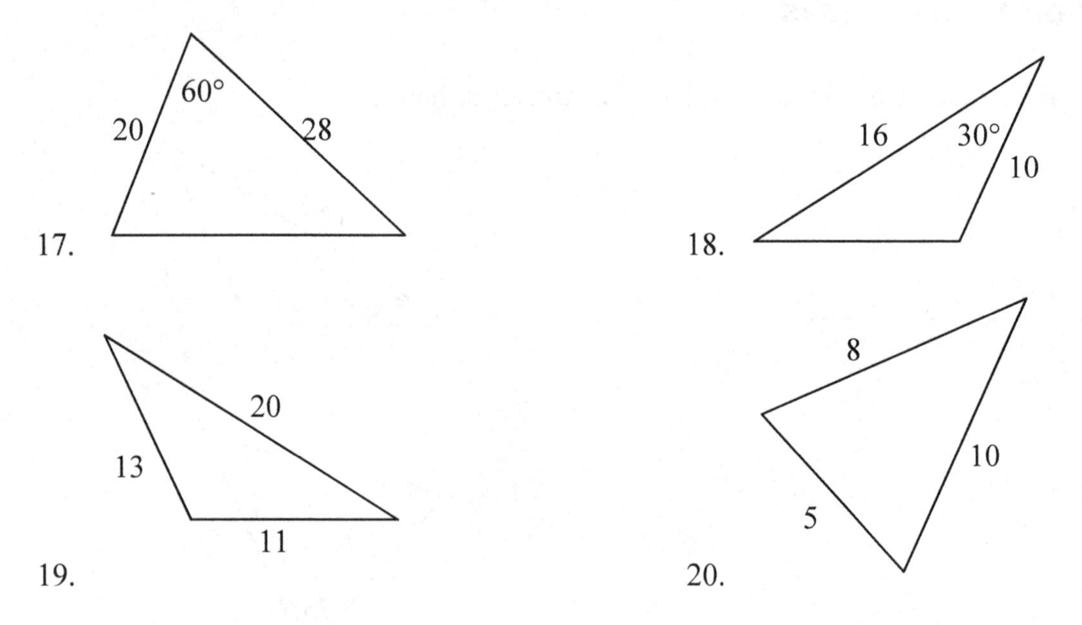

Assume α is opposite side a, β is opposite side b, and γ is opposite side c. Solve each triangle for the unknown sides and angles if possible. If there is more than one possible solution, give both.

21. $\gamma = 41.2°, a = 2.49, b = 3.13$

22. $\beta = 58.7°, a = 10.6, c = 15.7$

23. $\alpha = 120°, b = 6, c = 7$

24. $\gamma = 115°, a = 18, b = 23$

25. Find the area of a triangle with sides of length 18, 21, and 32.

26. Find the area of a triangle with sides of length 20, 26, and 37.

27. To find the distance across a small lake, a surveyor has taken the measurements shown. Find the distance across the lake.

28. To find the distance between two cities, a satellite calculates the distances and angle shown (*not to scale*). Find the distance between the cities.

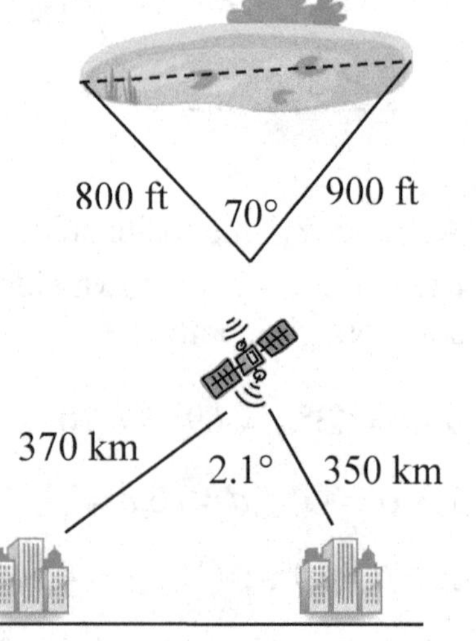

29. To determine how far a boat is from shore, two radar stations 500 feet apart determine the angles out to the boat, as shown. Find the distance of the boat from the station *A*, and the distance of the boat from shore.

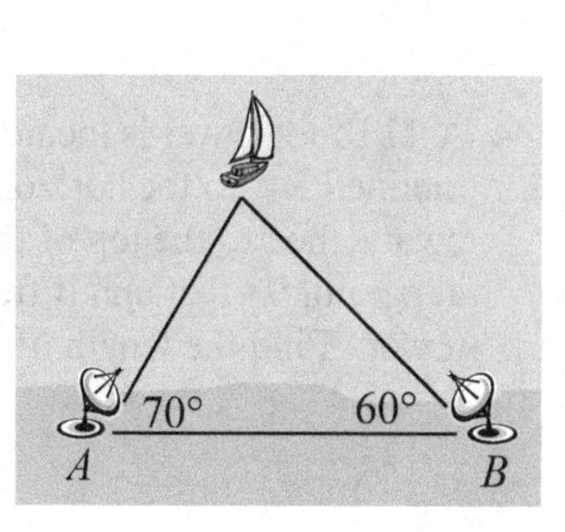

30. The path of a satellite orbiting the earth causes it to pass directly over two tracking stations *A* and *B*, which are 69 mi apart. When the satellite is on one side of the two stations, the angles of elevation at *A* and *B* are measured to be 86.2° and 83.9°, respectively. How far is the satellite from station *A* and how high is the satellite above the ground?

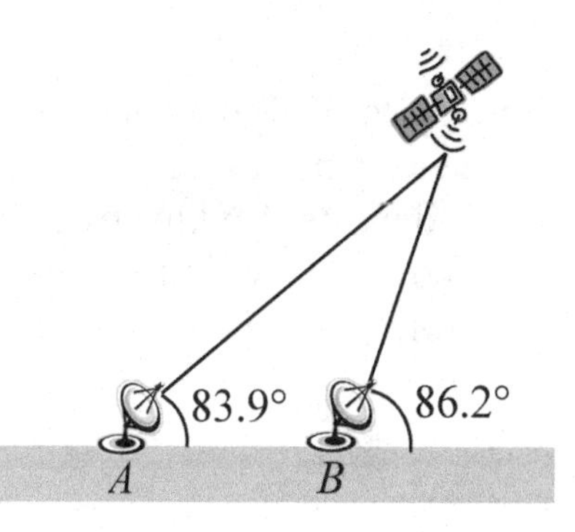

31. A communications tower is located at the top of a steep hill, as shown. The angle of inclination of the hill is 67°. A guy-wire is to be attached to the top of the tower and to the ground, 165 m downhill from the base of the tower. The angle formed by the guy-wire and the hill is 16°. Find the length of the cable required for the guy wire.

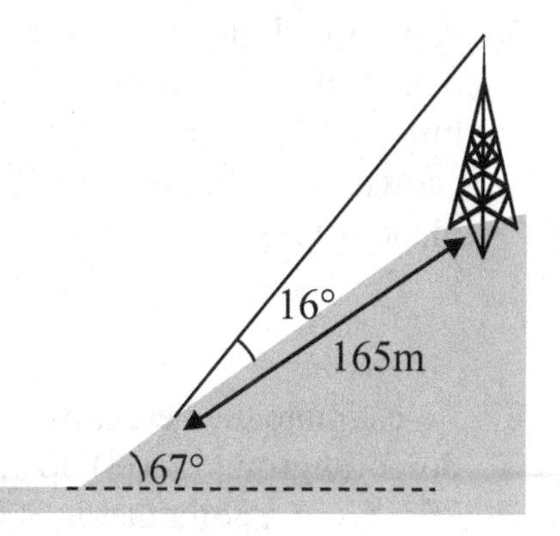

32. The roof of a house is at a 20° angle. An 8 foot solar panel is to be mounted on the roof, and should be angled 38° relative to the horizontal for optimal results. How long does the vertical support holding up the back of the panel need to be?

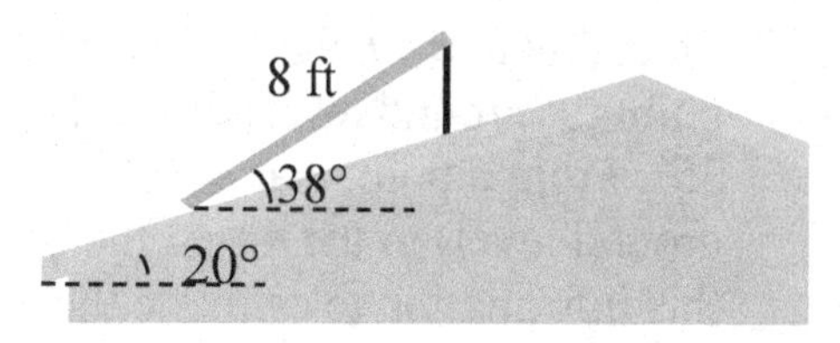

33. A 127 foot tower is located on a hill that is inclined 38° to the horizontal. A guy-wire is to be attached to the top of the tower and anchored at a point 64 feet downhill from the base of the tower. Find the length of wire needed.

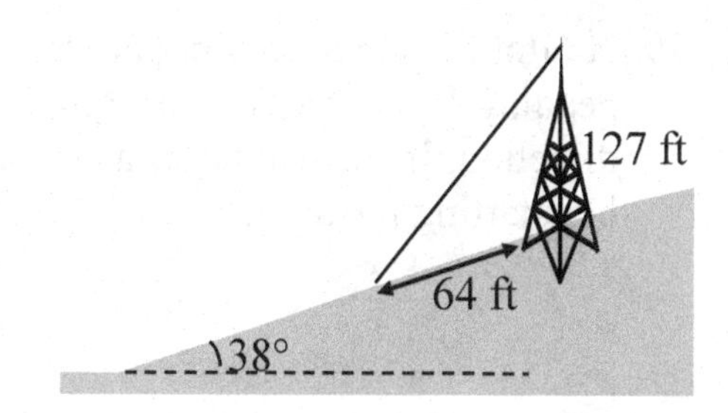

34. A 113 foot tower is located on a hill that is inclined 34° to the horizontal. A guy-wire is to be attached to the top of the tower and anchored at a point 98 feet uphill from the base of the tower. Find the length of wire needed.

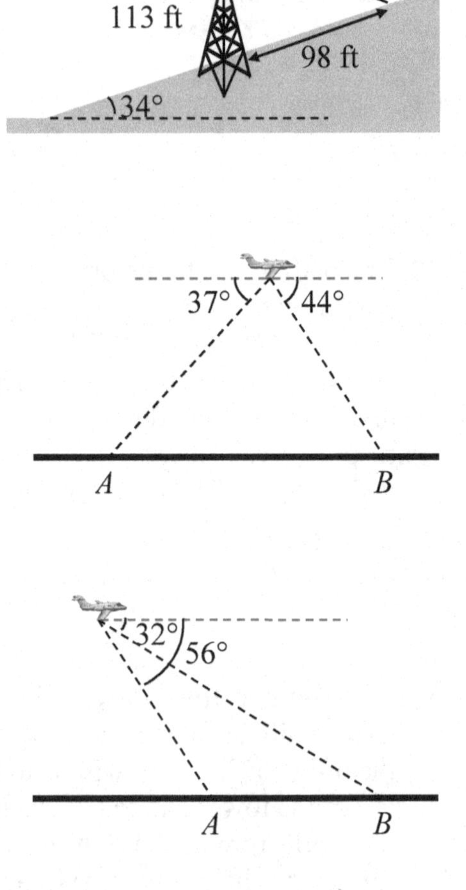

35. A pilot is flying over a straight highway. He determines the angles of depression to two mileposts, 6.6 km apart, to be 37° and 44°, as shown in the figure. Find the distance of the plane from point *A*, and the elevation of the plane.

36. A pilot is flying over a straight highway. He determines the angles of depression to two mileposts, 4.3 km apart, to be 32° and 56°, as shown in the figure. Find the distance of the plane from point *A*, and the elevation of the plane.

37. To estimate the height of a building, two students find the angle of elevation from a point (at ground level) down the street from the building to the top of the building is 39°. From a point that is 300 feet closer to the building, the angle of elevation (at ground level) to the top of the building is 50°. If we assume that the street is level, use this information to estimate the height of the building.

38. To estimate the height of a building, two students find the angle of elevation from a point (at ground level) down the street from the building to the top of the building is 35°. From a point that is 300 feet closer to the building, the angle of elevation (at ground level) to the top of the building is 53°. If we assume that the street is level, use this information to estimate the height of the building.

39. A pilot flies in a straight path for 1 hour 30 min. She then makes a course correction, heading 10 degrees to the right of her original course, and flies 2 hours in the new direction. If she maintains a constant speed of 680 miles per hour, how far is she from her starting position?

40. Two planes leave the same airport at the same time. One flies at 20 degrees east of north at 500 miles per hour. The second flies at 30 east of south at 600 miles per hour. How far apart are the planes after 2 hours?

41. The four sequential sides of a quadrilateral have lengths 4.5 cm, 7.9 cm, 9.4 cm, and 12.9 cm. The angle between the two smallest sides is 117°. What is the area of this quadrilateral?

42. The four sequential sides of a quadrilateral have lengths 5.7 cm, 7.2 cm, 9.4 cm, and 12.8 cm. The angle between the two smallest sides is 106°. What is the area of this quadrilateral?

43. Three circles with radii 6, 7, and 8, all touch as shown. Find the shaded area bounded by the three circles.

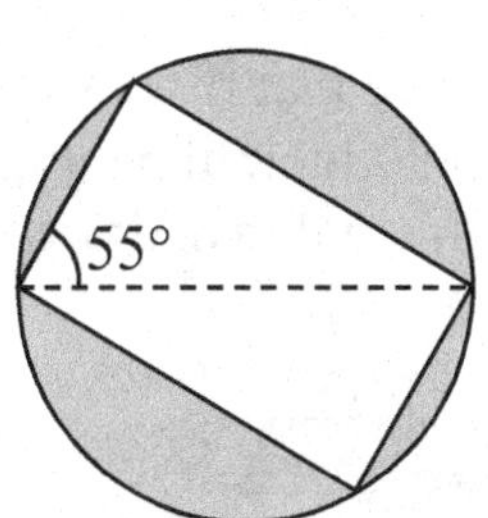

44. A rectangle is inscribed in a circle of radius 10 cm as shown. Find the shaded area, inside the circle but outside the rectangle.

55°

Section 8.2 Polar Coordinates

The coordinate system we are most familiar with is called the Cartesian coordinate system, a rectangular plane divided into four quadrants by horizontal and vertical axes.

In earlier chapters, we often found the Cartesian coordinates of a point on a circle at a given angle from the positive horizontal axis. Sometimes that angle, along with the point's distance from the origin, provides a more useful way of describing the point's location than conventional Cartesian coordinates.

Polar Coordinates
Polar coordinates of a point consist of an ordered pair, (r,θ), where r is the distance from the point to the origin, and θ is the angle measured in standard position.

Notice that if we were to "grid" the plane for polar coordinates, it would look like the graph to the right, with circles at incremental radii, and rays drawn at incremental angles.

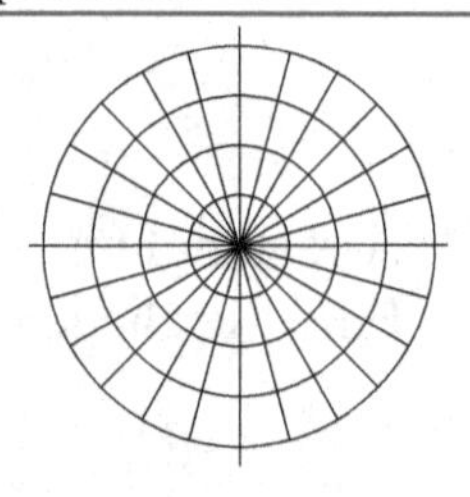

Example 1

Plot the polar point $\left(3,\dfrac{5\pi}{6}\right)$.

This point will be a distance of 3 from the origin, at an angle of $\dfrac{5\pi}{6}$. Plotting this,

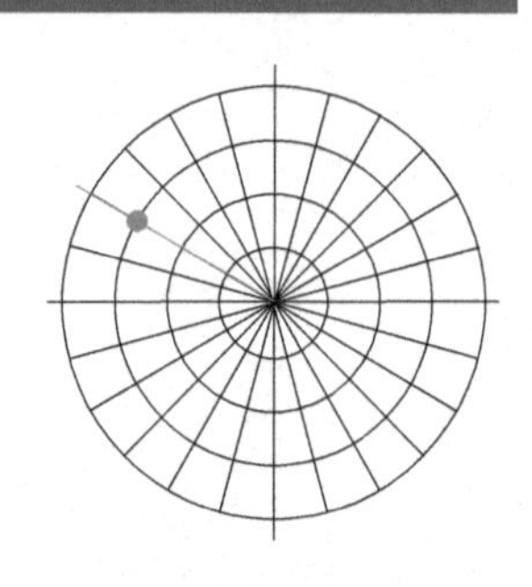

Example 2

Plot the polar point $\left(-2,\dfrac{\pi}{4}\right)$.

Typically we use positive r values, but occasionally we run into cases where r is negative. On a regular number line, we measure positive values to the right and negative values to the left. We will plot this point similarly. To start, we rotate to an angle of $\dfrac{\pi}{4}$. Moving this direction, into the first quadrant, would be positive r values. For negative r values, we move the opposite direction, into the third quadrant. Plotting this:

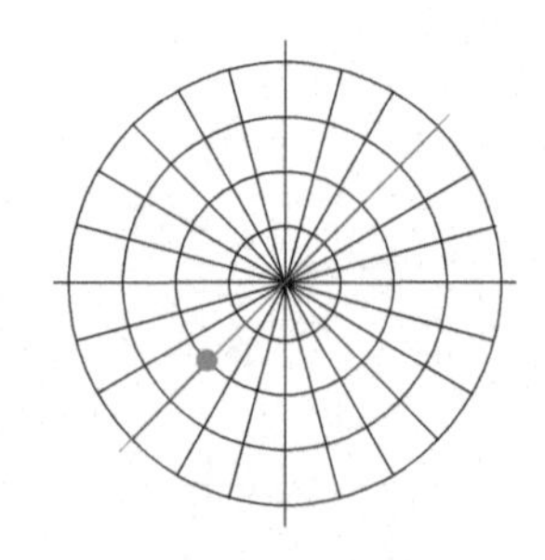

Note the resulting point is the same as the polar point $\left(2, \frac{5\pi}{4}\right)$. In fact, any Cartesian point can be represented by an infinite number of different polar coordinates by adding or subtracting full rotations to these points. For example, same point could also be represented as $\left(2, \frac{13\pi}{4}\right)$.

Try it Now

1. Plot the following points given in polar coordinates and label them.

   a. $A = \left(3, \frac{\pi}{6}\right)$ b. $B = \left(-2, \frac{\pi}{3}\right)$ c. $C = \left(4, \frac{3\pi}{4}\right)$

Converting Points

To convert between polar coordinates and Cartesian coordinates, we recall the relationships we developed back in Chapter 5.

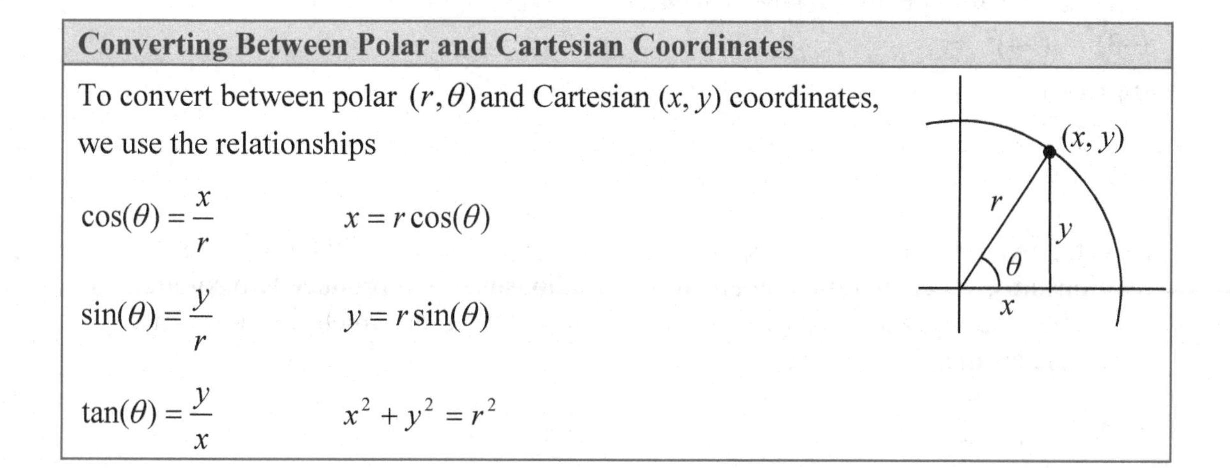

Converting Between Polar and Cartesian Coordinates

To convert between polar (r, θ) and Cartesian (x, y) coordinates, we use the relationships

$\cos(\theta) = \frac{x}{r}$ $\qquad x = r\cos(\theta)$

$\sin(\theta) = \frac{y}{r}$ $\qquad y = r\sin(\theta)$

$\tan(\theta) = \frac{y}{x}$ $\qquad x^2 + y^2 = r^2$

From these relationship and our knowledge of the unit circle, if $r = 1$ and $\theta = \frac{\pi}{3}$, the polar coordinates would be $(r, \theta) = \left(1, \frac{\pi}{3}\right)$, and the corresponding Cartesian coordinates $(x, y) = \left(\frac{1}{2}, \frac{\sqrt{3}}{2}\right)$.

Remembering your unit circle values will come in very handy as you convert between Cartesian and polar coordinates.

Example 3

Find the Cartesian coordinates of a point with polar coordinates $(r,\theta)=\left(5,\frac{2\pi}{3}\right)$.

To find the x and y coordinates of the point,

$$x=r\cos(\theta)=5\cos\left(\frac{2\pi}{3}\right)=5\left(-\frac{1}{2}\right)=-\frac{5}{2}$$

$$y=r\sin(\theta)=5\sin\left(\frac{2\pi}{3}\right)=5\left(\frac{\sqrt{3}}{2}\right)=\frac{5\sqrt{3}}{2}$$

The Cartesian coordinates are $\left(-\frac{5}{2},\frac{5\sqrt{3}}{2}\right)$.

Example 4

Find the polar coordinates of the point with Cartesian coordinates $(-3,-4)$.

We begin by finding the distance r using the Pythagorean relationship $x^2+y^2=r^2$

$$(-3)^2+(-4)^2=r^2$$
$$9+16=r^2$$
$$r^2=25$$
$$r=5$$

Now that we know the radius, we can find the angle using any of the three trig relationships. Keep in mind that any of the relationships will produce two solutions on the circle, and we need to consider the quadrant to determine which solution to accept. Using the cosine, for example:

$$\cos(\theta)=\frac{x}{r}=\frac{-3}{5}$$

$\theta=\cos^{-1}\left(\frac{-3}{5}\right)\approx 2.214$ By symmetry, there is a second possibility at

$$\theta=2\pi-2.214=4.069$$

Since the point (-3, -4) is located in the 3rd quadrant, we can determine that the second angle is the one we need. The polar coordinates of this point are $(r,\theta)=(5,4.069)$.

Try it Now

2. Convert the following.
   a. Convert polar coordinates $(r,\theta)=(2,\pi)$ to (x,y).
   b. Convert Cartesian coordinates $(x,y)=(0,-4)$ to (r,θ).

Polar Equations

Just as a Cartesian equation like $y=x^2$ describes a relationship between x and y values on a Cartesian grid, a polar equation can be written describing a relationship between r and θ values on the polar grid.

Example 5

Sketch a graph of the polar equation $r=\theta$.

The equation $r=\theta$ describes all the points for which the radius r is equal to the angle. To visualize this relationship, we can create a table of values.

θ	0	$\pi/4$	$\pi/2$	$3\pi/4$	π	$5\pi/4$	$3\pi/2$	$7\pi/4$	2π
r	0	$\pi/4$	$\pi/2$	$3\pi/4$	π	$5\pi/4$	$3\pi/2$	$7\pi/4$	2π

We can plot these points on the plane, and then sketch a curve that fits the points. The resulting graph is a spiral.

Notice that the resulting graph cannot be the result of a function of the form $y=f(x)$, as it does not pass the vertical line test, even though it resulted from a function giving r in terms of θ.

Although it is nice to see polar equations on polar grids, it is more common for polar graphs to be graphed on the Cartesian coordinate system, and so, the remainder of the polar equations will be graphed accordingly.

The spiral graph above on a Cartesian grid is shown here.

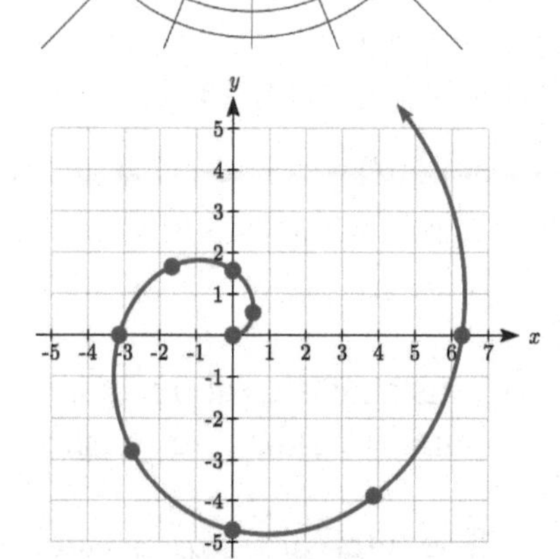

Example 6

Sketch a graph of the polar equation $r=3$.

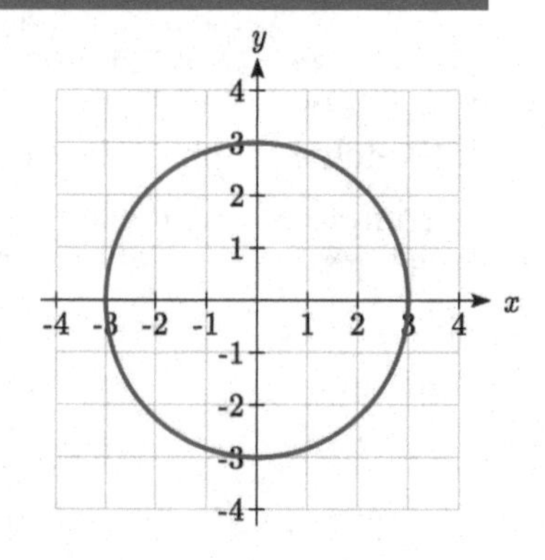

Recall that when a variable does not show up in the equation, it is saying that it does not matter what value that variable has; the output for the equation will remain the same. For example, the Cartesian equation $y=3$ describes all the points where $y=3$, no matter what the x values are, producing a horizontal line.

Likewise, this polar equation is describing all the points at a distance of 3 from the origin, no matter what the angle is, producing the graph of a **circle.**

The normal settings on graphing calculators and software graph on the Cartesian coordinate system with y being a function of x, where the graphing utility asks for $f(x)$, or simply $y =$.

To graph polar equations, you may need to change the mode of your calculator to Polar. You will know you have been successful in changing the mode if you now have r as a function of θ, where the graphing utility asks for $r(\theta)$, or simply $r =$.

Example 7

Sketch a graph of the polar equation $r = 4\cos(\theta)$, and find an interval on which it completes one cycle.

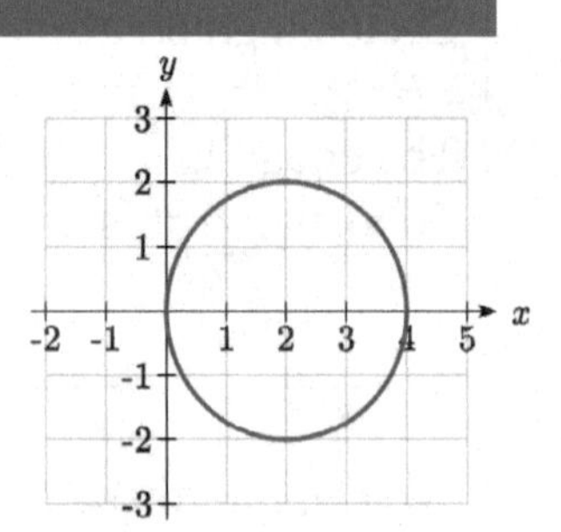

While we could again create a table, plot the corresponding points, and connect the dots, we can also turn to technology to directly graph it. Using technology, we produce the graph shown here, a circle passing through the origin.

Since this graph appears to close a loop and repeat itself, we might ask what interval of θ values yields the entire graph. At $\theta = 0$, $r = 4\cos(0) = 4$, yielding the point (4, 0). We want the next θ value when the graph returns to the point (4, 0). Solving for when $x = 4$ is equivalent to solving $r\cos(\theta) = 4$.

$r\cos(\theta) = 4$ Substituting the equation for r gives

$4\cos(\theta)\cos(\theta) = 4$ Dividing by 4 and simplifying

$\cos^2(\theta) = 1$ This has solutions when

$\cos(\theta) = 1$ or $\cos(\theta) = -1$ Solving these gives solutions

$\theta = 0$ or $\theta = \pi$

This shows us at 0 radians we are at the point (0, 4), and again at π radians we are at the point (0, 4) having finished one complete revolution.

The interval $0 \le \theta < \pi$ yields one complete iteration of the circle.

Try it Now

3. Sketch a graph of the polar equation $r = 3\sin(\theta)$, and find an interval on which it completes one cycle.

The last few examples have all been circles. Next, we will consider two other "named" polar equations, **limaçons** and **roses**.

Example 8

Sketch a graph of the polar equation $r = 4\sin(\theta) + 2$. What interval of θ values corresponds to the inner loop?

This type of graph is called a **limaçon**.
Using technology, we can draw the graph. The inner loop begins and ends at the origin, where $r = 0$. We can solve for the θ values for which $r = 0$.

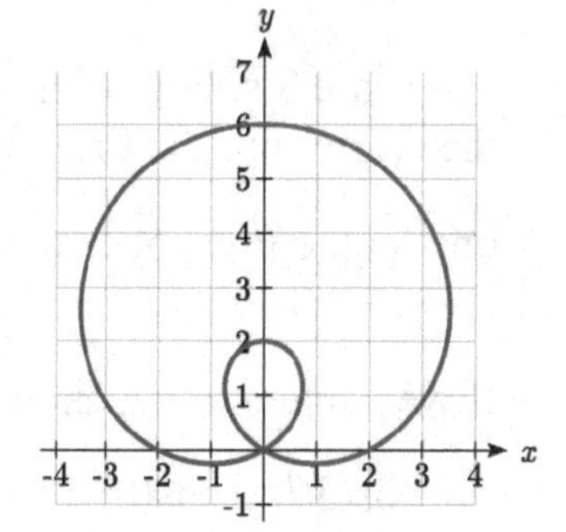

$$0 = 4\sin(\theta) + 2$$
$$-2 = 4\sin(\theta)$$
$$\sin(\theta) = -\frac{1}{2}$$
$$\theta = \frac{7\pi}{6} \text{ or } \theta = \frac{11\pi}{6}$$

This tells us that $r = 0$, so the graph passes through the origin, twice on the interval $[0, 2\pi)$.

The inner loop arises from the interval $\frac{7\pi}{6} \le \theta \le \frac{11\pi}{6}$.

This corresponds to where the function $r = 4\sin(\theta) + 2$ takes on negative values, as we could see if we graphed the function in the $r\theta$ plane.

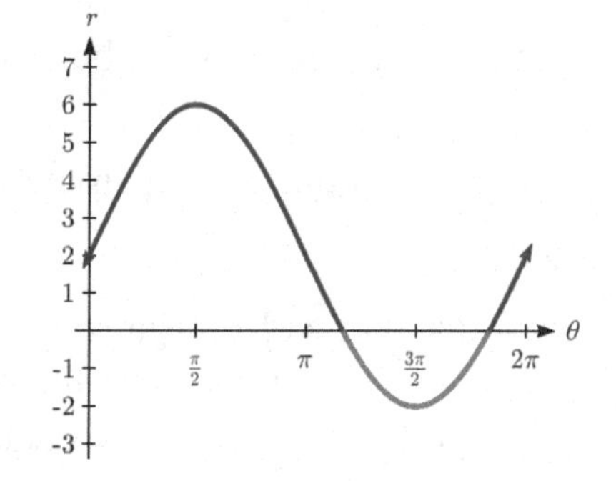

Example 9

Sketch a graph of the polar equation $r = \cos(3\theta)$. What interval of θ values describes one small loop of the graph?

This type of graph is called a 3 leaf **rose.**

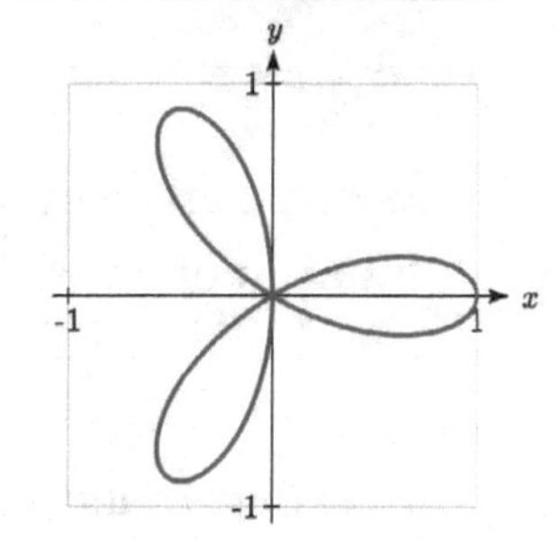

We can use technology to produce a graph. The interval $[0, \pi)$ yields one cycle of this function. As with the last problem, we can note that there is an interval on which one loop of this graph begins and ends at the origin, where $r = 0$. Solving for θ,

$$0 = \cos(3\theta) \qquad \text{Substitute } u = 3\theta$$
$$0 = \cos(u)$$
$$u = \frac{\pi}{2} \text{ or } u = \frac{3\pi}{2} \text{ or } u = \frac{5\pi}{2}$$

Undo the substitution,

$$3\theta = \frac{\pi}{2} \quad \text{or} \quad 3\theta = \frac{3\pi}{2} \quad \text{or} \quad 3\theta = \frac{5\pi}{2}$$

$$\theta = \frac{\pi}{6} \quad \text{or} \quad \theta = \frac{\pi}{2} \quad \text{or} \quad \theta = \frac{5\pi}{6}$$

There are 3 solutions on $0 \le \theta < \pi$ which correspond to the 3 times the graph returns to the origin, but the first two solutions we solved for above are enough to conclude that one loop corresponds to the interval $\frac{\pi}{6} \le \theta < \frac{\pi}{2}$.

If we wanted to get an idea of how the computer drew this graph, consider when $\theta = 0$. $r = \cos(3\theta) = \cos(0) = 1$, so the graph starts at (1,0). As we found above, at $\theta = \frac{\pi}{6}$ and $\theta = \frac{\pi}{2}$, the graph is at the origin. Looking at the equation, notice that any angle in between $\frac{\pi}{6}$ and $\frac{\pi}{2}$, for example at $\theta = \frac{\pi}{3}$, produces a negative r: $r = \cos\left(3 \cdot \frac{\pi}{3}\right) = \cos(\pi) = -1$.

θ	r	x	y
0	1	1	0
$\frac{\pi}{6}$	0	0	0
$\frac{\pi}{3}$	-1	$-\frac{1}{2}$	$-\frac{\sqrt{3}}{2}$
$\frac{\pi}{2}$	0	0	0

Notice that with a negative r value and an angle with terminal side in the first quadrant, the corresponding Cartesian point would be in the third quadrant. Since $r = \cos(3\theta)$ is negative on $\frac{\pi}{6} \le \theta < \frac{\pi}{2}$, this interval corresponds to the loop of the graph in the third quadrant.

Try it Now

4. Sketch a graph of the polar equation $r = \sin(2\theta)$. Would you call this function a **limaçon** or a **rose**?

Converting Equations

While many polar equations cannot be expressed nicely in Cartesian form (and vice versa), it can be beneficial to convert between the two forms, when possible. To do this we use the same relationships we used to convert points between coordinate systems.

Example 10

Rewrite the Cartesian equation $x^2 + y^2 = 6y$ as a polar equation.

We wish to eliminate *x* and *y* from the equation and introduce *r* and θ. Ideally, we would like to write the equation with *r* isolated, if possible, which represents *r* as a function of θ.

$x^2 + y^2 = 6y$	Remembering $x^2 + y^2 = r^2$ we substitute
$r^2 = 6y$	$y = r\sin(\theta)$ and so we substitute again
$r^2 = 6r\sin(\theta)$	Subtract $6r\sin(\theta)$ from both sides
$r^2 - 6r\sin(\theta) = 0$	Factor
$r(r - 6\sin(\theta)) = 0$	Use the zero factor theorem
$r = 6\sin(\theta)$ or $r = 0$	Since $r = 0$ is only a point, we reject that solution.

The solution $r = 6\sin(\theta)$ is fairly similar to the one we graphed in Example 7. In fact, this equation describes a circle with bottom at the origin and top at the point (0, 6).

Example 11

Rewrite the Cartesian equation $y = 3x + 2$ as a polar equation.

$y = 3x + 2$	Use $y = r\sin(\theta)$ and $x = r\cos(\theta)$
$r\sin(\theta) = 3r\cos(\theta) + 2$	Move all terms with *r* to one side
$r\sin(\theta) - 3r\cos(\theta) = 2$	Factor out *r*
$r(\sin(\theta) - 3\cos(\theta)) = 2$	Divide
$r = \dfrac{2}{\sin(\theta) - 3\cos(\theta)}$	

In this case, the polar equation is more unwieldy than the Cartesian equation, but there are still times when this equation might be useful.

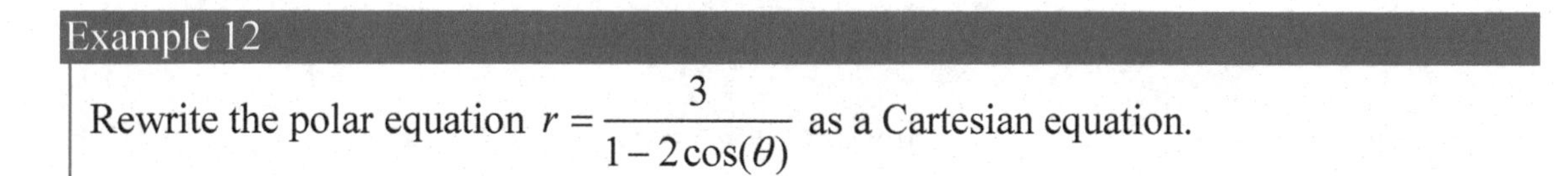

Example 12

Rewrite the polar equation $r = \dfrac{3}{1 - 2\cos(\theta)}$ as a Cartesian equation.

We want to eliminate θ and r and introduce x and y. It is usually easiest to start by clearing the fraction and looking to substitute values that will eliminate θ.

$r = \dfrac{3}{1-2\cos(\theta)}$ Clear the fraction

$r(1-2\cos(\theta)) = 3$ Use $\cos(\theta) = \dfrac{x}{r}$ to eliminate θ

$r\left(1-2\dfrac{x}{r}\right) = 3$ Distribute and simplify

$r - 2x = 3$ Isolate the r

$r = 3 + 2x$ Square both sides

$r^2 = (3+2x)^2$ Use $x^2 + y^2 = r^2$

$x^2 + y^2 = (3+2x)^2$

When our entire equation has been changed from r and θ to x and y we can stop unless asked to solve for y or simplify.

In this example, if desired, the right side of the equation could be expanded and the equation simplified further. However, the equation cannot be written as a function in Cartesian form.

Try it Now

5. a. Rewrite the Cartesian equation in polar form: $y = \pm\sqrt{3-x^2}$

   b. Rewrite the polar equation in Cartesian form: $r = 2\sin(\theta)$

Example 13

Rewrite the polar equation $r = \sin(2\theta)$ in Cartesian form.

$r = \sin(2\theta)$ Use the double angle identity for sine

$r = 2\sin(\theta)\cos(\theta)$ Use $\cos(\theta) = \dfrac{x}{r}$ and $\sin(\theta) = \dfrac{y}{r}$

$r = 2 \cdot \dfrac{x}{r} \cdot \dfrac{y}{r}$ Simplify

$r = \dfrac{2xy}{r^2}$ Multiply by r^2

$r^3 = 2xy$ Since $x^2 + y^2 = r^2$, $r = \sqrt{x^2 + y^2}$

$\left(\sqrt{x^2 + y^2}\right)^3 = 2xy$

This equation could also be written as

$$\left(x^2+y^2\right)^{3/2}=2xy \quad \text{or} \qquad x^2+y^2=\left(2xy\right)^{2/3}$$

Important Topics of This Section
Cartesian coordinate system
Polar coordinate system
Plotting points in polar coordinates
Converting coordinates between systems
Polar equations: Spirals, circles, limaçons and roses
Converting equations between systems

Try it Now Answers

1.

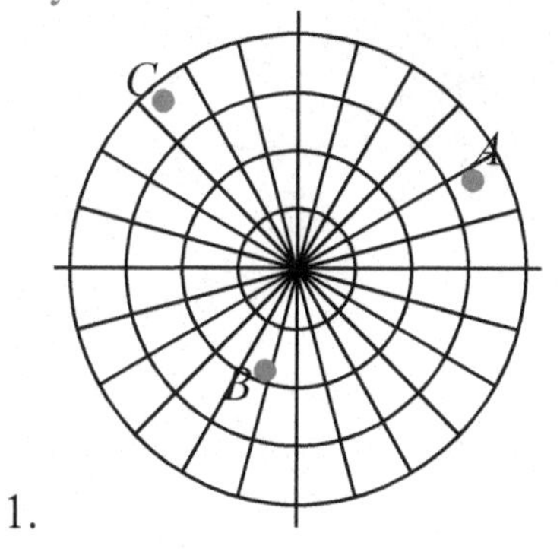

2. a. $(r,\theta)=\left(2,\pi\right)$ converts to $(x,y)=\left(2\cos(\pi),2\sin(\pi)\right)=(-2,0)$

b. $(x,y)=\left(0,-4\right)$ converts to $(r,\theta)=\left(4,\dfrac{3\pi}{2}\right) or \left(-4,\dfrac{\pi}{2}\right)$

3. $3\sin(\theta)=0$ at $\theta=0$ and $\theta=\pi$.
It completes one cycle on the interval $0\le\theta<\pi$.

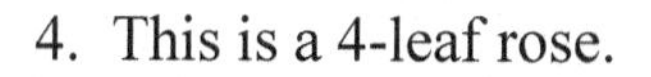

4. This is a 4-leaf rose.

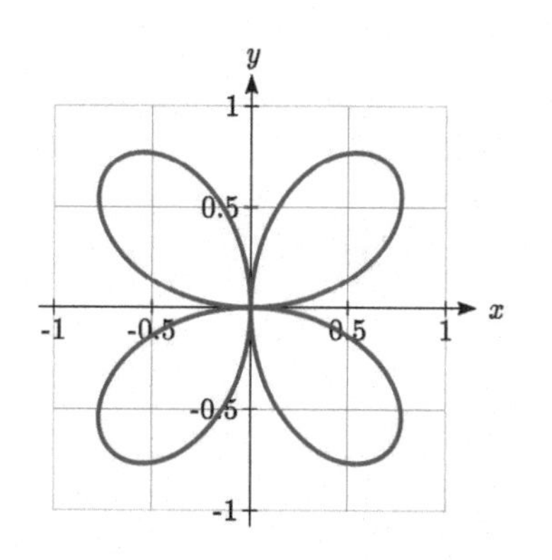

5. a. $y=\pm\sqrt{3-x^2}$ can be rewritten as $x^2+y^2=3$, and becomes $r=\sqrt{3}$

b. $r = 2\sin(\theta)$. $r = 2\frac{y}{r}$. $r^2 = 2y$. $x^2 + y^2 = 2y$

Section 8.2 Exercises

Convert the given polar coordinates to Cartesian coordinates.

1. $\left(7, \frac{7\pi}{6}\right)$
2. $\left(6, \frac{3\pi}{4}\right)$
3. $\left(4, \frac{7\pi}{4}\right)$
4. $\left(9, \frac{4\pi}{3}\right)$
5. $\left(6, -\frac{\pi}{4}\right)$
6. $\left(12, -\frac{\pi}{3}\right)$
7. $\left(3, \frac{\pi}{2}\right)$
8. $(5, \pi)$
9. $\left(-3, \frac{\pi}{6}\right)$
10. $\left(-2, \frac{2\pi}{3}\right)$
11. $(3, 2)$
12. $(7, 1)$

Convert the given Cartesian coordinates to polar coordinates.

13. $(4, 2)$
14. $(8, 8)$
15. $(-4, 6)$
16. $(-5, 1)$
17. $(3, -5)$
18. $(6, -5)$
19. $(-10, -13)$
20. $(-4, -7)$

Convert the given Cartesian equation to a polar equation.

21. $x = 3$
22. $y = 4$
23. $y = 4x^2$
24. $y = 2x^4$
25. $x^2 + y^2 = 4y$
26. $x^2 + y^2 = 3x$
27. $x^2 - y^2 = x$
28. $x^2 - y^2 = 3y$

Convert the given polar equation to a Cartesian equation.

29. $r = 3\sin(\theta)$
30. $r = 4\cos(\theta)$
31. $r = \dfrac{4}{\sin(\theta) + 7\cos(\theta)}$
32. $r = \dfrac{6}{\cos(\theta) + 3\sin(\theta)}$
33. $r = 2\sec(\theta)$
34. $r = 3\csc(\theta)$
35. $r = \sqrt{r\cos(\theta) + 2}$
36. $r^2 = 4\sec(\theta)\csc(\theta)$

Match each equation with one of the graphs shown.

37. $r = 2 + 2\cos(\theta)$
38. $r = 2 + 2\sin(\theta)$
39. $r = 4 + 3\cos(\theta)$
40. $r = 3 + 4\cos(\theta)$
41. $r = 5$
42. $r = 2\sin(\theta)$

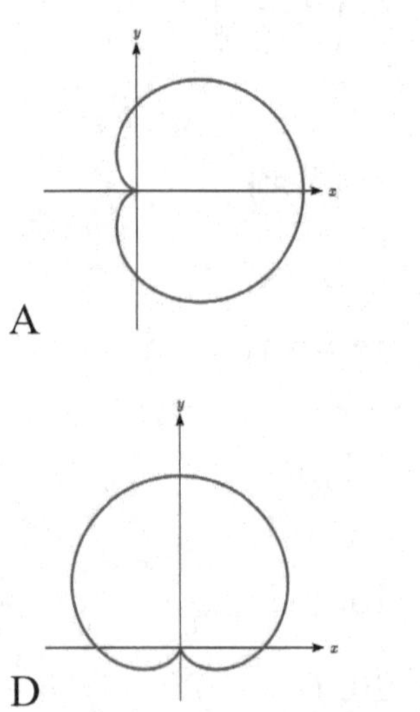

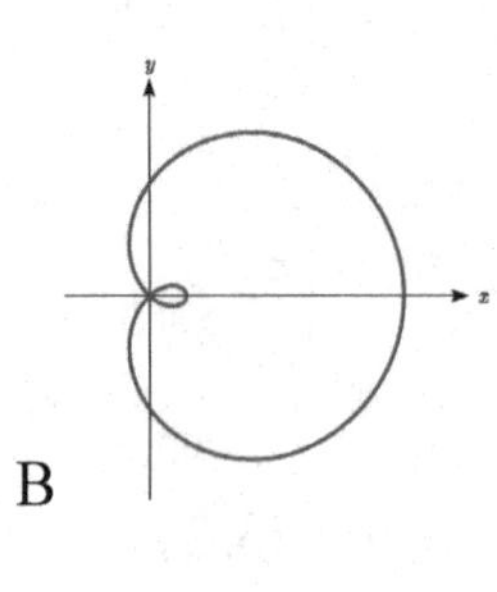

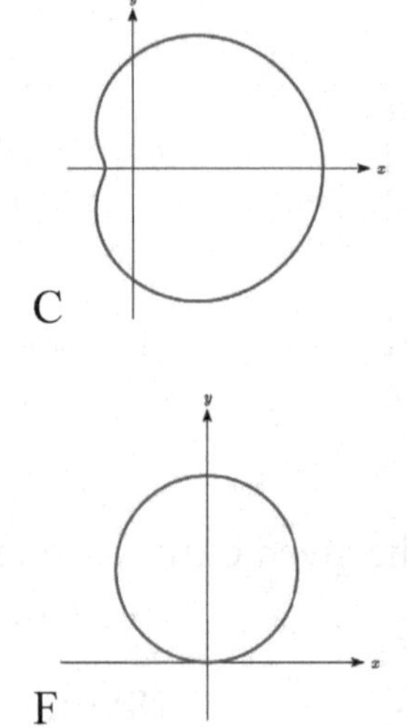

A B C

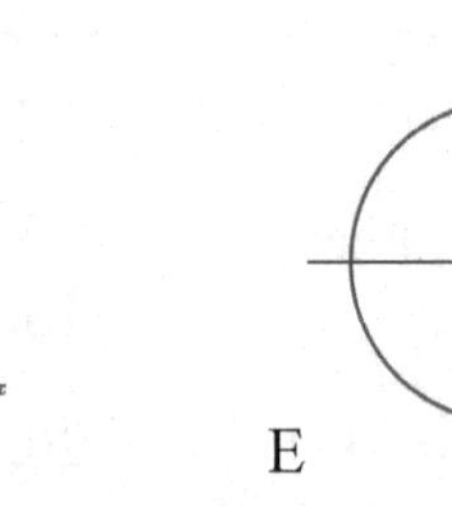

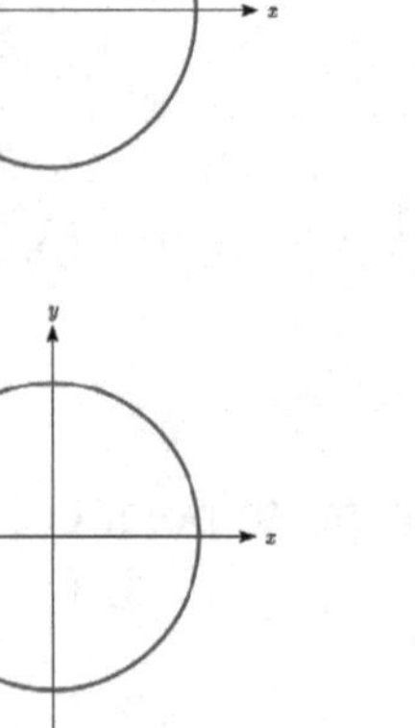

D E F

Match each equation with one of the graphs shown.

43. $r = \log(\theta)$
44. $r = \theta\cos(\theta)$
45. $r = \cos\left(\dfrac{\theta}{2}\right)$
46. $r = \sin(\theta)\cos^2(\theta)$
47. $r = 1 + 2\sin(3\theta)$
48. $r = 1 + \sin(2\theta)$

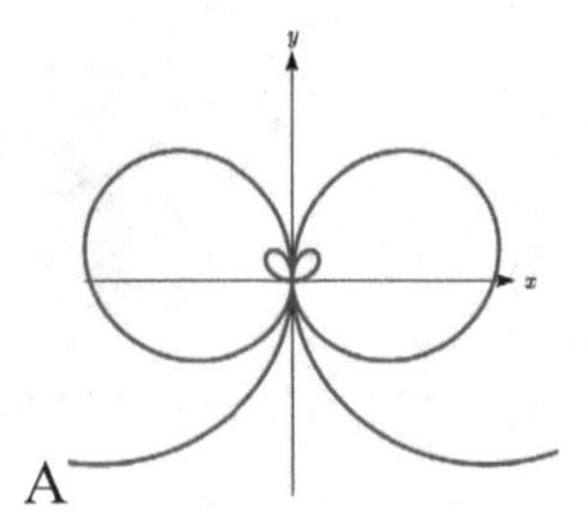

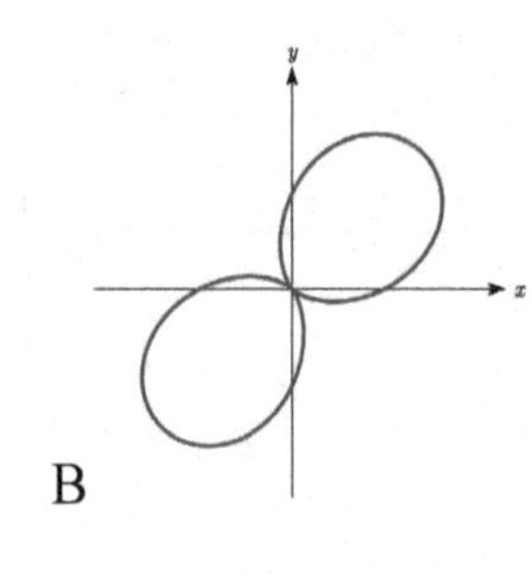

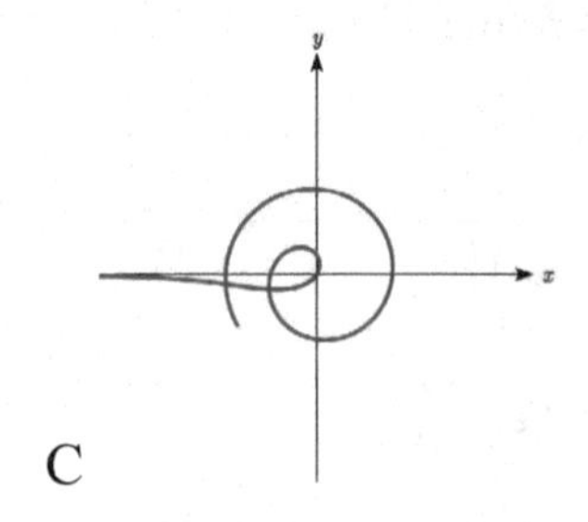

A B C

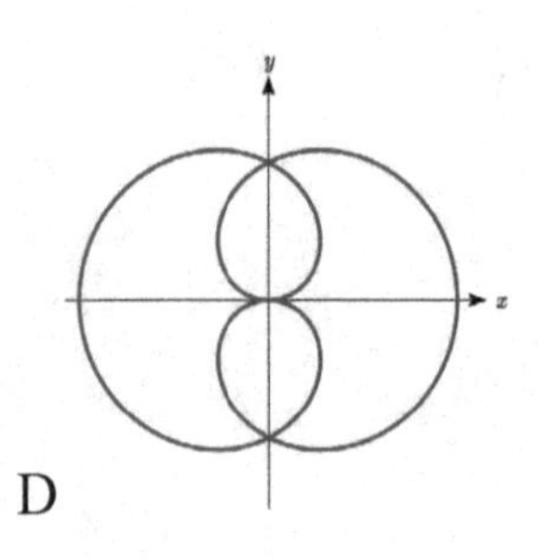

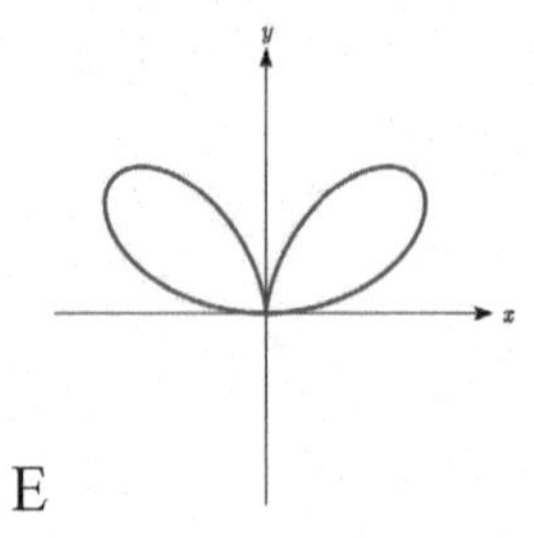

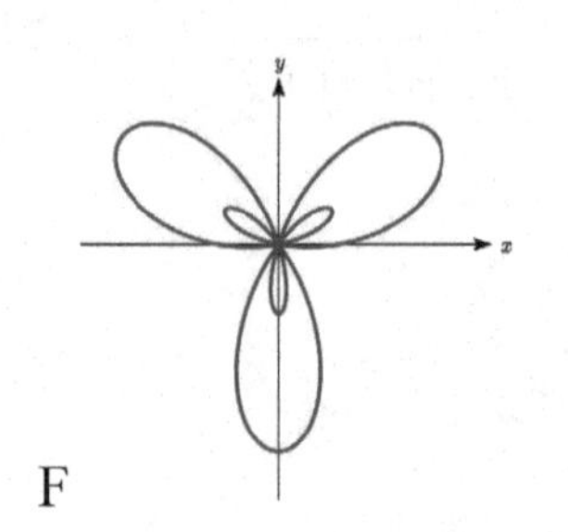

D E F

Sketch a graph of the polar equation.

49. $r = 3\cos(\theta)$

50. $r = 4\sin(\theta)$

51. $r = 3\sin(2\theta)$

52. $r = 4\sin(4\theta)$

53. $r = 5\sin(3\theta)$

54. $r = 4\sin(5\theta)$

55. $r = 3\cos(2\theta)$

56. $r = 4\cos(4\theta)$

57. $r = 2 + 2\cos(\theta)$

58. $r = 3 + 3\sin(\theta)$

59. $r = 1 + 3\sin(\theta)$

60. $r = 2 + 4\cos(\theta)$

61. $r = 2\theta$

62. $r = \dfrac{1}{\theta}$

63. $r = 3 + \sec(\theta)$, a conchoid

64. $r = \dfrac{1}{\sqrt{\theta}}$, a lituus[1]

65. $r = 2\sin(\theta)\tan(\theta)$, a cissoid

66. $r = 2\sqrt{1 - \sin^2(\theta)}$, a hippopede

---

[1] This curve was the inspiration for the artwork featured on the cover of this book.

Section 8.3 Polar Form of Complex Numbers

From previous classes, you may have encountered "imaginary numbers" – the square roots of negative numbers – and, more generally, complex numbers which are the sum of a real number and an imaginary number. While these are useful for expressing the solutions to quadratic equations, they have much richer applications in electrical engineering, signal analysis, and other fields. Most of these more advanced applications rely on properties that arise from looking at complex numbers from the perspective of polar coordinates.

We will begin with a review of the definition of complex numbers.

Imaginary Number *i*

The most basic complex number is i, defined to be $i = \sqrt{-1}$, commonly called an **imaginary number**. Any real multiple of i is also an imaginary number.

Example 1

Simplify $\sqrt{-9}$.

We can separate $\sqrt{-9}$ as $\sqrt{9}\sqrt{-1}$. We can take the square root of 9, and write the square root of -1 as i.

$\sqrt{-9} = \sqrt{9}\sqrt{-1} = 3i$

A complex number is the sum of a real number and an imaginary number.

Complex Number

A **complex number** is a number $z = a + bi$, where a and b are real numbers

a is the real part of the complex number

b is the imaginary part of the complex number

$i = \sqrt{-1}$

Plotting a complex number

We can plot real numbers on a number line. For example, if we wanted to show the number 3, we plot a point:

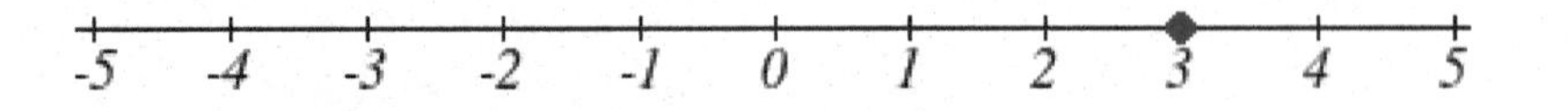

To plot a complex number like $3-4i$, we need more than just a number line since there are two components to the number. To plot this number, we need two number lines, crossed to form a complex plane.

Complex Plane

In the **complex plane**, the horizontal axis is the real axis and the vertical axis is the imaginary axis.

Example 2

Plot the number $3-4i$ on the complex plane.

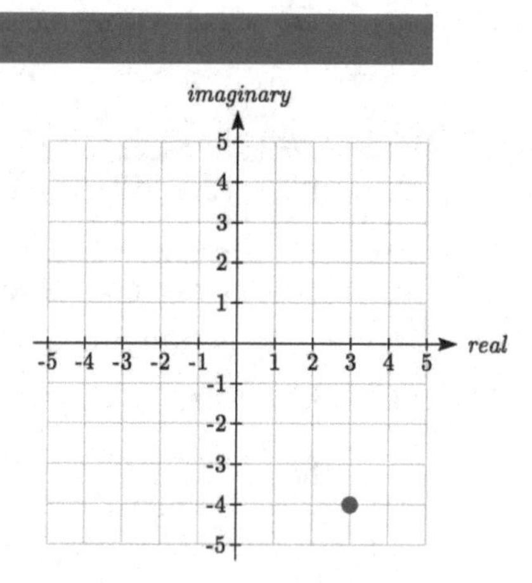

The real part of this number is 3, and the imaginary part is -4. To plot this, we draw a point 3 units to the right of the origin in the horizontal direction and 4 units down in the vertical direction.

Because this is analogous to the Cartesian coordinate system for plotting points, we can think about plotting our complex number $z=a+bi$ as if we were plotting the point (a, b) in Cartesian coordinates. Sometimes people write complex numbers as $z=x+yi$ to highlight this relation.

Arithmetic on Complex Numbers

Before we dive into the more complicated uses of complex numbers, let's make sure we remember the basic arithmetic involved. To add or subtract complex numbers, we simply add the like terms, combining the real parts and combining the imaginary parts.

Example 3

Add $3-4i$ and $2+5i$.

Adding $(3-4i)+(2+5i)$, we add the real parts and the imaginary parts

$3+2-4i+5i$

$5+i$

Try it Now

1. Subtract $2+5i$ from $3-4i$.

We can also multiply and divide complex numbers.

Example 4

Multiply: $4(2+5i)$.

To multiply the complex number by a real number, we simply distribute as we would when multiplying polynomials.

$4(2+5i)$	Distribute
$=4\cdot 2+4\cdot 5i$	Simplify
$=8+20i$	

Example 5

Multiply: $(2-3i)(1+4i)$.

To multiply two complex numbers, we expand the product as we would with polynomials (the process commonly called FOIL – "first outer inner last").

$(2-3i)(1+4i)$	Expand
$=2+8i-3i-12i^2$	Since $i=\sqrt{-1}$, $i^2=-1$
$=2+8i-3i-12(-1)$	Simplify
$=14+5i$	

Example 6

Divide $\dfrac{(2+5i)}{(4-i)}$.

To divide two complex numbers, we have to devise a way to write this as a complex number with a real part and an imaginary part.

We start this process by eliminating the complex number in the denominator. To do this, we multiply the numerator and denominator by a special complex number so that the result in the denominator is a real number. The number we need to multiply by is called the **complex conjugate**, in which the sign of the imaginary part is changed. Here, $4+i$ is the complex conjugate of $4-i$. Of course, obeying our algebraic rules, we must multiply by $4+i$ on both the top and bottom.

$$\frac{(2+5i)}{(4-i)}\cdot\frac{(4+i)}{(4+i)}$$

In the numerator,

$(2+5i)(4+i)$	Expand
$=8+20i+2i+5i^2$	Since $i=\sqrt{-1}$, $i^2=-1$
$=8+20i+2i+5(-1)$	Simplify
$=3+22i$	

Multiplying the denominator

$(4-i)(4+i)$	Expand
$(16-4i+4i-i^2)$	Since $i=\sqrt{-1}$, $i^2=-1$
$(16-(-1))$	
$=17$	

Combining this we get $\dfrac{3+22i}{17}=\dfrac{3}{17}+\dfrac{22i}{17}$

Try it Now

2. Multiply $3-4i$ and $2+3i$.

With the interpretation of complex numbers as points in a plane, which can be related to the Cartesian coordinate system, you might be starting to guess our next step – to refer to this point not by its horizontal and vertical components, but using its polar location, given by the distance from the origin and an angle.

Polar Form of Complex Numbers

Remember, because the complex plane is analogous to the Cartesian plane that we can think of a complex number $z=x+yi$ as analogous to the Cartesian point (x, y) and recall how we converted from (x, y) to polar (r, θ) coordinates in the last section.

Bringing in all of our old rules we remember the following:

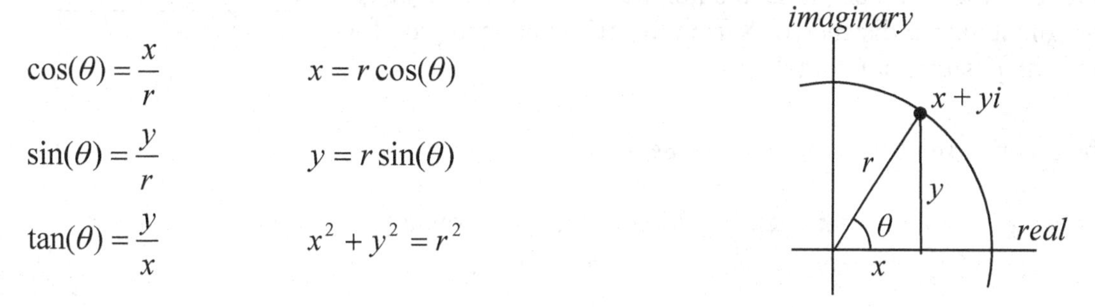

With this in mind, we can write $z=x+yi=r\cos(\theta)+ir\sin(\theta)$.

Example 7

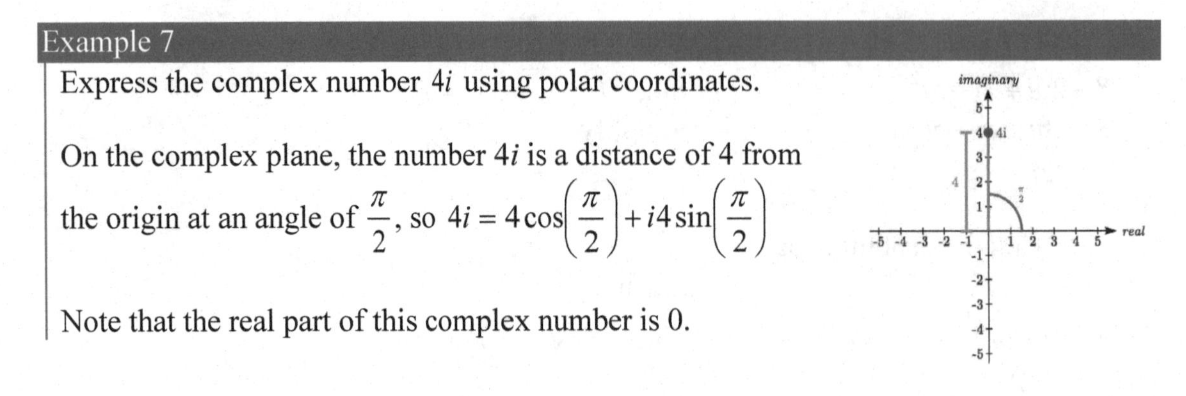

Express the complex number $4i$ using polar coordinates.

On the complex plane, the number $4i$ is a distance of 4 from the origin at an angle of $\frac{\pi}{2}$, so $4i = 4\cos\left(\frac{\pi}{2}\right) + i4\sin\left(\frac{\pi}{2}\right)$

Note that the real part of this complex number is 0.

In the 18th century, Leonhard Euler demonstrated a relationship between exponential and trigonometric functions that allows the use of complex numbers to greatly simplify some trigonometric calculations. While the proof is beyond the scope of this class, you will likely see it in a later calculus class.

Polar Form of a Complex Number and Euler's Formula

The **polar form of a complex number** is $z = r\cos(\theta) + ir\sin(\theta)$.

An alternate form, which will be the primary one used, is $z = re^{i\theta}$

Euler's Formula states $re^{i\theta} = r\cos(\theta) + ir\sin(\theta)$

Similar to plotting a point in the polar coordinate system we need r and θ to find the polar form of a complex number.

Example 8

Find the polar form of the complex number -8.

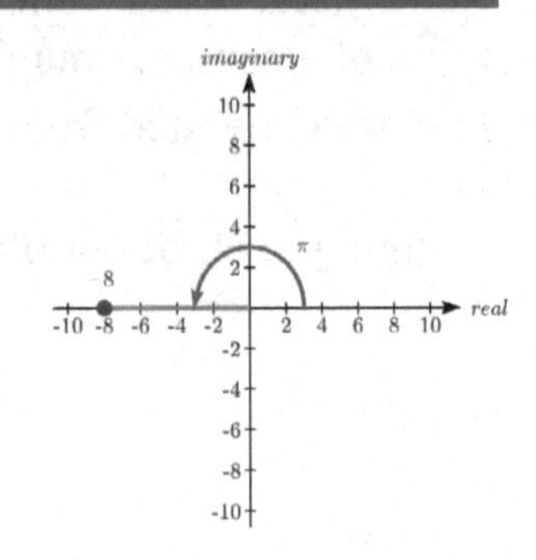

Treating this is a complex number, we can write it as $-8+0i$.

Plotted in the complex plane, the number -8 is on the negative horizontal axis, a distance of 8 from the origin at an angle of π from the positive horizontal axis.

The polar form of the number -8 is $8e^{i\pi}$.

Plugging $r = 8$ and $\theta = \pi$ back into Euler's formula, we have:
$8e^{i\pi} = 8\cos(\pi) + 8i\sin(\pi) = -8 + 0i = -8$ as desired.

Example 9

Find the polar form of $-4+4i$.

On the complex plane, this complex number would correspond to the point (-4, 4) on a Cartesian plane. We can find the distance r and angle θ as we did in the last section.

$$r^2 = x^2 + y^2$$
$$r^2 = (-4)^2 + 4^2$$
$$r = \sqrt{32} = 4\sqrt{2}$$

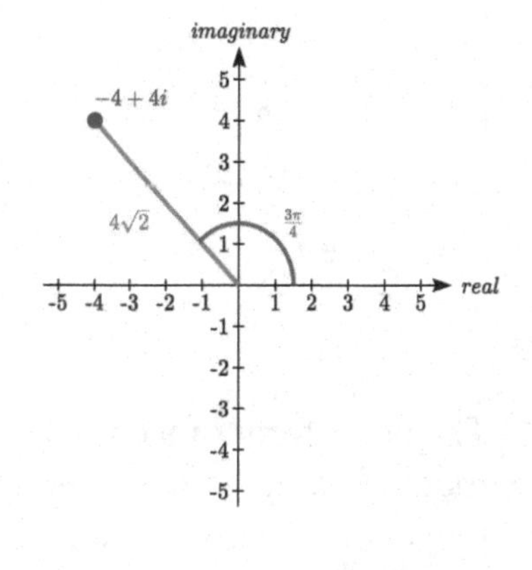

To find θ, we can use $\cos(\theta) = \frac{x}{r}$

$$\cos(\theta) = \frac{-4}{4\sqrt{2}} = -\frac{\sqrt{2}}{2}$$

This is one of known cosine values, and since the point is in the second quadrant, we can conclude that $\theta = \frac{3\pi}{4}$.

The polar form of this complex number is $4\sqrt{2}e^{\frac{3\pi}{4}i}$.

Example 10

Find the polar form of $-3-5i$.

On the complex plane, this complex number would correspond to the point (-3, -5) on a Cartesian plane. First, we find r.

$$r^2 = x^2 + y^2$$
$$r^2 = (-3)^2 + (-5)^2$$
$$r = \sqrt{34}$$

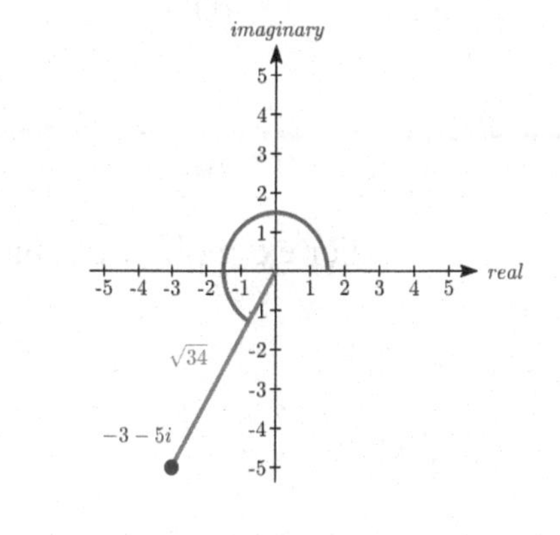

To find θ, we might use $\tan(\theta) = \frac{y}{x}$

$$\tan(\theta) = \frac{-5}{-3}$$

$$\theta = \tan^{-1}\left(\frac{5}{3}\right) = 1.0304$$

This angle is in the wrong quadrant, so we need to find a second solution. For tangent, we can find that by adding π.

$$\theta = 1.0304 + \pi = 4.1720$$

The polar form of this complex number is $\sqrt{34}e^{4.1720i}$.

Try it Now

3. Write $\sqrt{3}+i$ in polar form.

Example 11

Write $3e^{\frac{\pi}{6}i}$ in complex $a+bi$ form.

$3e^{\frac{\pi}{6}i} = 3\cos\left(\frac{\pi}{6}\right)+i3\sin\left(\frac{\pi}{6}\right)$ — Evaluate the trig functions

$= 3\cdot\frac{\sqrt{3}}{2}+i3\cdot\frac{1}{2}$ — Simplify

$= \frac{3\sqrt{3}}{2}+i\frac{3}{2}$

The polar form of a complex number provides a powerful way to compute powers and roots of complex numbers by using exponent rules you learned in algebra. To compute a power of a complex number, we:

1) Convert to polar form
2) Raise to the power, using exponent rules to simplify
3) Convert back to $a + bi$ form, if needed

Example 12

Evaluate $(-4+4i)^6$.

While we could multiply this number by itself five times, that would be very tedious. To compute this more efficiently, we can utilize the polar form of the complex number.

In an earlier example, we found that $-4+4i = 4\sqrt{2}e^{\frac{3\pi}{4}i}$. Using this,

$(-4+4i)^6$ — Write the complex number in polar form

$= \left(4\sqrt{2}e^{\frac{3\pi}{4}i}\right)^6$ — Utilize the exponent rule $(ab)^m = a^m b^m$

$= \left(4\sqrt{2}\right)^6\left(e^{\frac{3\pi}{4}i}\right)^6$ — On the second factor, use the rule $(a^m)^n = a^{mn}$

$= \left(4\sqrt{2}\right)^6 e^{\frac{3\pi}{4}i\cdot 6}$ — Simplify

$= 32768e^{\frac{9\pi}{2}i}$

At this point, we have found the power as a complex number in polar form. If we want the answer in standard $a + bi$ form, we can utilize Euler's formula.

$$32768e^{\frac{9\pi}{2}i} = 32768\cos\left(\frac{9\pi}{2}\right) + i32768\sin\left(\frac{9\pi}{2}\right)$$

Since $\frac{9\pi}{2}$ is coterminal with $\frac{\pi}{2}$, we can use our special angle knowledge to evaluate the sine and cosine.

$$32768\cos\left(\frac{9\pi}{2}\right) + i32768\sin\left(\frac{9\pi}{2}\right) = 32768(0) + i32768(1) = 32768i$$

We have found that $(-4+4i)^6 = 32768i$.

The result of the process can be summarized by DeMoivre's Theorem. This is a shorthand to using exponent rules.

DeMoivre's Theorem
If $z = r(\cos(\theta) + i\sin(\theta))$, then for any integer n, $z^n = r^n(\cos(n\theta) + i\sin(n\theta))$

We omit the proof, but note we can easily verify it holds in one case using Example 12:

$$(-4+4i)^6 = \left(4\sqrt{2}\right)^6\left(\cos\left(6\cdot\frac{3\pi}{4}\right) + i\sin\left(6\cdot\frac{3\pi}{4}\right)\right) = 32768\left(\cos\left(\frac{9\pi}{2}\right) + i\sin\left(\frac{9\pi}{2}\right)\right) = 32768i$$

Example 13

Evaluate $\sqrt{9i}$.

To evaluate the square root of a complex number, we can first note that the square root is the same as having an exponent of $\frac{1}{2}$: $\sqrt{9i} = (9i)^{1/2}$.

To evaluate the power, we first write the complex number in polar form. Since $9i$ has no real part, we know that this value would be plotted along the vertical axis, a distance of 9 from the origin at an angle of $\frac{\pi}{2}$. This gives the polar form: $9i = 9e^{\frac{\pi}{2}i}$.

$\sqrt{9i} = (9i)^{1/2}$ Use the polar form

$= \left(9e^{\frac{\pi}{2}i}\right)^{1/2}$ Use exponent rules to simplify

$= 9^{1/2}\left(e^{\frac{\pi}{2}i}\right)^{1/2}$

$= 9^{1/2}e^{\frac{\pi}{2}i\cdot\frac{1}{2}}$ Simplify

$= 3e^{\frac{\pi}{4}i}$ Rewrite using Euler's formula if desired

$= 3\cos\left(\frac{\pi}{4}\right) + i3\sin\left(\frac{\pi}{4}\right)$ Evaluate the sine and cosine

$= 3\frac{\sqrt{2}}{2} + i3\frac{\sqrt{2}}{2}$

Using the polar form, we were able to find a square root of a complex number.

$$\sqrt{9i} = \frac{3\sqrt{2}}{2} + \frac{3\sqrt{2}}{2}i$$

Alternatively, using DeMoivre's Theorem we could write

$\left(9e^{\frac{\pi}{2}i}\right)^{1/2} = 9^{1/2}\left(\cos\left(\frac{1}{2}\cdot\frac{\pi}{2}\right) + i\sin\left(\frac{1}{2}\cdot\frac{\pi}{2}\right)\right) = 3\left(\cos\left(\frac{\pi}{4}\right) + i\sin\left(\frac{\pi}{4}\right)\right)$ and simplify

Try it Now

4. Evaluate $\left(\sqrt{3}+i\right)^6$ using polar form.

You may remember that equations like $x^2 = 4$ have two solutions, 2 and -2 in this case, though the square root $\sqrt{4}$ only gives one of those solutions. Likewise, the square root we found in Example 11 is only one of two complex numbers whose square is $9i$. Similarly, the equation $z^3 = 8$ would have three solutions where only one is given by the cube root. In this case, however, only one of those solutions, $z = 2$, is a real value. To find the others, we can use the fact that complex numbers have multiple representations in polar form.

Example 14

Find all complex solutions to $z^3 = 8$.

Since we are trying to solve $z^3 = 8$, we can solve for z as $z = 8^{1/3}$. Certainly one of these solutions is the basic cube root, giving z = 2. To find others, we can turn to the polar representation of 8.

Since 8 is a real number, is would sit in the complex plane on the horizontal axis at an angle of 0, giving the polar form $8e^{0i}$. Taking the 1/3 power of this gives the real solution:

$$\left(8e^{0i}\right)^{1/3} = 8^{1/3}\left(e^{0i}\right)^{1/3} = 2e^0 = 2\cos(0) + i2\sin(0) = 2$$

However, since the angle 2π is coterminal with the angle of 0, we could also represent the number 8 as $8e^{2\pi i}$. Taking the 1/3 power of this gives a first complex solution:

$$\left(8e^{2\pi i}\right)^{1/3} = 8^{1/3}\left(e^{2\pi i}\right)^{1/3} = 2e^{\frac{2\pi}{3}i} = 2\cos\left(\frac{2\pi}{3}\right) + i2\sin\left(\frac{2\pi}{3}\right) = 2\left(-\frac{1}{2}\right) + i2\left(\frac{\sqrt{3}}{2}\right) = -1 + \sqrt{3}i$$

For the third root, we use the angle of 4π, which is also coterminal with an angle of 0.

$$\left(8e^{4\pi i}\right)^{1/3} = 8^{1/3}\left(e^{4\pi i}\right)^{1/3} = 2e^{\frac{4\pi}{3}i} = 2\cos\left(\frac{4\pi}{3}\right) + i2\sin\left(\frac{4\pi}{3}\right) = 2\left(-\frac{1}{2}\right) + i2\left(-\frac{\sqrt{3}}{2}\right) = -1 - \sqrt{3}i$$

Altogether, we found all three complex solutions to $z^3 = 8$,

$z = 2, \quad -1+\sqrt{3}i, \quad -1-\sqrt{3}i$

Graphed, these three numbers would be equally spaced on a circle about the origin at a radius of 2.

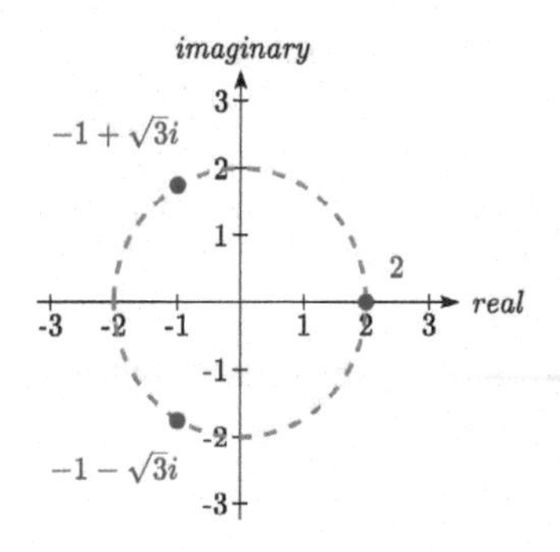

Important Topics of This Section
Complex numbers
Imaginary numbers
Plotting points in the complex coordinate system
Basic operations with complex numbers
Euler's Formula
DeMoivre's Theorem
Finding complex solutions to equations

Try it Now Answers

1. $(3-4i)-(2+5i)=1-9i$

2. $(3-4i)(2+3i)=18+i$

3. $\sqrt{3}+i$ would correspond with the point $\left(\sqrt{3},1\right)$ in the first quadrant.

$$r=\sqrt{\sqrt{3}^2+1^2}=\sqrt{4}=2$$

$$\sin\left(\theta\right)=\frac{1}{2}, \text{ so } \theta=\frac{\pi}{6}$$

$\sqrt{3}+i$ in polar form is $2e^{i\pi/6}$

4. $\left(\sqrt{3}+i\right)^6=\left(2e^{i\pi/6}\right)^6=2^6e^{i\pi}=64\cos(\pi)+i64\sin(\pi)=-64$

Section 8.3 Exercises

Simplify each expression to a single complex number.

1. $\sqrt{-9}$
2. $\sqrt{-16}$
3. $\sqrt{-6}\sqrt{-24}$
4. $\sqrt{-3}\sqrt{-75}$
5. $\dfrac{2+\sqrt{-12}}{2}$
6. $\dfrac{4+\sqrt{-20}}{2}$

Simplify each expression to a single complex number.

7. $(3+2i)+(5-3i)$
8. $(-2-4i)+(1+6i)$
9. $(-5+3i)-(6-i)$
10. $(2-3i)-(3+2i)$
11. $(2+3i)(4i)$
12. $(5-2i)(3i)$
13. $(6-2i)(5)$
14. $(-2+4i)(8)$
15. $(2+3i)(4-i)$
16. $(-1+2i)(-2+3i)$
17. $(4-2i)(4+2i)$
18. $(3+4i)(3-4i)$
19. $\dfrac{3+4i}{2}$
20. $\dfrac{6-2i}{3}$
21. $\dfrac{-5+3i}{2i}$
22. $\dfrac{6+4i}{i}$
23. $\dfrac{2-3i}{4+3i}$
24. $\dfrac{3+4i}{2-i}$
25. i^6
26. i^{11}
27. i^{17}
28. i^{24}

Rewrite each complex number from polar form into $a+bi$ form.

29. $3e^{2i}$
30. $4e^{4i}$
31. $6e^{\frac{\pi}{6}i}$
32. $8e^{\frac{\pi}{3}i}$
33. $3e^{\frac{5\pi}{4}i}$
34. $5e^{\frac{7\pi}{4}i}$

Rewrite each complex number into polar $re^{i\theta}$ form.

35. 6
36. -8
37. $-4i$
38. $6i$
39. $2+2i$
40. $4+4i$
41. $-3+3i$
42. $-4-4i$
43. $5+3i$
44. $4+7i$
45. $-3+i$
46. $-2+3i$
47. $-1-4i$
48. $-3-6i$
49. $5-i$
50. $1-3i$

Compute each of the following, leaving the result in polar $re^{i\theta}$ form.

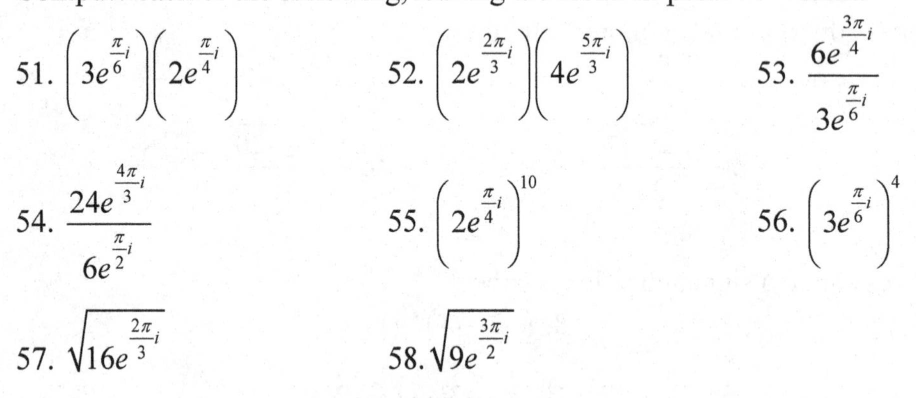

51. $\left(3e^{\frac{\pi}{6}i}\right)\left(2e^{\frac{\pi}{4}i}\right)$
52. $\left(2e^{\frac{2\pi}{3}i}\right)\left(4e^{\frac{5\pi}{3}i}\right)$
53. $\dfrac{6e^{\frac{3\pi}{4}i}}{3e^{\frac{\pi}{6}i}}$

54. $\dfrac{24e^{\frac{4\pi}{3}i}}{6e^{\frac{\pi}{2}i}}$
55. $\left(2e^{\frac{\pi}{4}i}\right)^{10}$
56. $\left(3e^{\frac{\pi}{6}i}\right)^{4}$

57. $\sqrt{16e^{\frac{2\pi}{3}i}}$
58. $\sqrt{9e^{\frac{3\pi}{2}i}}$

Compute each of the following, simplifying the result into $a+bi$ form.

59. $(2+2i)^8$
60. $(4+4i)^6$
61. $\sqrt{-3+3i}$

62. $\sqrt{-4-4i}$
63. $\sqrt[3]{5+3i}$
64. $\sqrt[4]{4+7i}$

Solve each of the following equations for all complex solutions.

65. $z^5=2$
66. $z^7=3$
67. $z^6=1$
68. $z^8=1$

Section 8.4 Vectors

A woman leaves home, walks 3 miles north, then 2 miles southeast. How far is she from home, and in which direction would she need to walk to return home? How far has she walked by the time she gets home?

This question may seem familiar – indeed we did a similar problem with a boat in the first section of this chapter. In that section, we solved the problem using triangles. In this section, we will investigate another way to approach the problem using vectors, a geometric entity that indicates both a distance and a direction. We will begin our investigation using a purely geometric view of vectors.

A Geometric View of Vectors

Vector

A **vector** is an object that has both a length and a direction.

Geometrically, a vector can be represented by an arrow that has a fixed length and indicates a direction. If, starting at the point A, a vector, which means "carrier" in Latin, moves toward point B, we write $\overrightarrow{AB}$ to represent the vector.

A vector may also be indicated using a single letter in boldface type, like **u**, or by capping the letter representing the vector with an arrow, like $\vec{u}$.

Example 1

Draw a vector that represents the movement from the point $P(-1, 2)$ to the point $Q(3,3)$

By drawing an arrow from the first point to the second, we can construct a vector $\overrightarrow{PQ}$.

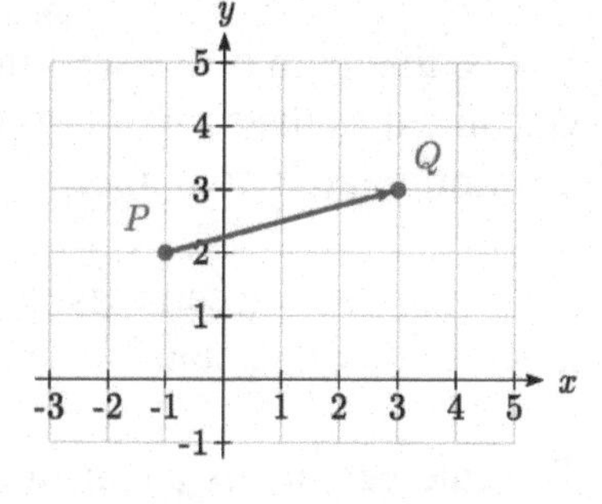

Try it Now

1. Draw a vector, $\vec{v}$, that travels from the origin to the point (3, 5).

Using this geometric representation of vectors, we can visualize the addition and scaling of vectors.

To add vectors, we envision a sum of two movements. To find $\vec{u}+\vec{v}$, we first draw the vector $\vec{u}$, then from the end of $\vec{u}$ we drawn the vector $\vec{v}$. This corresponds to the notion that first we move along the first vector, and then from that end position we move along the second vector. The sum $\vec{u}+\vec{v}$ is the new vector that travels directly from the beginning of $\vec{u}$ to the end of $\vec{v}$ in a straight path.

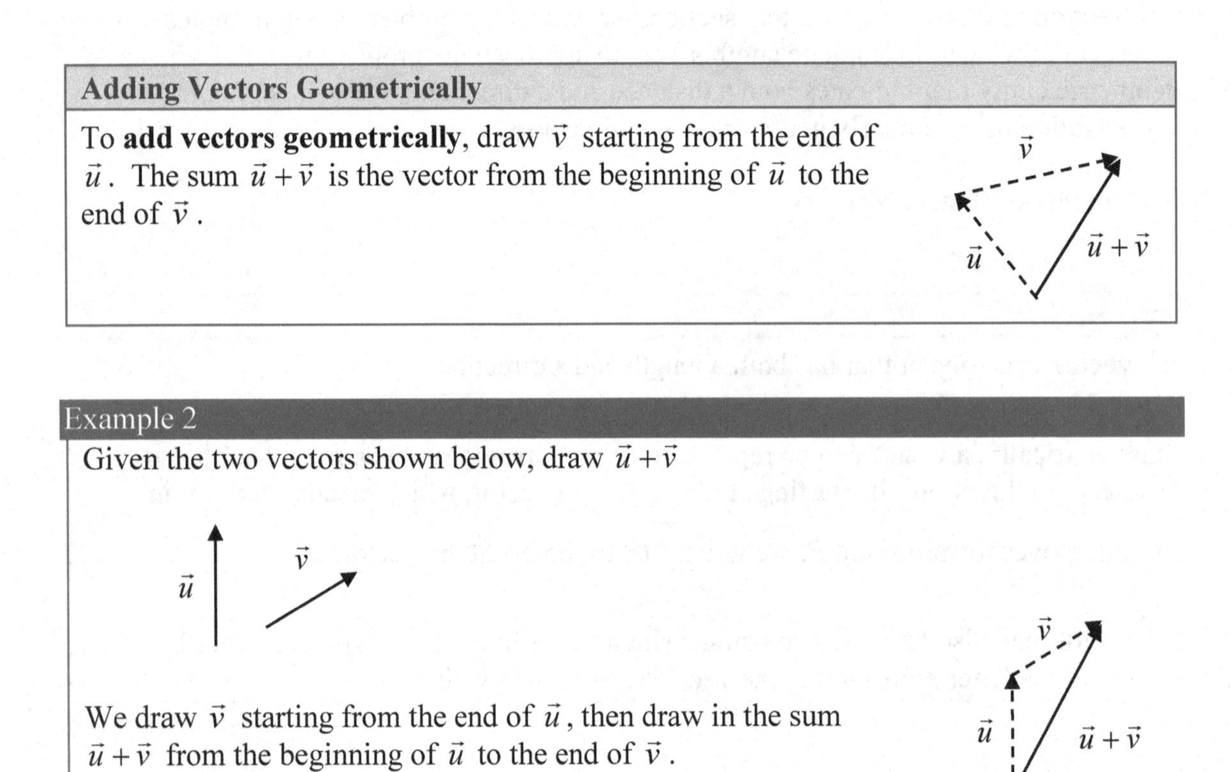

Adding Vectors Geometrically

To **add vectors geometrically**, draw $\vec{v}$ starting from the end of $\vec{u}$. The sum $\vec{u}+\vec{v}$ is the vector from the beginning of $\vec{u}$ to the end of $\vec{v}$.

Example 2

Given the two vectors shown below, draw $\vec{u}+\vec{v}$

We draw $\vec{v}$ starting from the end of $\vec{u}$, then draw in the sum $\vec{u}+\vec{v}$ from the beginning of $\vec{u}$ to the end of $\vec{v}$.

Notice that path of the walking woman from the beginning of the section could be visualized as the sum of two vectors. The resulting sum vector would indicate her end position relative to home.

Although vectors can exist anywhere in the plane, if we put the starting point at the origin it is easy to understand its size and direction relative to other vectors.

To scale vectors by a constant, such as $3\vec{u}$, we can imagine adding $\vec{u}+\vec{u}+\vec{u}$. The result will be a vector three times as long in the same direction as the original vector. If we were to scale a vector by a negative number, such as $-\vec{u}$, we can envision this as the opposite of $\vec{u}$; the vector so that $\vec{u}+(-\vec{u})$ returns us to the starting point. This vector $-\vec{u}$ would point in the opposite direction as $\vec{u}$ but have the same length.

Another way to think about scaling a vector is to maintain its direction and multiply its length by a constant, so that $3\vec{u}$ would point in the same direction but will be 3 times as long.

Scaling a Vector Geometrically
To **geometrically scale a vector** by a constant, scale the length of the vector by the constant. Scaling a vector by a negative constant will reverse the direction of the vector.

Example 3

Given the vector shown, draw $3\vec{u}$, $-\vec{u}$, and $-2\vec{u}$.

The vector $3\vec{u}$ will be three times as long. The vector $-\vec{u}$ will have the same length but point in the opposite direction. The vector $-2\vec{u}$ will point in the opposite direction and be twice as long.

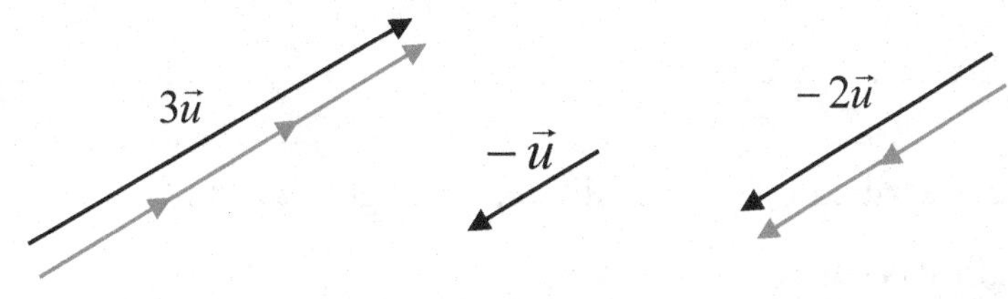

By combining scaling and addition, we can find the difference between vectors geometrically as well, since $\vec{u}-\vec{v}=\vec{u}+(-\vec{v})$.

Example 4

Given the vectors shown, draw $\vec{u}-\vec{v}$

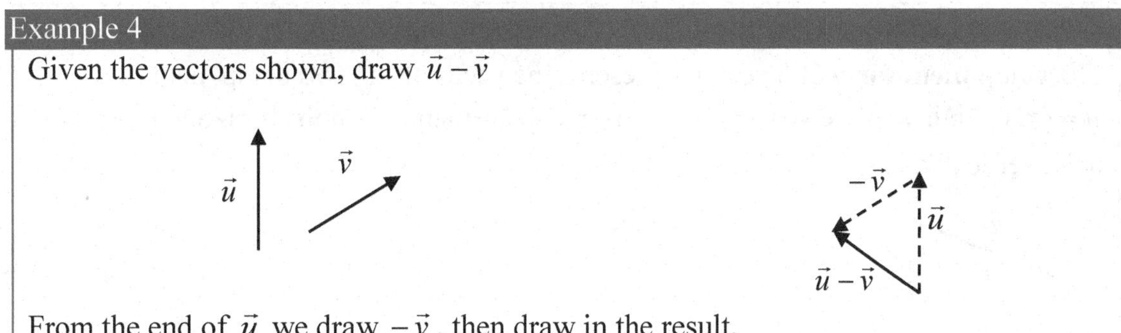

From the end of $\vec{u}$ we draw $-\vec{v}$, then draw in the result.

Notice that the sum and difference of two vectors are the two diagonals of a parallelogram with the vectors $\vec{u}$ and $\vec{v}$ as edges.

Try it Now

2. Using vector $\vec{v}$ from Try it Now #1, draw $-2\vec{v}$.

Component Form of Vectors

While the geometric interpretation of vectors gives us an intuitive understanding of vectors, it does not provide us a convenient way to do calculations. For that, we need a handy way to represent vectors. Since a vector involves a length and direction, it would be logical to want to represent a vector using a length and an angle θ, usually measured from standard position.

Magnitude and Direction of a Vector

A vector $\vec{u}$ can be described by its magnitude, or length, $|\vec{u}|$, and an angle θ.

A vector with length 1 is called **unit vector**.

While this is very reasonable, and a common way to describe vectors, it is often more convenient for calculations to represent a vector by horizontal and vertical components.

Component Form of a Vector

The **component form** of a vector represents the vector using two components. $\vec{u} = \langle x, y \rangle$ indicates the vector represents a displacement of x units horizontally and y units vertically.

Notice how we can see the magnitude of the vector as the length of the hypotenuse of a right triangle, or in polar form as the radius, r.

Alternate Notation for Vector Components

Sometimes you may see vectors written as the combination of unit vectors $\vec{i}$ and $\vec{j}$, where $\vec{i}$ points the right and $\vec{j}$ points up. In other words, $\vec{i} = \langle 1,0 \rangle$ and $\vec{j} = \langle 0,1 \rangle$.

In this notation, the vector $\vec{u} = \langle 3,-4 \rangle$ would be written as $\vec{u} = 3\vec{i} - 4\vec{j}$ since both indicate a displacement of 3 units to the right, and 4 units down.

While it can be convenient to think of the vector $\vec{u} = \langle x, y \rangle$ as an arrow from the origin to the point (x, y), be sure to remember that most vectors can be situated anywhere in the plane, and simply indicate a displacement (change in position) rather than a position.
It is common to need to convert from a magnitude and angle to the component form of the vector and vice versa. Happily, this process is identical to converting from polar coordinates to Cartesian coordinates, or from the polar form of complex numbers to the $a+bi$ form.

Example 5

Find the component form of a vector with length 7 at an angle of 135 degrees.

Using the conversion formulas $x = r\cos(\theta)$ and $y = r\sin(\theta)$, we can find the components

$$x = 7\cos(135°) = -\frac{7\sqrt{2}}{2}$$

$$y = 7\sin(135°) = \frac{7\sqrt{2}}{2}$$

This vector can be written in component form as $\left\langle -\frac{7\sqrt{2}}{2}, \frac{7\sqrt{2}}{2} \right\rangle$.

Example 6

Find the magnitude and angle θ representing the vector $\vec{u} = \langle 3,-2 \rangle$.

First we can find the magnitude by remembering the relationship between x, y and r:

$$r^2 = 3^2 + (-2)^2 = 13$$

$$r = \sqrt{13}$$

Second we can find the angle. Using the tangent,

$\tan(\theta) = \dfrac{-2}{3}$

$\theta = \tan^{-1}\left(-\dfrac{2}{3}\right) \approx -33.69°$, or written as a coterminal positive angle, 326.31°. This angle is in the 4th quadrant as desired.

Try it Now

3. Using vector $\vec{v}$ from Try it Now #1, the vector that travels from the origin to the point (3, 5), find the components, magnitude and angle θ that represent this vector.

In addition to representing distance movements, vectors are commonly used in physics and engineering to represent any quantity that has both direction and magnitude, including velocities and forces.

Example 7

An object is launched with initial velocity 200 meters per second at an angle of 30 degrees. Find the initial horizontal and vertical velocities.

By viewing the initial velocity as a vector, we can resolve the vector into horizontal and vertical components.

$x = 200\cos(30°) = 200 \cdot \dfrac{\sqrt{3}}{2} \approx 173.205$ m/sec

$y = 200\sin(30°) = 200 \cdot \dfrac{1}{2} = 100$ m/sec

200 m/s
100 m/s
30°
173 m/s

This tells us that, absent wind resistance, the object will travel horizontally at about 173 meters each second. Gravity will cause the vertical velocity to change over time – we'll leave a discussion of that to physics or calculus classes.

Adding and Scaling Vectors in Component Form

To add vectors in component form, we can simply add the corresponding components. To scale a vector by a constant, we scale each component by that constant.

Combining Vectors in Component Form

To **add, subtract, or scale vectors** in component form

If $\vec{u} = \langle u_1, u_2 \rangle$, $\vec{v} = \langle v_1, v_2 \rangle$, and c is any constant, then

$$\vec{u} + \vec{v} = \langle u_1 + v_1, u_2 + v_2 \rangle$$

$$\vec{u} - \vec{v} = \langle u_1 - v_1, u_2 - v_2 \rangle$$

$$c\vec{u} = \langle cu_1, cu_2 \rangle$$

Example 8

Given $\vec{u} = \langle 3, -2 \rangle$ and $\vec{v} = \langle -1, 4 \rangle$, find a new vector $\vec{w} = 3\vec{u} - 2\vec{v}$

Using the vectors given,

$\vec{w} = 3\vec{u} - 2\vec{v}$

$= 3\langle 3, -2 \rangle - 2\langle -1, 4 \rangle$ Scale each vector

$= \langle 9, -6 \rangle - \langle -2, 8 \rangle$ Subtract corresponding components

$= \langle 11, -14 \rangle$

By representing vectors in component form, we can find the resulting displacement vector after a multitude of movements without needing to draw a lot of complicated non-right triangles. For a simple example, we revisit the problem from the opening of the section. The general procedure we will follow is:

1) Convert vectors to component form
2) Add the components of the vectors
3) Convert back to length and direction if needed to suit the context of the question

Example 9

A woman leaves home, walks 3 miles north, then 2 miles southeast. How far is she from home, and what direction would she need to walk to return home? How far has she walked by the time she gets home?

Let's begin by understanding the question in a little more depth. When we use vectors to describe a traveling direction, we often position things so north points in the upward direction, east points to the right, and so on, as pictured here.

Consequently, travelling NW, SW, NE or SE, means we are travelling through the quadrant bordered by the given directions at a 45 degree angle.

With this in mind, we begin by converting each vector to components.

A walk 3 miles north would, in components, be $\langle 0,3 \rangle$.

A walk of 2 miles southeast would be at an angle of 45° South of East. Measuring from standard position the angle would be 315°.

Converting to components, we choose to use the standard position angle so that we do not have to worry about whether the signs are negative or positive; they will work out automatically.

$$\langle 2\cos(315°), 2\sin(315°) \rangle = \left\langle 2 \cdot \frac{\sqrt{2}}{2}, 2 \cdot \frac{-\sqrt{2}}{2} \right\rangle \approx \langle 1.414, -1.414 \rangle$$

Adding these vectors gives the sum of the movements in component form

$$\langle 0,3 \rangle + \langle 1.414, -1.414 \rangle = \langle 1.414, 1.586 \rangle$$

To find how far she is from home and the direction she would need to walk to return home, we could find the magnitude and angle of this vector.

$$\text{Length} = \sqrt{1.414^2 + 1.586^2} = 2.125$$

To find the angle, we can use the tangent

$$\tan(\theta) = \frac{1.586}{1.414}$$

$$\theta = \tan^{-1}\left(\frac{1.586}{1.414}\right) = 48.273° \text{ north of east}$$

Of course, this is the angle from her starting point to her ending point. To return home, she would need to head the opposite direction, which we could either describe as $180° + 48.273° = 228.273°$ measured in standard position, or as 48.273° south of west (or 41.727° west of south).

She has walked a total distance of $3 + 2 + 2.125 = 7.125$ miles.

Keep in mind that total distance travelled is not the same as the length of the resulting displacement vector or the "return" vector.

Try it Now

4. In a scavenger hunt, directions are given to find a buried treasure. From a starting point at a flag pole you must walk 30 feet east, turn 30 degrees to the north and travel 50 feet, and then turn due south and travel 75 feet. Sketch a picture of these vectors, find their components, and calculate how far and in what direction you must travel to go directly to the treasure from the flag pole without following the map.

While using vectors is not much faster than using law of cosines with only two movements, when combining three or more movements, forces, or other vector quantities, using vectors quickly becomes much more efficient than trying to use triangles.

Example 10

Three forces are acting on an object as shown below, each measured in Newtons (N). What force must be exerted to keep the object in equilibrium (where the sum of the forces is zero)?

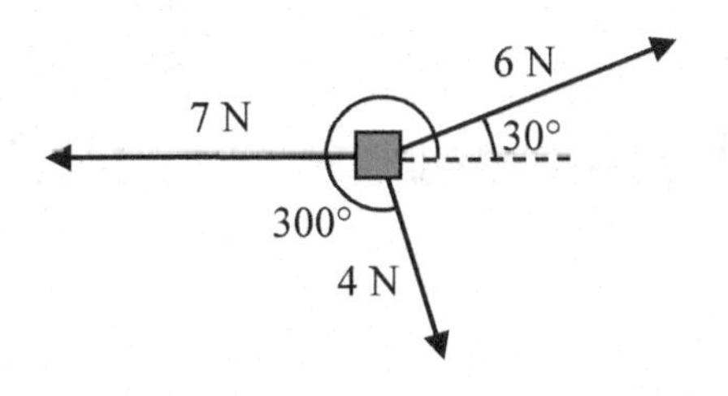

We start by resolving each vector into components.

The first vector with magnitude 6 Newtons at an angle of 30 degrees will have components

$$\langle 6\cos(30°), 6\sin(30°)\rangle = \left\langle 6\cdot\frac{\sqrt{3}}{2}, 6\cdot\frac{1}{2}\right\rangle = \langle 3\sqrt{3}, 3\rangle$$

The second vector is only in the horizontal direction, so can be written as $\langle -7, 0\rangle$.

The third vector with magnitude 4 Newtons at an angle of 300 degrees will have components

$$\langle 4\cos(300°), 4\sin(300°)\rangle = \left\langle 4\cdot\frac{1}{2}, 4\cdot\frac{-\sqrt{3}}{2}\right\rangle = \langle 2, -2\sqrt{3}\rangle$$

To keep the object in equilibrium, we need to find a force vector $\langle x, y\rangle$ so the sum of the four vectors is the zero vector, $\langle 0,0\rangle$.

$\langle 3\sqrt{3}, 3\rangle + \langle -7, 0\rangle + \langle 2, -2\sqrt{3}\rangle + \langle x, y\rangle = \langle 0, 0\rangle$ Add component-wise

$\langle 3\sqrt{3} - 7 + 2, 3 + 0 - 2\sqrt{3}\rangle + \langle x, y\rangle = \langle 0, 0\rangle$ Simplify

$\langle 3\sqrt{3} - 5, 3 - 2\sqrt{3}\rangle + \langle x, y\rangle = \langle 0, 0\rangle$ Solve

$\langle x, y\rangle = \langle 0, 0\rangle - \langle 3\sqrt{3} - 5, 3 - 2\sqrt{3}\rangle$

$\langle x, y\rangle = \langle -3\sqrt{3} + 5, -3 + 2\sqrt{3}\rangle \approx \langle -0.196, 0.464\rangle$

This vector gives in components the force that would need to be applied to keep the object in equilibrium. If desired, we could find the magnitude of this force and direction it would need to be applied in.

Magnitude = $\sqrt{(-0.196)^2 + 0.464^2} = 0.504$ N

Angle:

$$\tan(\theta) = \frac{0.464}{-0.196}$$

$$\theta = \tan^{-1}\left(\frac{0.464}{-0.196}\right) = -67.089°.$$

This is in the wrong quadrant, so we adjust by finding the next angle with the same tangent value by adding a full period of tangent:

$\theta = -67.089° + 180° = 112.911°$

To keep the object in equilibrium, a force of 0.504 Newtons would need to be applied at an angle of 112.911°.

Important Topics of This Section
Vectors, magnitude (length) & direction
Addition of vectors
Scaling of vectors
Components of vectors
Vectors as velocity
Vectors as forces
Adding & Scaling vectors in component form
Total distance travelled vs. total displacement

Try it Now Answers

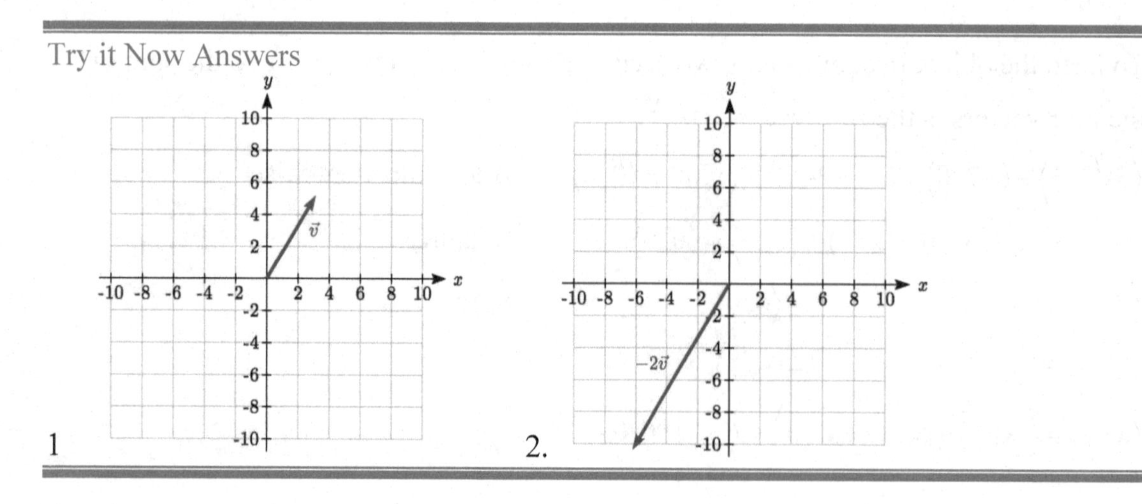

2. $\vec{v} = \langle 3, 5 \rangle$ magnitude $= \sqrt{34}$ $\theta = \tan^{-1}\left(\frac{5}{3}\right) = 59.04°$

3.

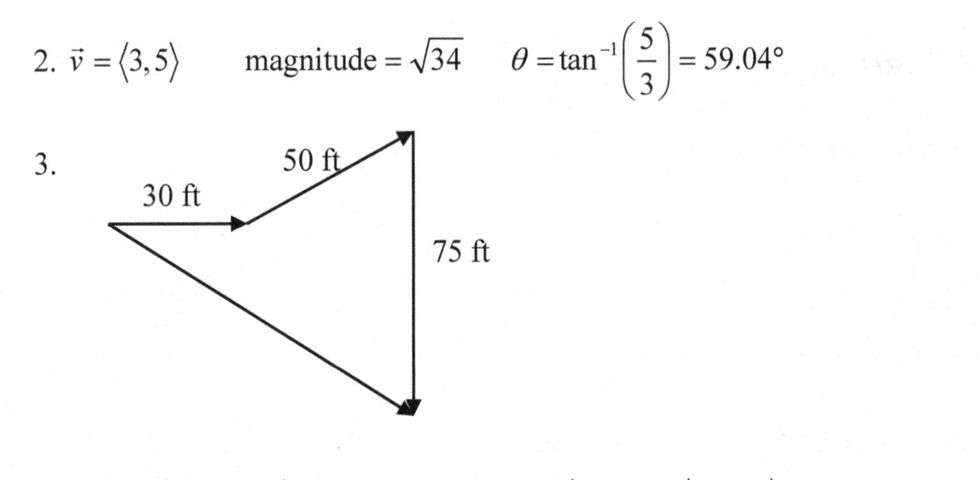

$\vec{v}_1 = \langle 30, 0 \rangle$ $\vec{v}_2 = \langle 50\cos(30°), 50\sin(30°) \rangle$ $\vec{v}_3 = \langle 0, -75 \rangle$

$\vec{v}_f = \langle 30 + 50\cos(30°), 50\sin(30°) - 75 \rangle = \langle 73.301, -50 \rangle$

Magnitude = 88.73 feet at an angle of 34.3° south of east.

Section 8.4 Exercises

Write the vector shown in component form.

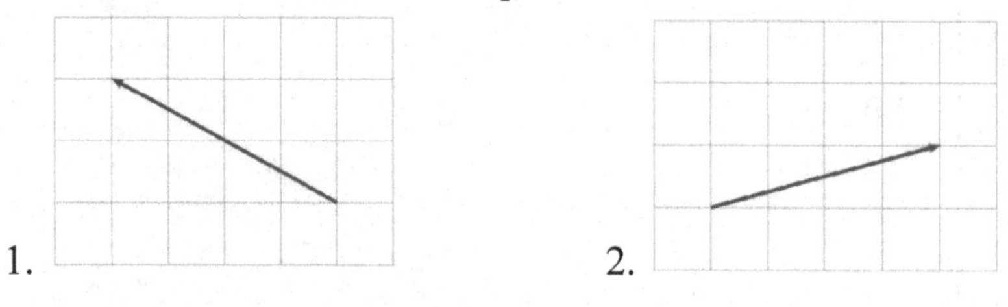

1. 2.

Given the vectors shown, sketch $\vec{u}+\vec{v}$, $\vec{u}-\vec{v}$, and $2\vec{u}$.

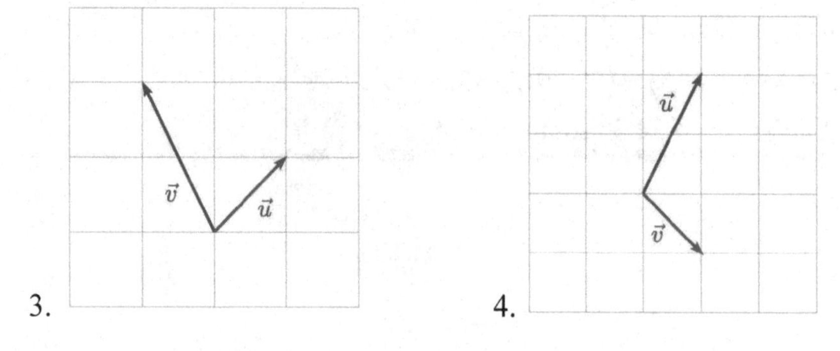

3. 4.

Write each vector below as a combination of the vectors $\vec{u}$ and $\vec{v}$ from question #3.

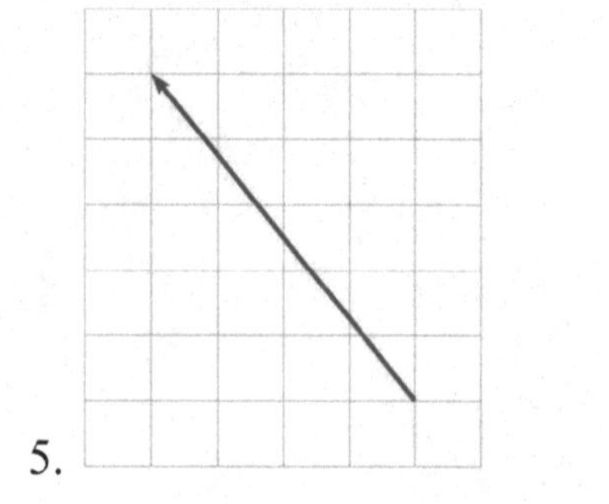

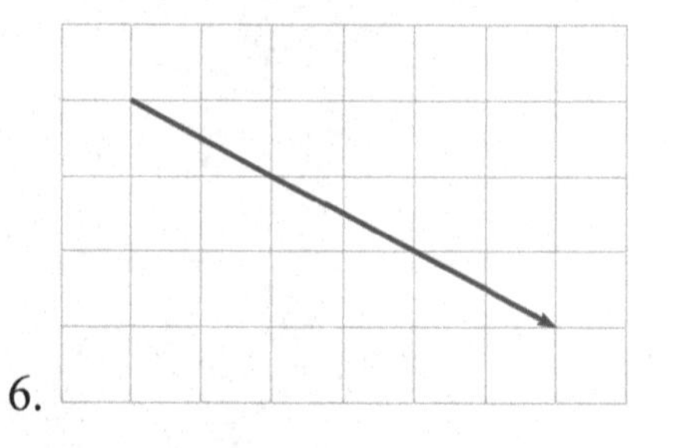

5. 6.

From the given magnitude and direction in standard position, write the vector in component form.

7. Magnitude: 6, Direction: 45°
8. Magnitude: 10, Direction: 120°
9. Magnitude: 8, Direction: 220°
10. Magnitude: 7, Direction: 305°

Find the magnitude and direction of the vector.

11. $\langle 0,4\rangle$
12. $\langle -3,0\rangle$
13. $\langle 6,5\rangle$
14. $\langle 3,7\rangle$
15. $\langle -2,1\rangle$
16. $\langle -10,13\rangle$
17. $\langle 2,-5\rangle$
18. $\langle 8,-4\rangle$
19. $\langle -4,-6\rangle$
20. $\langle -1,9\rangle$

Using the vectors given, compute $\vec{u}+\vec{v}$, $\vec{u}-\vec{v}$, and $2\vec{u}-3\vec{v}$.

21. $\vec{u}=\langle 2,-3\rangle$, $\vec{v}=\langle 1,5\rangle$
22. $\vec{u}=\langle -3,4\rangle$, $\vec{v}=\langle -2,1\rangle$

23. A woman leaves home and walks 3 miles west, then 2 miles southwest. How far from home is she, and in what direction must she walk to head directly home?

24. A boat leaves the marina and sails 6 miles north, then 2 miles northeast. How far from the marina is the boat, and in what direction must it sail to head directly back to the marina?

25. A person starts walking from home and walks 4 miles east, 2 miles southeast, 5 miles south, 4 miles southwest, and 2 miles east. How far have they walked? If they walked straight home, how far would they have to walk?

26. A person starts walking from home and walks 4 miles east, 7 miles southeast, 6 miles south, 5 miles southwest, and 3 miles east. How far have they walked? If they walked straight home, how far would they have to walk?

27. Three forces act on an object: $\vec{F_1} = \langle -8, -5 \rangle$, $\vec{F_2} = \langle 0, 1 \rangle$, $\vec{F_3} = \langle 4, -7 \rangle$. Find the net force acting on the object.

28. Three forces act on an object: $\vec{F_1} = \langle 2, 5 \rangle$, $\vec{F_2} = \langle 8, 3 \rangle$, $\vec{F_3} = \langle 0, -7 \rangle$. Find the net force acting on the object.

29. A person starts walking from home and walks 3 miles at 20° north of west, then 5 miles at 10° west of south, then 4 miles at 15° north of east. If they walked straight home, how far would they have to walk, and in what direction?

30. A person starts walking from home and walks 6 miles at 40° north of east, then 2 miles at 15° east of south, then 5 miles at 30° south of west. If they walked straight home, how far would they have to walk, and in what direction?

31. An airplane is heading north at an airspeed of 600 km/hr, but there is a wind blowing from the southwest at 80 km/hr. How many degrees off course will the plane end up flying, and what is the plane's speed relative to the ground?

32. An airplane is heading north at an airspeed of 500 km/hr, but there is a wind blowing from the northwest at 50 km/hr. How many degrees off course will the plane end up flying, and what is the plane's speed relative to the ground?

33. An airplane needs to head due north, but there is a wind blowing from the southwest at 60 km/hr. The plane flies with an airspeed of 550 km/hr. To end up flying due north, the pilot will need to fly the plane how many degrees west of north?

34. An airplane needs to head due north, but there is a wind blowing from the northwest at 80 km/hr. The plane flies with an airspeed of 500 km/hr. To end up flying due north, the pilot will need to fly the plane how many degrees west of north?

35. As part of a video game, the point (5, 7) is rotated counterclockwise about the origin through an angle of 35 degrees. Find the new coordinates of this point.

36. As part of a video game, the point (7, 3) is rotated counterclockwise about the origin through an angle of 40 degrees. Find the new coordinates of this point.

37. Two children are throwing a ball back and forth straight across the back seat of a car. The ball is being thrown 10 mph relative to the car, and the car is travelling 25 mph down the road. If one child doesn't catch the ball and it flies out the window, in what direction does the ball fly (ignoring wind resistance)?

38. Two children are throwing a ball back and forth straight across the back seat of a car. The ball is being thrown 8 mph relative to the car, and the car is travelling 45 mph down the road. If one child doesn't catch the ball and it flies out the window, in what direction does the ball fly (ignoring wind resistance)?

Section 8.5 Dot Product

Now that we can add, subtract, and scale vectors, you might be wondering whether we can multiply vectors. It turns out there are two different ways to multiply vectors, one which results in a number, and one which results in a vector. In this section, we'll focus on the first, called the **dot product** or **scalar product**, since it produces a single numeric value (a scalar). We'll begin with some motivation.

In physics, we often want to know how much of a force is acting in the direction of motion. To determine this, we need to know the angle between direction of force and the direction of motion. Likewise, in computer graphics, the lighting system determines how bright a triangle on the object should be based on the angle between object and the direction of the light. In both applications, we're interested in the angle between the vectors, so let's start there.

Suppose we have two vectors, $\vec{a} = \langle a_1, a_2 \rangle$ and $\vec{b} = \langle b_1, b_2 \rangle$. Using our polar coordinate conversions, we could write $\vec{a} = \langle |\vec{a}|\cos(\alpha), |\vec{a}|\sin(\alpha) \rangle$ and $\vec{b} = \left\langle |\vec{b}|\cos(\beta), |\vec{b}|\sin(\beta) \right\rangle$.

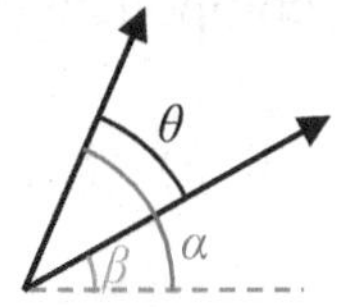

Now, if we knew the angles α and β, we wouldn't have much work to do - the angle between the vectors would be $\theta = \alpha - \beta$. While we certainly could use some inverse tangents to find the two angles, it would be great if we could find a way to determine the angle between the vector just from the vector components.

To help us manipulate $\theta = \alpha - \beta$, we might try introducing a trigonometric function:

$$\cos(\theta) = \cos(\alpha - \beta)$$

Now we can apply the difference of angles identity

$$\cos(\theta) = \cos(\alpha)\cos(\beta) + \sin(\alpha)\sin(\beta)$$

Now $a_1 = |\vec{a}|\cos(\alpha)$, so $\cos(\alpha) = \dfrac{a_1}{|\vec{a}|}$, and likewise for the other three components.

Making those substitutions,

$$\cos(\theta) = \frac{a_1}{|\vec{a}|}\frac{b_1}{|\vec{b}|} + \frac{a_2}{|\vec{a}|}\frac{b_2}{|\vec{b}|} = \frac{a_1 b_1 + a_2 b_2}{|\vec{a}||\vec{b}|}$$

$$|\vec{a}||\vec{b}|\cos(\theta) = a_1 b_1 + a_2 b_2$$

Notice the expression on the right is a very simple calculation based on the components of the vectors. It comes up so frequently we define it to be the **dot product** of the two vectors, notated by a dot. This gives us two definitions of the dot product.

Definitions of the Dot Product	
$\vec{a} \cdot \vec{b} = a_1 b_1 + a_2 b_2$	Component definition
$\vec{a} \cdot \vec{b} = \lvert\vec{a}\rvert\lvert\vec{b}\rvert\cos(\theta)$	Geometric definition

The first definition, $\vec{a} \cdot \vec{b} = a_1 b_1 + a_2 b_2$, gives us a simple way to calculate the dot product from components. The second definition, $\vec{a} \cdot \vec{b} = \lvert\vec{a}\rvert\lvert\vec{b}\rvert\cos(\theta)$, gives us a geometric interpretation of the dot product, and gives us a way to find the angle between two vectors, as we desired.

Example 1

Find the dot product $\langle 3,-2\rangle \cdot \langle 5,1\rangle$.

Using the first definition, we can calculate the dot product by multiplying the x components and adding that to the product of the y components.

$$\langle 3,-2\rangle \cdot \langle 5,1\rangle = (3)(5) + (-2)(1) = 15 - 2 = 13$$

Example 2

Find the dot product of the two vectors shown.

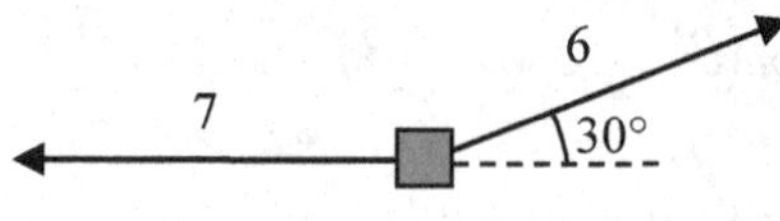

We can immediately see that the magnitudes of the two vectors are 7 and 6. We can quickly calculate that the angle between the vectors is 150°. Using the geometric definition of the dot product,

$$\vec{a} \cdot \vec{b} = \lvert\vec{a}\rvert\lvert\vec{b}\rvert\cos(\theta) = (6)(7)\cos(150°) = 42 \cdot \frac{\sqrt{3}}{2} = 21\sqrt{3}.$$

Try it Now

1. Calculate the dot product $\langle -7,3\rangle \cdot \langle -2,-6\rangle$

Now we can return to our goal of finding the angle between vectors.

Example 3

An object is being pulled up a ramp in the direction $\langle 5,1\rangle$ by a rope pulling in the direction $\langle 4,2\rangle$. What is the angle between the rope and the ramp?

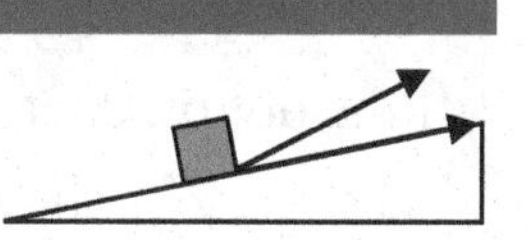

Using the component form, we can easily calculate the dot product.
$\vec{a}\cdot\vec{b} = \langle 5,1\rangle\cdot\langle 4,2\rangle = (5)(4)+(1)(2) = 20+2 = 22$

We can also calculate the magnitude of each vector.
$|\vec{a}| = \sqrt{5^2+1^2} = \sqrt{26}$, $|\vec{b}| = \sqrt{4^2+2^2} = \sqrt{20}$

Substituting these values into the geometric definition, we can solve for the angle between the vectors.

$$\vec{a}\cdot\vec{b} = |\vec{a}||\vec{b}|\cos(\theta)$$

$$22 = \sqrt{26}\sqrt{20}\cos(\theta)$$

$$\theta = \cos^{-1}\left(\frac{22}{\sqrt{26}\sqrt{20}}\right) \approx 15.255°.$$

Example 4

Calculate the angle between the vectors $\langle 6,4\rangle$ and $\langle -2,3\rangle$.

Calculating the dot product, $\langle 6,4\rangle\cdot\langle -2,3\rangle = (6)(-2)+(4)(3) = -12+12 = 0$

We don't even need to calculate the magnitudes in this case since the dot product is 0.

$$\vec{a}\cdot\vec{b} = |\vec{a}||\vec{b}|\cos(\theta)$$

$$0 = |\vec{a}||\vec{b}|\cos(\theta)$$

$$\theta = \cos^{-1}\left(\frac{0}{|\vec{a}||\vec{b}|}\right) = \cos^{-1}(0) = 90°$$

With the dot product equaling zero, as in the last example, the angle between the vectors will always be 90°, indicating that the vectors are **orthogonal**, a more general way of saying perpendicular. This gives us a quick way to check if vectors are orthogonal. Also, if the dot product is positive, then the inside of the inverse cosine will be positive, giving an angle less than 90°. A negative dot product will then lead to an angle larger than 90°

Sign of the Dot Product	
If the dot product is:	
Zero	The vectors are **orthogonal** (perpendicular).
Positive	The angle between the vectors is less than 90°
Negative	The angle between the vectors is greater than 90°

Try it Now

2. Are the vectors $\langle -7,3\rangle$ and $\langle -2,-6\rangle$ orthogonal? If not, find the angle between them.

Projections

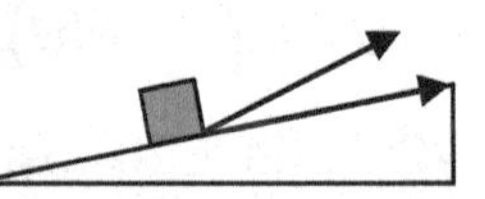

In addition to finding the angle between vectors, sometimes we want to know how much one vector points in the direction of another. For example, when pulling an object up a ramp, we might want to know how much of the force is exerted in the direction of motion. To determine this we can use the idea of a **projection**.

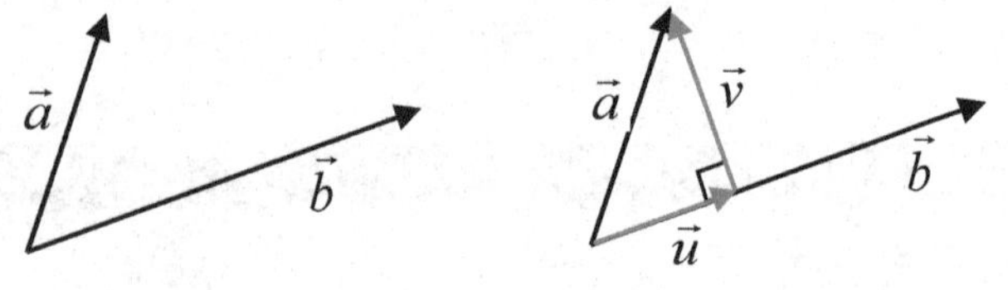

In the picture above, $\vec{u}$ is a projection of $\vec{a}$ onto $\vec{b}$. In other words, it is the portion of $\vec{a}$ that points in the same direction as $\vec{b}$.

To find the length of $\vec{u}$, we could notice that it is one side of a right triangle. If we define θ to be the angle between $\vec{a}$ and $\vec{u}$, then $\cos(\theta) = \frac{|\vec{u}|}{|\vec{a}|}$, so $|\vec{a}|\cos(\theta) = |\vec{u}|$.

While we could find the angle between the vectors to determine this magnitude, we could skip some steps by using the dot product directly. Since $\vec{a}\cdot\vec{b} = |\vec{a}||\vec{b}|\cos(\theta)$,

$|\vec{a}|\cos(\theta) = \frac{\vec{a}\cdot\vec{b}}{|\vec{b}|}$. Using this, we can rewrite $|\vec{u}| = |\vec{a}|\cos(\theta)$ as $|\vec{u}| = \frac{\vec{a}\cdot\vec{b}}{|\vec{b}|}$. This gives us

the length of the projection, sometimes denoted as $comp_{\vec{b}}\vec{a} = |\vec{u}| = \frac{\vec{a}\cdot\vec{b}}{|\vec{b}|}$.

To find the vector $\vec{u}$ itself, we could first scale $\vec{b}$ to a unit vector with length 1: $\dfrac{\vec{b}}{|\vec{b}|}$.

Multiplying this by the length of the projection will give a vector in the direction of $\vec{b}$ but with the correct length.

$$proj_{\vec{b}}\vec{a} = |\vec{u}|\frac{\vec{b}}{|\vec{b}|} = \left(\frac{\vec{a}\cdot\vec{b}}{|\vec{b}|}\right)\frac{\vec{b}}{|\vec{b}|} = \left(\frac{\vec{a}\cdot\vec{b}}{|\vec{b}|^2}\right)\vec{b}$$

Projection Vector

The projection of vector $\vec{a}$ onto $\vec{b}$ is $proj_{\vec{b}}\vec{a} = \left(\dfrac{\vec{a}\cdot\vec{b}}{|\vec{b}|^2}\right)\vec{b}$

The magnitude of the projection is $comp_{\vec{b}}\vec{a} = \dfrac{\vec{a}\cdot\vec{b}}{|\vec{b}|}$

Example 5

Find the projection of the vector $\langle 3,-2\rangle$ onto the vector $\langle 8,6\rangle$.

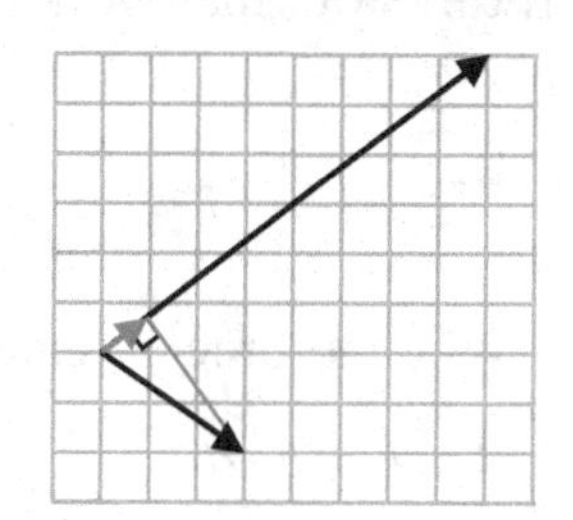

We will need to know the dot product of the vectors and the magnitude of the vector we are projecting onto.

$\langle 3,-2\rangle\cdot\langle 8,6\rangle = (3)(8)+(-2)(6) = 24-12 = 12$

$|\langle 8,6\rangle| = \sqrt{8^2+6^2} = \sqrt{64+36} = \sqrt{100} = 10$

The magnitude of the projection will be $\dfrac{\langle 3,-2\rangle\cdot\langle 8,6\rangle}{|\langle 8,6\rangle|} = \dfrac{12}{10} = \dfrac{6}{5}$.

To find the projection vector itself, we would multiply that magnitude by $\langle 8,6\rangle$ scaled to a unit vector.

$$\frac{6}{5}\frac{\langle 8,6\rangle}{|\langle 8,6\rangle|} = \frac{6}{5}\frac{\langle 8,6\rangle}{10} = \frac{6}{50}\langle 8,6\rangle = \left\langle\frac{48}{50},\frac{36}{50}\right\rangle = \left\langle\frac{24}{25},\frac{18}{25}\right\rangle.$$

Based on the sketch above, this answer seems reasonable.

Try it Now

3. Find the component of the vector $\langle -3,4\rangle$ that is *orthogonal* to the vector $\langle -8,4\rangle$

Work

In physics, when a constant force causes an object to move, the mechanical **work** done by that force is the product of the force and the distance the object is moved. However, we only consider the portion of force that is acting in the direction of motion.

This is simply the magnitude of the projection of the force vector onto the distance vector, $\dfrac{\vec{F}\cdot\vec{d}}{\left|\vec{d}\right|}$. The work done is the product of that component of force times the distance moved, the magnitude of the distance vector.

$$Work = \left(\frac{\vec{F}\cdot\vec{d}}{\left|\vec{d}\right|}\right)\left|\vec{d}\right| = \vec{F}\cdot\vec{d}$$

It turns out that work is simply the dot product of the force vector and the distance vector.

Work

When a force $\vec{F}$ causes an object to move some distance $\vec{d}$, the work done is

$Work = \vec{F}\cdot\vec{d}$

Example 6

A cart is pulled 20 feet by applying a force of 30 pounds on a rope held at a 30 degree angle. How much work is done?

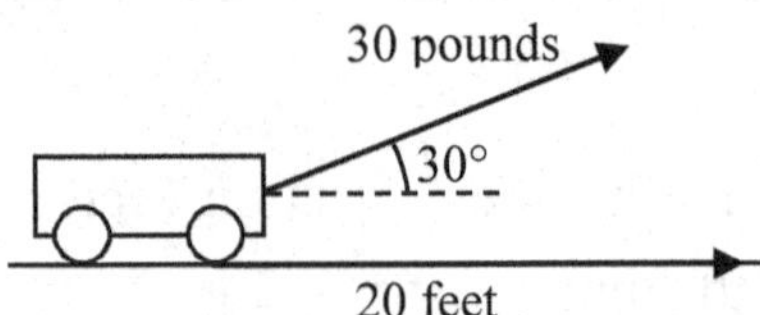

Since work is simply the dot product, we can take advantage of the geometric definition of the dot product in this case.

$Work = \vec{F}\cdot\vec{d} = \left|\vec{F}\right|\cdot\left|\vec{d}\right|\cos(\theta) = (30)(20)\cos(30°) \approx 519.615$ ft-lbs.

Try it Now

4. Find the work down moving an object from the point (1, 5) to (9, 14) by the force vector $\vec{F} = \langle 3,2\rangle$

Important Topics of This Section
Calculate Dot Product
Using component definition
Using geometric definition
Find the angle between two vectors
Sign of the dot product
Projections
Work

Try it Now Answers

1. $\langle -7,3\rangle \cdot \langle -2,-6\rangle = (-7)(-2)+(3)(-6) = 14-18 = -4$

2. In the previous Try it Now, we found the dot product was -4, so the vectors are not orthogonal. The magnitudes of the vectors are $\sqrt{(-7)^2+3^2} = \sqrt{58}$ and $\sqrt{(-2)^2+6^2} = \sqrt{40}$. The angle between the vectors will be
$$\theta = \cos^{-1}\left(\frac{-4}{\sqrt{58}\sqrt{40}}\right) \approx 94.764°$$

3. We want to find the component of $\langle -3,4\rangle$ that is *orthogonal* to the vector $\langle -8,4\rangle$. In the picture to the right, that component is vector $\vec{v}$. Notice that $\vec{u}+\vec{v}=\vec{a}$, so if we can find the projection vector, we can find $\vec{v}$.

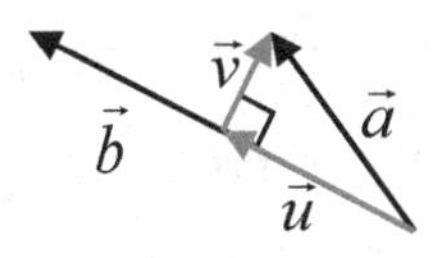

$$\vec{u} = proj_{\vec{b}}\vec{a} = \left(\frac{\vec{a}\cdot\vec{b}}{\left|\vec{b}\right|^2}\right)\vec{b} = \left(\frac{\langle -3,4\rangle\cdot\langle -8,4\rangle}{\left(\sqrt{(-8)^2+4^2}\right)^2}\right)\langle -8,4\rangle = \frac{40}{80}\langle -8,4\rangle = \langle -4,2\rangle.$$

Now we can solve $\vec{u}+\vec{v}=\vec{a}$ for $\vec{v}$.
$\vec{v} = \vec{a}-\vec{u} = \langle -3,4\rangle - \langle -4,2\rangle = \langle 1,2\rangle$

4. The distance vector is $\langle 9-1, 14-5\rangle = \langle 8,9\rangle$.
The work is the dot product: $Work = \vec{F}\cdot\vec{d} = \langle 3,2\rangle\cdot\langle 8,9\rangle = 24+18 = 42$

Section 8.5 Exercises

Two vectors are described by their magnitude and direction in standard position. Find the dot product of the vectors.

1. Magnitude: 6, Direction: 45°; Magnitude: 10, Direction: 120°
2. Magnitude: 8, Direction: 220°; Magnitude: 7, Direction: 305°

Find the dot product of each pair of vectors.

3. $\langle 0,4\rangle$; $\langle -3,0\rangle$
4. $\langle 6,5\rangle$; $\langle 3,7\rangle$
5. $\langle -2,1\rangle$; $\langle -10,13\rangle$
6. $\langle 2,-5\rangle$; $\langle 8,-4\rangle$

Find the angle between the vectors

7. $\langle 0,4\rangle$; $\langle -3,0\rangle$
8. $\langle 6,5\rangle$; $\langle 3,7\rangle$
9. $\langle 2,4\rangle$; $\langle 1,-3\rangle$
10. $\langle -4,1\rangle$; $\langle 8,-2\rangle$
11. $\langle 4,2\rangle$; $\langle 8,4\rangle$
12. $\langle 5,3\rangle$; $\langle -6,10\rangle$

13. Find a value for k so that $\langle 2,7\rangle$ and $\langle k,4\rangle$ will be orthogonal.
14. Find a value for k so that $\langle -3,5\rangle$ and $\langle 2,k\rangle$ will be orthogonal.

15. Find the magnitude of the projection of $\langle 8,-4\rangle$ onto $\langle 1,-3\rangle$.
16. Find the magnitude of the projection of $\langle 2,7\rangle$ onto $\langle 4,5\rangle$.
17. Find the projection of $\langle -6,10\rangle$ onto $\langle 1,-3\rangle$.
18. Find the projection of $\langle 0,4\rangle$ onto $\langle 3,7\rangle$.

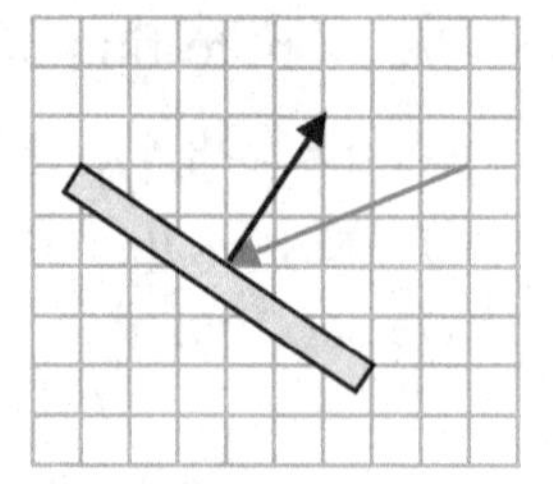

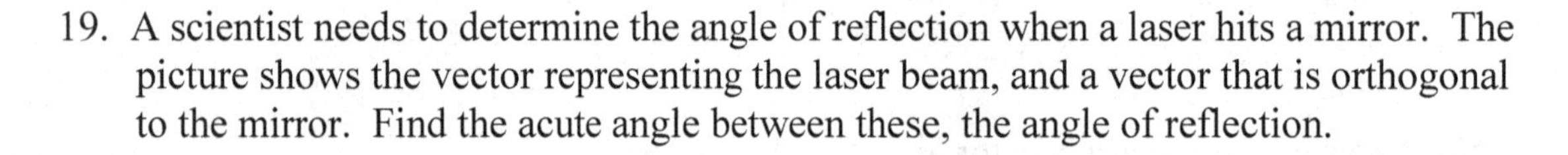

19. A scientist needs to determine the angle of reflection when a laser hits a mirror. The picture shows the vector representing the laser beam, and a vector that is orthogonal to the mirror. Find the acute angle between these, the angle of reflection.

20. A triangle has coordinates at A: (1,4), B: (2,7), and C: (4,2). Find the angle at point B.

21. A boat is trapped behind a log lying parallel to the dock. It only requires 10 pounds of force to pull the boat directly towards you, but because of the log, you'll have to pull at a 45° angle. How much force will you have to pull with? (We're going to assume that the log is very slimy and doesn't contribute any additional resistance.)

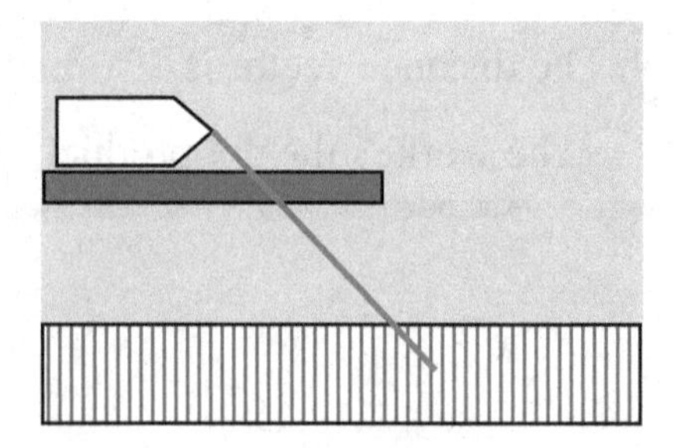

22. A large boulder needs to be dragged to a new position. If pulled directly horizontally, the boulder would require 400 pounds of pulling force to move. We need to pull the boulder using a rope tied to the back of a large truck, forming a 15° angle from the ground. How much force will the truck need to pull with?

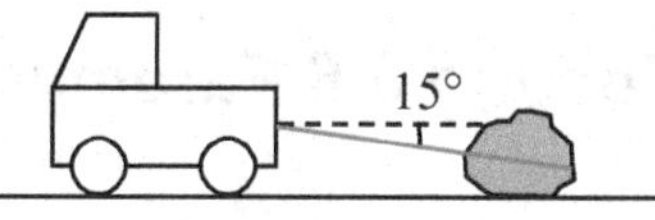

23. Find the work done against gravity by pushing a 20 pound cart 10 feet up a ramp that is 10° above horizontal. Assume there is no friction, so the only force is 20 pounds downwards due to gravity.

24. Find the work done against gravity by pushing a 30 pound cart 15 feet up a ramp that is 8° above horizontal. Assume there is no friction, so the only force is 30 pounds downwards due to gravity.

25. An object is pulled to the top of a 40 foot ramp that forms a 10° angle with the ground. It is pulled by rope exerting a force of 120 pounds at a 35° angle relative to the ground. Find the work done.

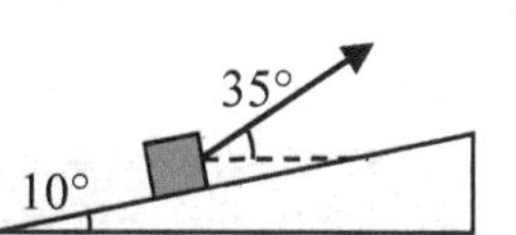

26. An object is pulled to the top of a 30 foot ramp that forms a 20° angle with the ground. It is pulled by rope exerting a force of 80 pounds at a 30° angle relative to the ground. Find the work done.

Section 8.6 Parametric Equations

Many shapes, even ones as simple as circles, cannot be represented as an equation where y is a function of x. Consider, for example, the path a moon follows as it orbits around a planet, which simultaneously rotates around a sun. In some cases, polar equations provide a way to represent such a path. In others, we need a more versatile approach that allows us to represent both the x and y coordinates in terms of a third variable, or parameter.

Parametric Equations

A system of **parametric equations** is a pair of functions $x(t)$ and $y(t)$ in which the x and y coordinates are the output, represented in terms of a third input parameter, t.

Example 1

Moving at a constant speed, an object moves at a steady rate along a straight path from coordinates (-5, 3) to the coordinates (3, -1) in 4 seconds, where the coordinates are measured in meters. Find parametric equations for the position of the object.

The x coordinate of the object starts at -5 meters, and goes to +3 meters, this means the x direction has changed by 8 meters in 4 seconds, giving us a rate of 2 meters per second. We can now write the x coordinate as a linear function with respect to time, t, $x(t) = -5 + 2t$. Similarly, the y value starts at 3 and goes to -1, giving a change in y value of 4 meters, meaning the y values have decreased by 4 meters in 4 seconds, for a rate of -1 meter per second, giving equation $y(t) = 3 - t$. Together, these are the parametric equations for the position of the object:

$x(t) = -5 + 2t$

$y(t) = 3 - t$

t	x	y
0	-5	3
1	-3	2
2	-1	1
3	1	0
4	3	-1

Using these equations, we can build a table of t, x, and y values. Because of the context, we limited ourselves to non-negative t values for this example, but in general you can use any values.

From this table, we could create three possible graphs: a graph of x vs. t, which would show the horizontal position over time, a graph of y vs. t, which would show the vertical position over time, or a graph of y vs. x, showing the position of the object in the plane.

Position of x as a function of time

Position of y as a function of time

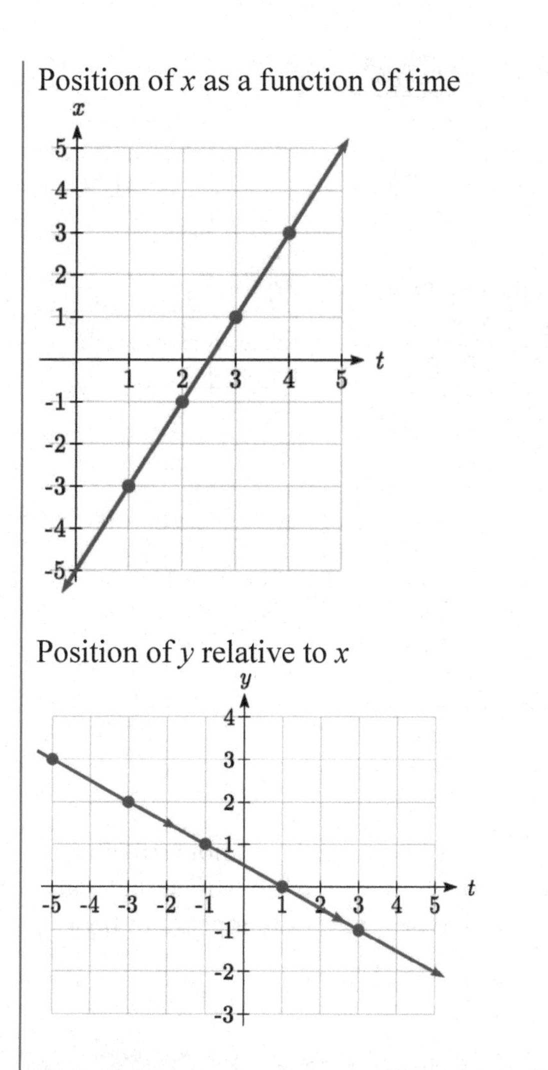

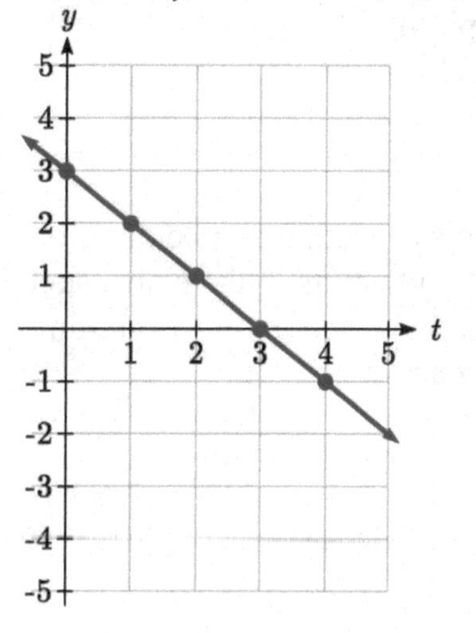

Position of y relative to x

Notice that the parameter t does not explicitly show up in this third graph. Sometimes, when the parameter t does represent a quantity like time, we might indicate the direction of movement on the graph using an arrow, as shown above.

There is often no single parametric representation for a curve. In Example 1 we assumed the object was moving at a steady rate along a straight line. If we kept the assumption about the path (straight line) but did not assume the speed was constant, we might get a system like:

$x(t) = -5 + 2t^2$

$y(t) = 3 - t^2$

This starts at (-5, 3) when $t = 0$ and ends up at (3, -1) when $t = 2$. If we graph the $x(t)$ and $y(t)$ function separately, we can see that those are no longer linear, but if we graph x vs. y we will see that we still get a straight-line path.

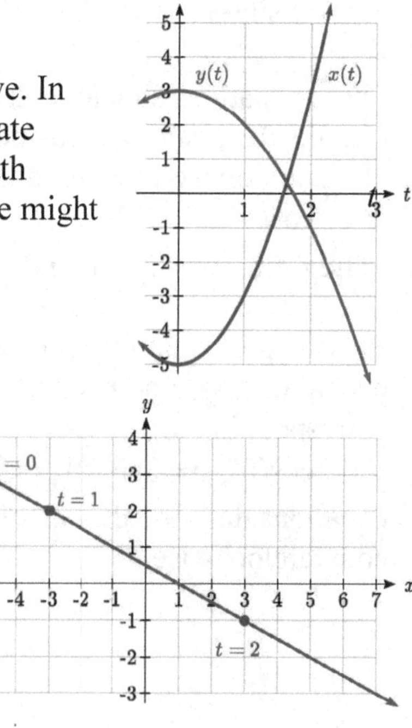

Example 2

Sketch a graph of

$x(t) = t^2 + 1$

$y(t) = 2 + t$

t	x	y
-3	10	-1
-2	5	0
-1	2	1
0	1	2
1	2	3
2	5	4

We can begin by creating a table of values. From this table, we can plot the (x, y) points in the plane, sketch in a rough graph of the curve, and indicate the direction of motion with respect to time by using arrows.

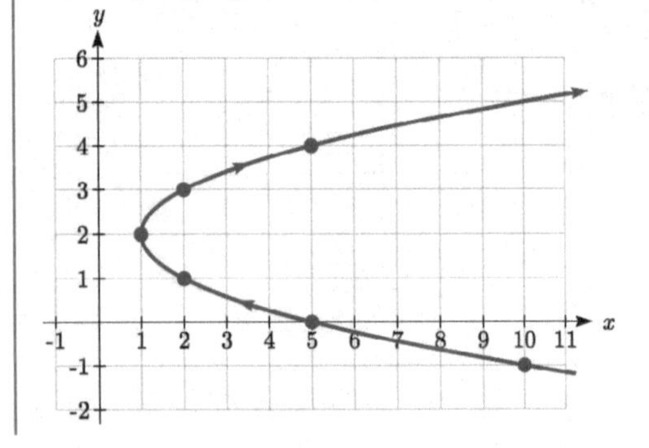

Notice that here the parametric equations describe a shape for which y is not a function of x. This is an example of why using parametric equations can be useful – since they can represent such a graph as a set of functions. This particular graph also appears to be a parabola where x is a function of y, which we will soon verify.

Example 3

Sketch a graph of

$x(t) = 3\cos(t)$

$y(t) = 3\sin(t)$

These equations should look familiar. Back when we first learned about sine and cosine we found that the coordinates of a point on a circle of radius r at an angle of θ will be $x = r\cos(\theta),\ y = r\sin(\theta)$. The equations above are in the same form, with $r = 3$, and t used in place of θ.

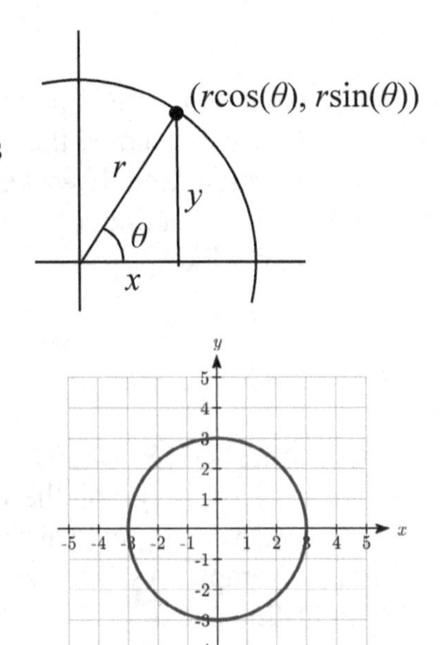

This suggests that for each value of t, these parametric equations give a point on a circle of radius 3 at the angle corresponding to t. At $t = 0$, the graph would be at $x = 3\cos(0),\ y = 3\sin(0)$, the point (3,0). Indeed, these equations describe the equation of a circle, drawn counterclockwise.

Interestingly, these similar parametric equations also describe the circle of radius 3:

$x(t) = 3\sin(t)$

$y(t) = 3\cos(t)$

The difference with these equations it the graph would start at $x = 3\sin(0),\ y = 3\cos(0)$, the point (0,3). As t increases from 0, the x value will increase, indicating these equations would draw the graph in a clockwise direction.

While creating a t-x-y table, plotting points and connecting the dots with a smooth curve usually works to give us a rough idea of what the graph of a system of parametric equations looks like, it's generally easier to use technology to create these tables and (simultaneously) much nicer-looking graphs.

Example 4

Sketch a graph of $\begin{matrix} x(t) = 2\cos(t) \\ y(t) = 3\sin(t) \end{matrix}$.

Notice first that this equation looks very similar to the ones from the previous example, except the coefficients are not equal. You might guess that the pairing of cos and sin will still produce rotation, but now x will vary from -2 to 2 while y will vary from -3 to 3, creating an ellipse.

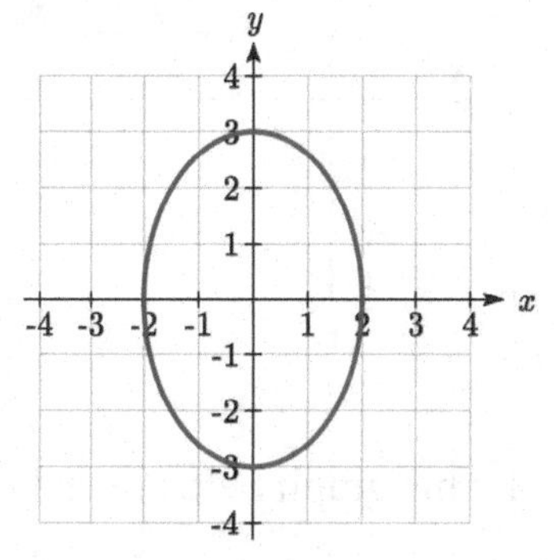

Using technology we can generate a graph of this equation, verifying it is indeed an ellipse.

Similar to graphing polar equations, you must change the MODE on your calculator (or select parametric equations on your graphing technology) before graphing a system of parametric equations. You will know you have successfully entered parametric mode when the equation input has changed to ask for a *x(t)=* and *y(t)=* pair of equations.

Try it Now

1. Sketch a graph of $\begin{matrix} x(t) = 4\cos(3t) \\ y(t) = 3\sin(2t) \end{matrix}$. This is an example of a **Lissajous** figure.

Example 5

The populations of rabbits and wolves on an island over time are given by the graphs below. Use these graphs to sketch a graph in the *r-w* plane showing the relationship between the number of rabbits and number of wolves.

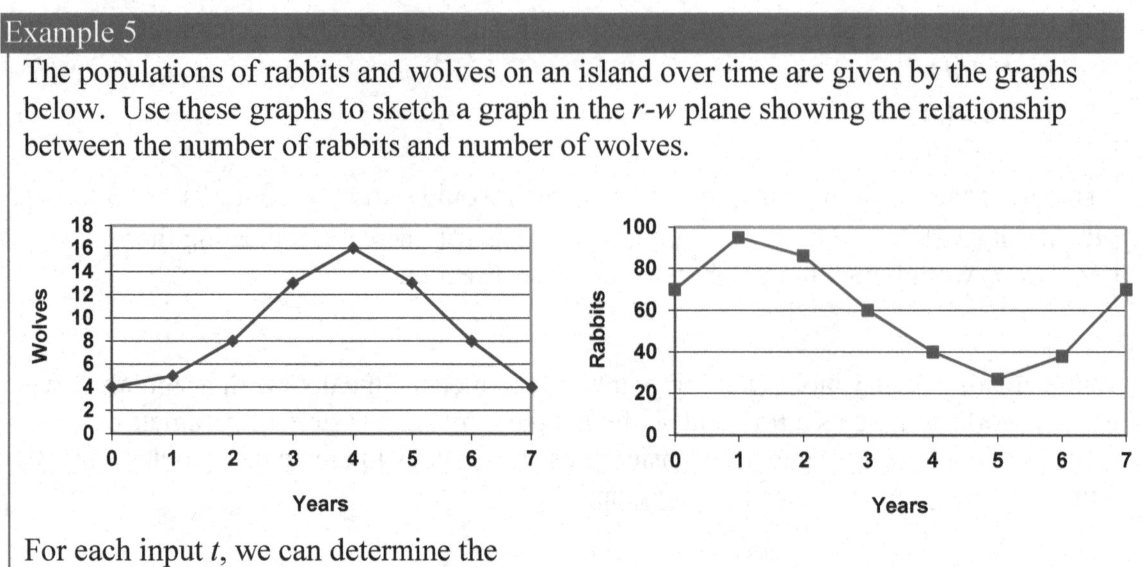

For each input *t*, we can determine the number of rabbits, *r*, and the number of wolves, *w*, from the respective graphs, and then plot the corresponding point in the *r-w* plane.

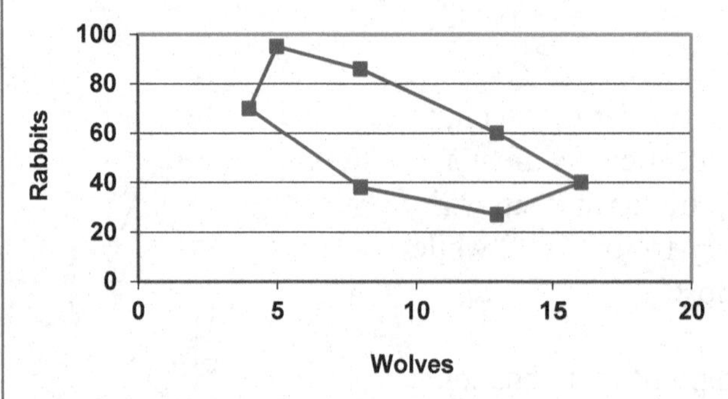

This graph helps reveal the cyclical interaction between the two populations.

Converting from Parametric to Cartesian

In some cases, it is possible to eliminate the parameter *t*, allowing you to write a pair of parametric equations as a Cartesian equation.

It is easiest to do this if one of the *x(t)* or *y(t)* functions can easily be solved for *t*, allowing you to then substitute the remaining expression into the second part.

Example 6

Write $\begin{aligned} x(t) &= t^2 + 1 \\ y(t) &= 2 + t \end{aligned}$ as a Cartesian equation, if possible.

Here, the equation for *y* is linear, so is relatively easy to solve for *t*. Since the resulting Cartesian equation will likely not be a function, and for convenience, we drop the function notation.

$y = 2 + t$ Solve for t

$y - 2 = t$ Substitute this for t in the x equation

$x = (y-2)^2 + 1$

Notice that this is the equation of a parabola with x as a function of y, with vertex at (1,2), opening to the right. Comparing this with the graph from Example 2, we see (unsurprisingly) that it yields the same graph in the *x-y* plane as did the original parametric equations.

Try it Now

2. Write $\begin{array}{l} x(t) = t^3 \\ y(t) = t^6 \end{array}$ as a Cartesian equation, if possible.

Example 7

Write $\begin{array}{l} x(t) = \sqrt{t} + 2 \\ y(t) = \log(t) \end{array}$ as a Cartesian equation, if possible.

We could solve either the first or second equation for t. Solving the first,

$x = \sqrt{t} + 2$

$x - 2 = \sqrt{t}$ Square both sides

$(x-2)^2 = t$ Substitute into the y equation

$y = \log\left((x-2)^2\right)$

Since the parametric equation is only defined for $t > 0$, this Cartesian equation is equivalent to the parametric equation on the corresponding domain. The parametric equations show that when $t > 0$, $x > 2$ and $y > 0$, so the domain of the Cartesian equation should be limited to $x > 2$.

To ensure that the Cartesian equation is as equivalent as possible to the original parametric equation, we try to avoid using domain-restricted inverse functions, such as the inverse trig functions, when possible. For equations involving trig functions, we often try to find an identity to utilize to avoid the inverse functions.

Example 8

Write $\begin{matrix} x(t) = 2\cos(t) \\ y(t) = 3\sin(t) \end{matrix}$ as a Cartesian equation, if possible.

To rewrite this, we can utilize the Pythagorean identity $\cos^2(t) + \sin^2(t) = 1$.

$x = 2\cos(t)$ so $\frac{x}{2} = \cos(t)$

$y = 3\sin(t)$ so $\frac{y}{3} = \sin(t)$

Starting with the Pythagorean Identity,

$\cos^2(t) + \sin^2(t) = 1$ Substitute in the expressions from the parametric form

$\left(\frac{x}{2}\right)^2 + \left(\frac{y}{3}\right)^2 = 1$ Simplify

$\frac{x^2}{4} + \frac{y^2}{9} = 1$

This is a Cartesian equation for the ellipse we graphed earlier.

Parameterizing Curves

While converting from parametric form to Cartesian can be useful, it is often more useful to parameterize a Cartesian equation – converting it into parametric form.

If the Cartesian equation gives one variable as a function of the other, then parameterization is trivial – the independent variable in the function can simply be defined as t.

Example 9

Parameterize the equation $x = y^3 - 2y$.

In this equation, x is expressed as a function of y. By defining $y = t$ we can then substitute that into the Cartesian equation, yielding $x = t^3 - 2t$. Together, this produces the parametric form:

$x(t) = t^3 - 2t$

$y(t) = t$

Try it Now

3. Write $x^2 + y^2 = 3$ in parametric form, if possible.

In addition to parameterizing Cartesian equations, we also can parameterize behaviors and movements.

Example 10

A robot follows the path shown. Create a table of values for the *x(t)* and *y(t)* functions, assuming the robot takes one second to make each movement.

Since we know the direction of motion, we can introduce consecutive values for t along the path of the robot. Using these values with the x and y coordinates of the robot, we can create the tables. For example, we designate the starting point, at (1, 1), as the position at $t = 0$, the next point at (3, 1) as the position at $t = 1$, and so on.

t	0	1	2	3	4	5	6
x	1	3	3	2	4	1	1

t	0	1	2	3	4	5	6
y	1	1	2	2	4	5	4

Notice how this also ties back to vectors. The journey of the robot as it moves through the Cartesian plane could also be displayed as vectors and total distance traveled and displacement could be calculated.

Example 11

A light is placed on the edge of a bicycle tire as shown and the bicycle starts rolling down the street. Find a parametric equation for the position of the light after the wheel has rotated through an angle of θ.

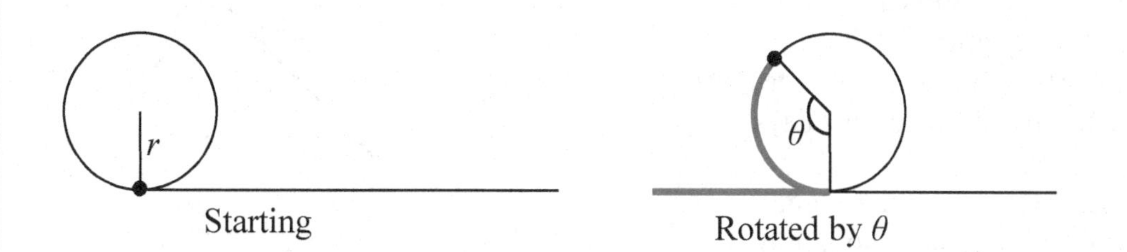

Relative to the center of the wheel, the position of the light can be found as the coordinates of a point on a circle, but since the x coordinate begins at 0 and moves in the negative direction, while the y coordinate starts at the lowest value, the coordinates of the point will be given by:

$x = -r\sin(\theta)$

$y = -r\cos(\theta)$

The center of the wheel, meanwhile, is moving horizontally. It remains at a constant height of r, but the horizontal position will move a distance equivalent to the arclength of the circle drawn out by the angle, $s = r\theta$. The position of the center of the circle is then

$x = r\theta$

$y = r$

Combining the position of the center of the wheel with the position of the light on the wheel relative to the center, we get the following parametric equationw, with θ as the parameter:

$x = r\theta - r\sin(\theta) = r(\theta - \sin(\theta))$

$y = r - r\cos(\theta) = r(1 - \cos(\theta))$

The result graph is called a cycloid.

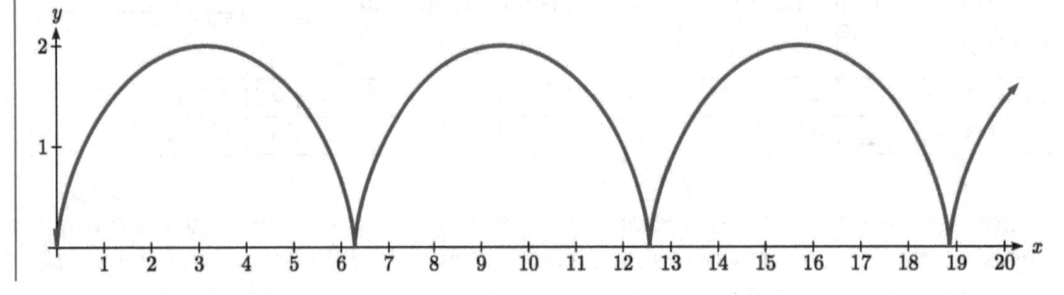

Example 12

A moon travels around a planet as shown, orbiting once every 10 days. The planet travels around a sun as shown, orbiting once every 100 days. Find a parametric equation for the position of the moon, relative to the center of the sun, after t days.

For this example, we'll assume the orbits are circular, though in real life they're actually elliptical.

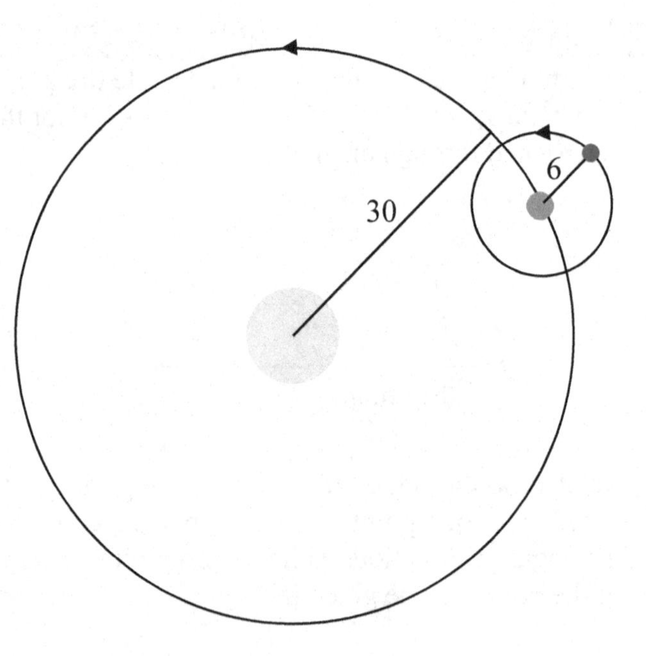

The coordinates of a point on a circle can always be written in the form
$x = r\cos(\theta)$
$y = r\sin(\theta)$

Since the orbit of the moon around the planet has a period of 10 days, the equation for the position of the *moon relative to the planet* will be
$$x(t) = 6\cos\left(\frac{2\pi}{10}t\right) = 6\cos\left(\frac{\pi}{5}t\right)$$
$$y(t) = 6\sin\left(\frac{2\pi}{10}t\right) = 6\sin\left(\frac{\pi}{5}t\right)$$

With a period of 100 days, the equation for the position of the *planet relative to the sun* will be
$$x(t) = 30\cos\left(\frac{2\pi}{100}t\right) = 30\cos\left(\frac{\pi}{50}t\right)$$
$$y(t) = 30\sin\left(\frac{2\pi}{100}t\right) = 30\sin\left(\frac{\pi}{50}t\right)$$

Combining these together, we can find the position of the *moon relative to the sun* as the sum of the components.
$$x(t) = 6\cos\left(\frac{\pi}{5}t\right) + 30\cos\left(\frac{\pi}{50}t\right)$$
$$y(t) = 6\sin\left(\frac{\pi}{5}t\right) + 30\sin\left(\frac{\pi}{50}t\right)$$

The resulting graph is shown here.

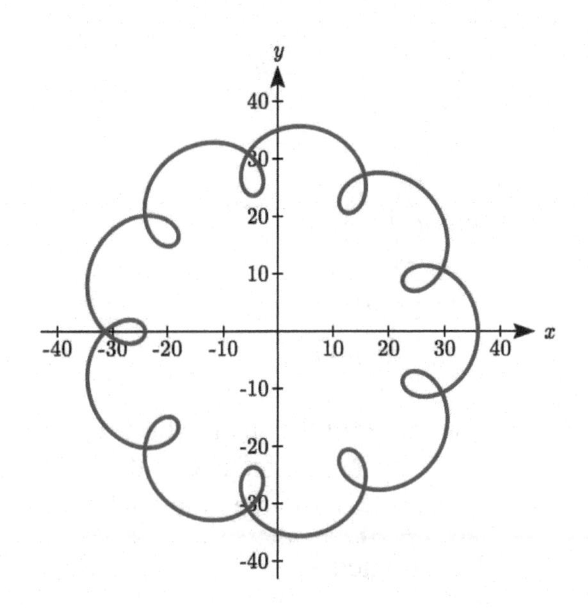

Try it Now

4. A wheel of radius 4 is rolled around the outside of a circle of radius 7. Find a parametric equation for the position of a point on the boundary of the smaller wheel. This shape is called an epicycloid.

Important Topics of This Section
Parametric equations
Graphing *x(t)* , *y(t)* and the corresponding *x-y* graph
Sketching graphs and building a table of values
Converting parametric to Cartesian
Converting Cartesian to parametric (parameterizing curves)

Try it Now Answers

1. 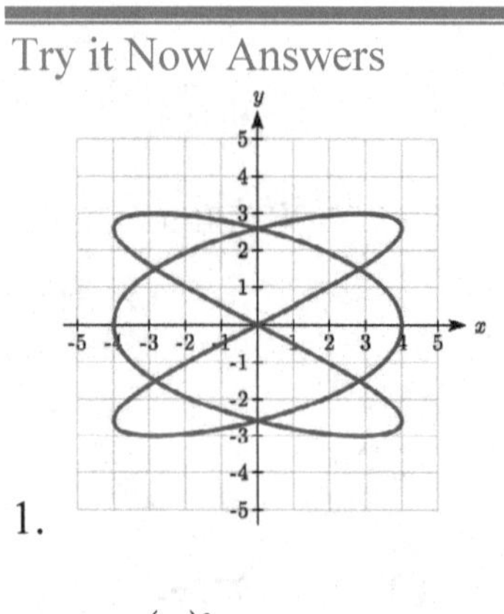

2. $y = (t^3)^2$, so $y = x^2$

3. $x(t) = 3\cos(t)$
$y(t) = 3\sin(t)$

4. The center of the small wheel rotates in circle with radius 7+4=11.
Since the circumference of the small circle is 8π and the circumference of the large circle is 22π, in the time it takes to roll around the large circle, the small circle will have rotated $\dfrac{22\pi}{8\pi} = \dfrac{11}{4}$ rotations. We use this as the stretch factor. The position of a point on the small circle will be the combination of the position of the center of the small wheel around the center of the large wheel, and the position of the point around the small wheel:

$$x(t) = 11\cos(t) - 4\cos\left(\frac{11}{4}t\right)$$

$$y(t) = 11\sin(t) - 4\sin\left(\frac{11}{4}t\right)$$

Section 8.6 Exercises

Match each set of equations with one of the graphs below.

1. $\begin{cases} x(t) = t \\ y(t) = t^2 - 1 \end{cases}$

2. $\begin{cases} x(t) = t - 1 \\ y(t) = t^2 \end{cases}$

3. $\begin{cases} x(t) = 4\sin(t) \\ y(t) = 2\cos(t) \end{cases}$

4. $\begin{cases} x(t) = 2\sin(t) \\ y(t) = 4\cos(t) \end{cases}$

5. $\begin{cases} x(t) = 2 + t \\ y(t) = 3 - 2t \end{cases}$

6. $\begin{cases} x(t) = -2 - 2t \\ y(t) = 3 + t \end{cases}$

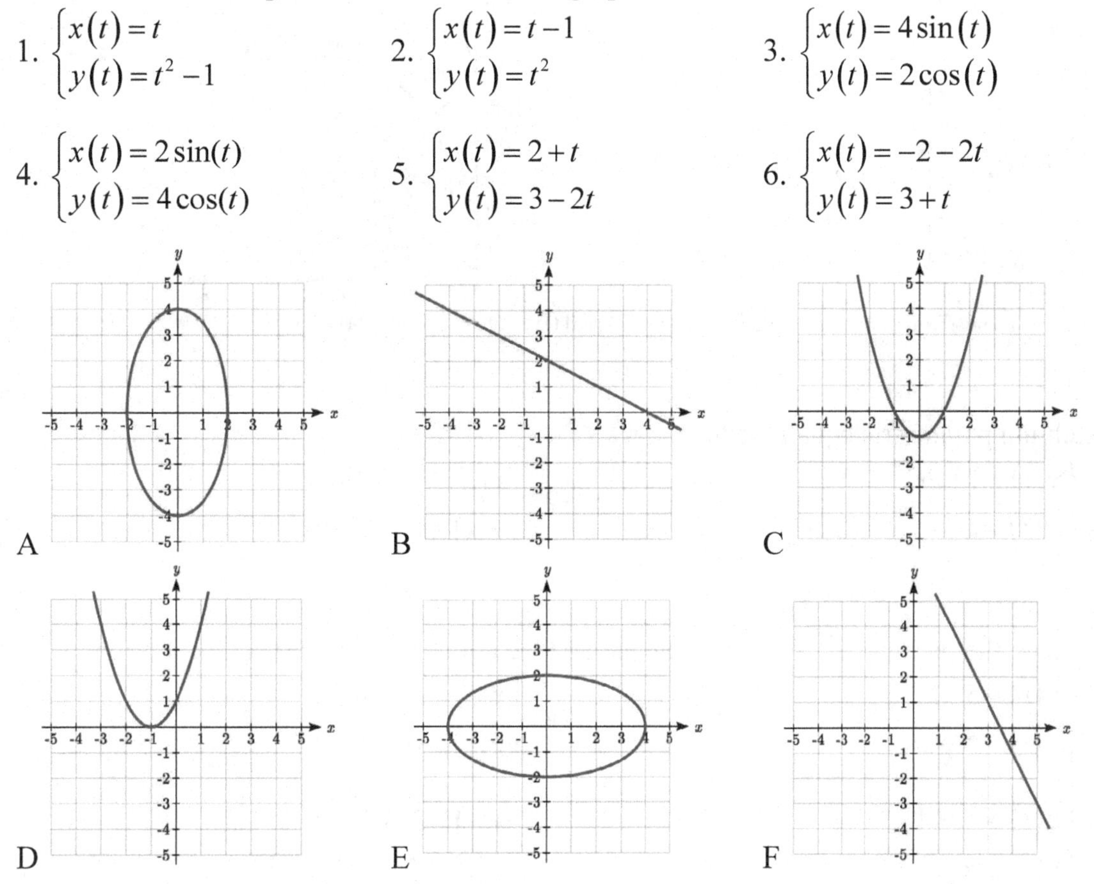

From each pair of graphs in the *t*-*x* and *t*-*y* planes shown, sketch a graph in the *x*-*y* plane.

7.

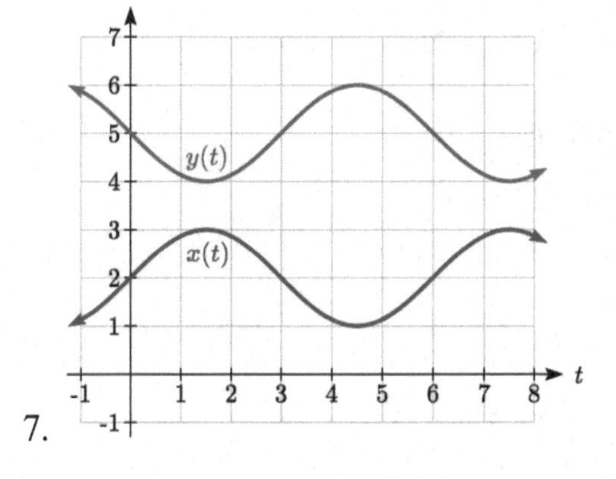

8.

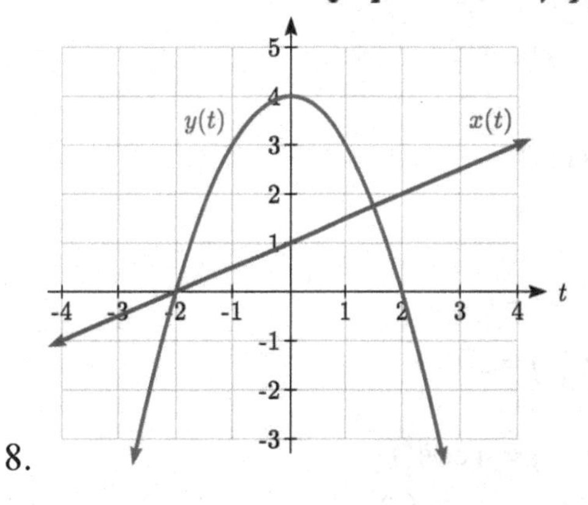

From each graph in the *x*-*y* plane shown, sketch a graph of the parameter functions in the *t*-*x* and *t*-*y* planes.

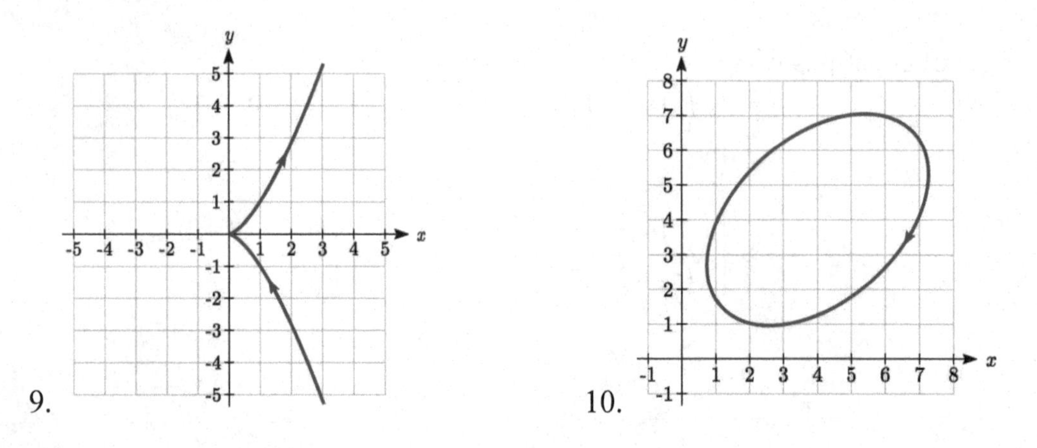

9. 10.

Sketch the parametric equations for $-2 \le t \le 2$.

11. $\begin{cases} x(t) = 1 + 2t \\ y(t) = t^2 \end{cases}$

12. $\begin{cases} x(t) = 2t - 2 \\ y(t) = t^3 \end{cases}$

Eliminate the parameter *t* to rewrite the parametric equation as a Cartesian equation

13. $\begin{cases} x(t) = 5 - t \\ y(t) = 8 - 2t \end{cases}$

14. $\begin{cases} x(t) = 6 - 3t \\ y(t) = 10 - t \end{cases}$

15. $\begin{cases} x(t) = 2t + 1 \\ y(t) = 3\sqrt{t} \end{cases}$

16. $\begin{cases} x(t) = 3t - 1 \\ y(t) = 2t^2 \end{cases}$

17. $\begin{cases} x(t) = 2e^t \\ y(t) = 1 - 5t \end{cases}$

18. $\begin{cases} x(t) = 4\log(t) \\ y(t) = 3 + 2t \end{cases}$

19. $\begin{cases} x(t) = t^3 - t \\ y(t) = 2t \end{cases}$

20. $\begin{cases} x(t) = t - t^4 \\ y(t) = t + 2 \end{cases}$

21. $\begin{cases} x(t) = e^{2t} \\ y(t) = e^{6t} \end{cases}$

22. $\begin{cases} x(t) = t^5 \\ y(t) = t^{10} \end{cases}$

23. $\begin{cases} x(t) = 4\cos(t) \\ y(t) = 5\sin(t) \end{cases}$

24. $\begin{cases} x(t) = 3\sin(t) \\ y(t) = 6\cos(t) \end{cases}$

Parameterize (write a parametric equation for) each Cartesian equation

25. $y(x) = 3x^2 + 3$

26. $y(x) = 2\sin(x) + 1$

27. $x(y) = 3\log(y) + y$

28. $x(y) = \sqrt{y} + 2y$

29. $\dfrac{x^2}{4} + \dfrac{y^2}{9} = 1$

30. $\dfrac{x^2}{16} + \dfrac{y^2}{36} = 1$

Parameterize the graphs shown.

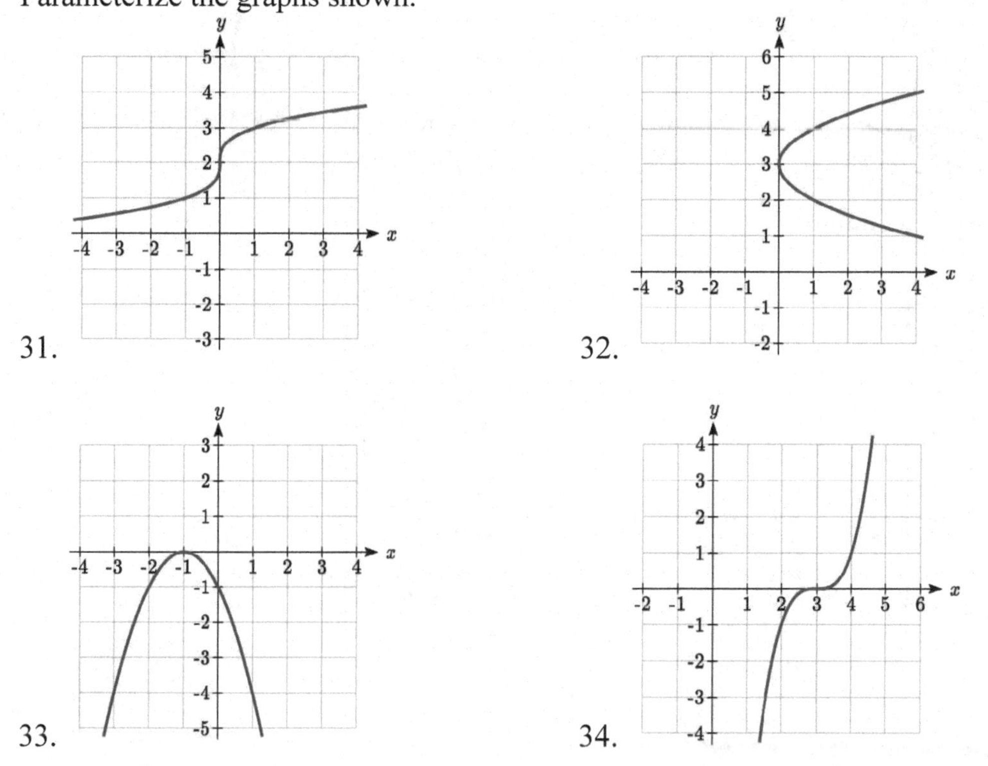

35. Parameterize the line from $(-1,5)$ to $(2,3)$ so that the line is at $(-1,5)$ at $t = 0$, and at $(2,3)$ at $t = 1$.

36. Parameterize the line from $(4,1)$ to $(6,-2)$ so that the line is at $(4,1)$ at $t = 0$, and at $(6,-2)$ at $t = 1$.

The graphs below are created by parameteric equations of the form $\begin{cases} x(t) = a\cos(bt) \\ y(t) = c\sin(dt) \end{cases}$.

Find the values of *a, b, c,* and *d* to achieve each graph.

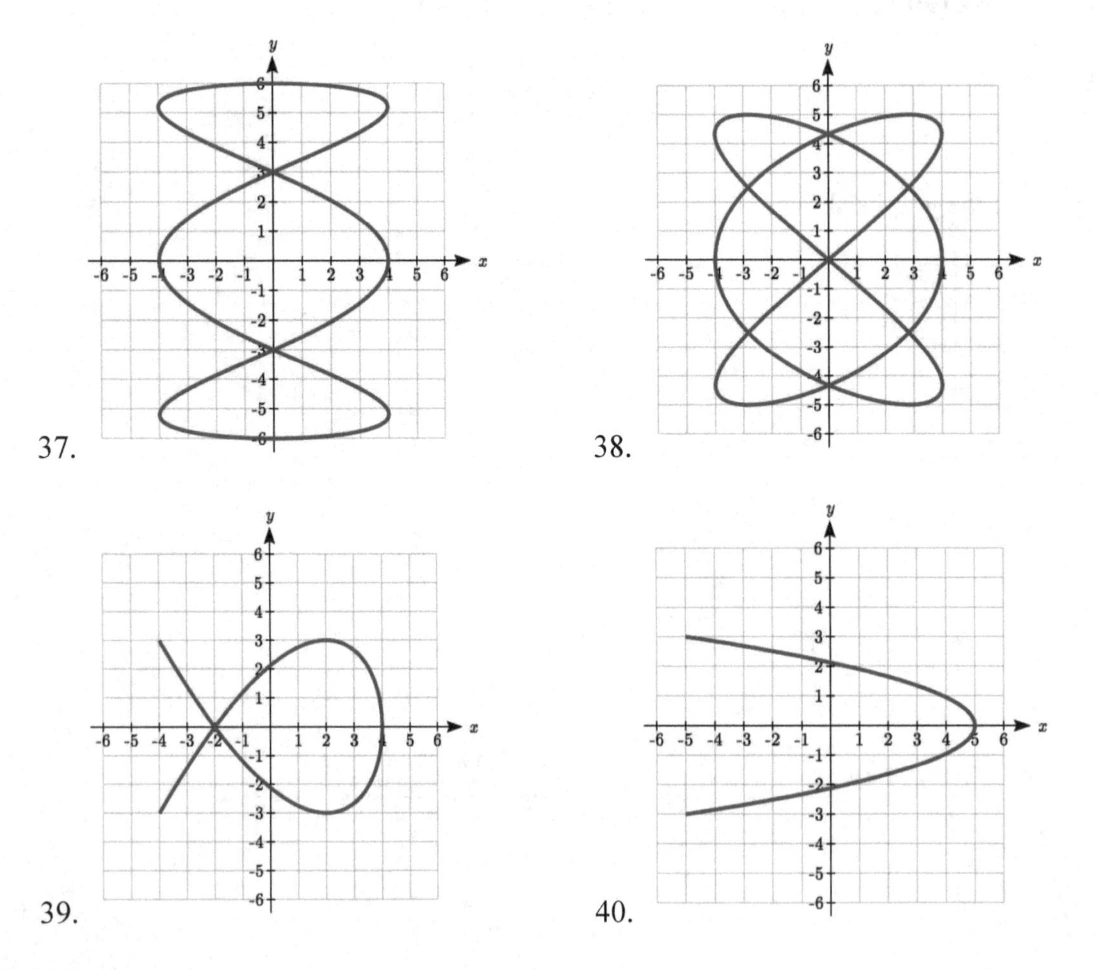

37.

38.

39.

40.

41. An object is thrown in the air with vertical velocity 20 ft/s and horizontal velocity 15 ft/s. The object's height can be described by the equation $y(t) = -16t^2 + 20t$, while the object moves horizontally with constant velocity 15 ft/s. Write parametric equations for the object's position, then eliminate time to write height as a function of horizontal position.

42. A skateboarder riding on a level surface at a constant speed of 9 ft/s throws a ball in the air, the height of which can be described by the equation $y(t) = -16t^2 + 10t + 5$. Write parametric equations for the ball's position, then eliminate time to write height as a function of horizontal position.

43. A carnival ride has a large rotating arm with diameter 40 feet centered 35 feet off the ground. At each end of the large arm are two smaller rotating arms with diameter 16 feet each. The larger arm rotates once every 5 seconds, while the smaller arms rotate once every 2 seconds. If you board the ride when the point P is closest to the ground, find parametric equations for your position over time.

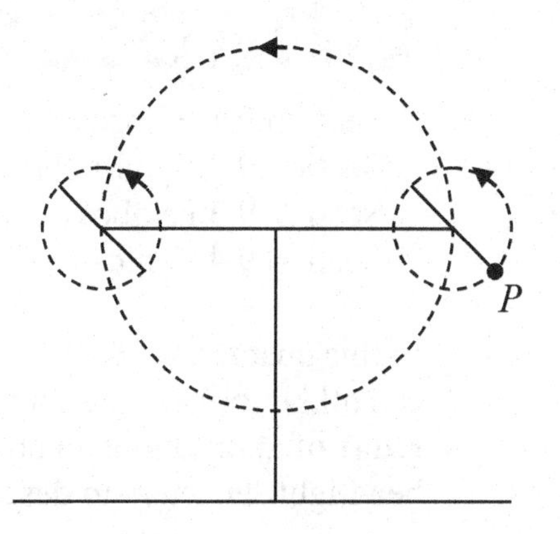

44. A hypocycloid is a shape generated by tracking a fixed point on a small circle as it rolls around the inside of a larger circle. If the smaller circle has radius 1 and the large circle has radius 6, find parametric equations for the position of the point P as the smaller wheel rolls in the direction indicated.

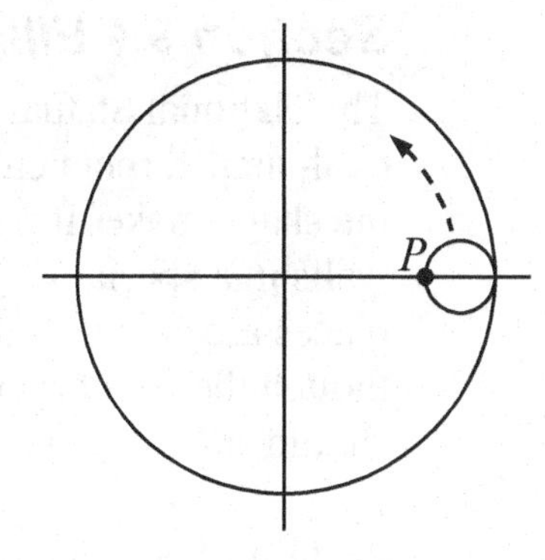

Chapter 9: Conics

In this chapter, we will explore a set of shapes defined by a common characteristic: they can all be formed by slicing a cone with a plane. These families of curves have a broad range of applications in physics and astronomy, from describing the shape of your car headlight reflectors to describing the orbits of planets and comets.

Section 9.1 Ellipses

The National Statuary Hall[1] in Washington, D.C. is an oval-shaped room called a whispering chamber because the shape makes it possible for sound to reflect from the walls in a special way. Two people standing in specific places are able to hear each other whispering even though they are far apart. To determine where they should stand, we will need to better understand ellipses.

An ellipse is a type of **conic section**, a shape resulting from intersecting a plane with a cone and looking at the curve where they intersect. They were discovered by the Greek mathematician Menaechmus over two millennia ago.

The figure below[2] shows two types of conic sections. When a plane is perpendicular to the axis of the cone, the shape of the intersection is a circle. A slightly titled plane creates an oval-shaped conic section called an **ellipse**.

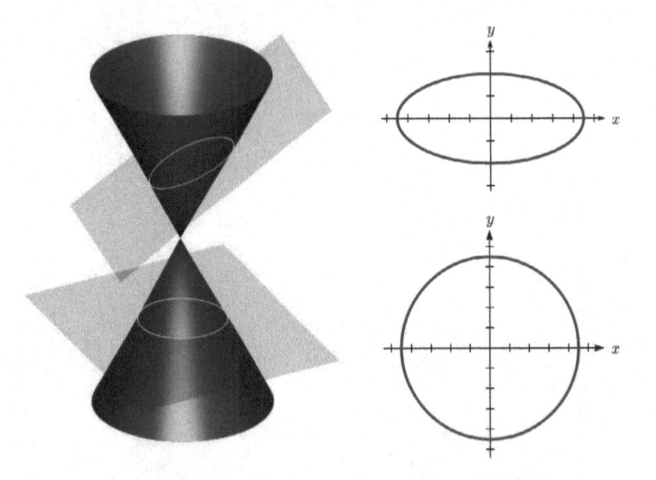

[1] Photo by Gary Palmer, Flickr, CC-BY, https://www.flickr.com/photos/gregpalmer/2157517950
[2] Pbroks13 (https://commons.wikimedia.org/wiki/File:Conic_sections_with_plane.svg), "Conic sections with plane", cropped to show only ellipse and circle by L Michaels, CC BY 3.0

An ellipse can be drawn by placing two thumbtacks in a piece of cardboard then cutting a piece of string longer than the distance between the thumbtacks. Tack each end of the string to the cardboard, and trace a curve with a pencil held taught against the string. An ellipse is the set of all points where the sum of the distances from two fixed points is constant. The length of the string is the constant, and the two thumbtacks are the fixed points, called foci.

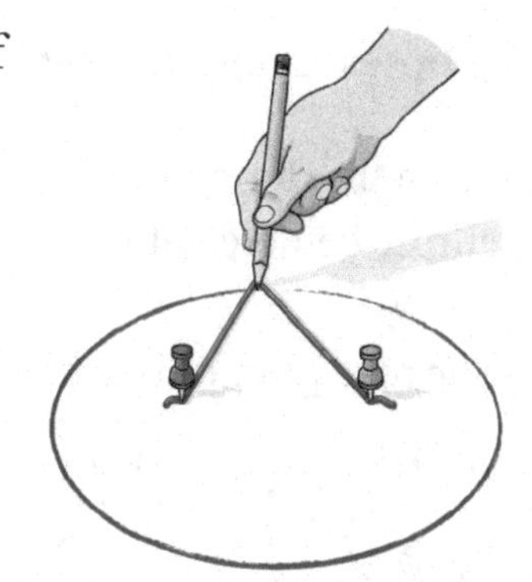

Ellipse Definition and Vocabulary

An **ellipse** is the set of all points $Q(x, y)$ for which the sum of the distance to two fixed points $F_1(x_1, y_1)$ and $F_2(x_2, y_2)$, called the **foci** (plural of focus), is a constant *k*: $d(Q, F_1) + d(Q, F_2) = k$.

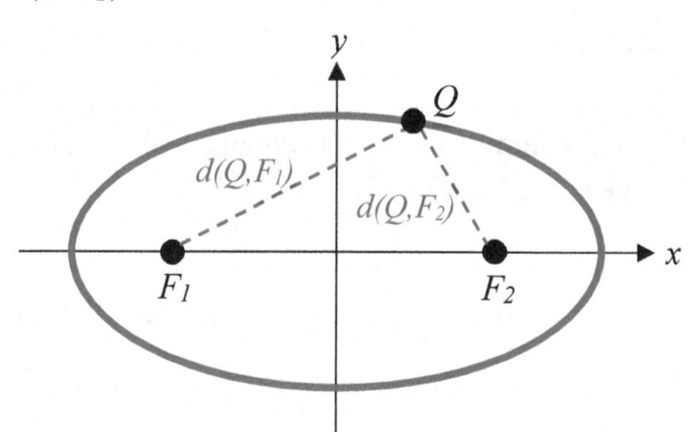

The **major axis** is the line passing through the foci.
Vertices are the points on the ellipse which intersect the major axis.
The **major axis length** is the length of the line segment between the vertices.
The **center** is the midpoint between the vertices (or the midpoint between the foci).
The **minor axis** is the line perpendicular to the minor axis passing through the center.
Minor axis endpoints are the points on the ellipse which intersect the minor axis.
The minor axis endpoints are also sometimes called **co-vertices**.
The **minor axis length** is the length of the line segment between minor axis endpoints.

Note that which axis is major and which is minor will depend on the orientation of the ellipse. In the ellipse shown at right, the foci lie on the *y* axis, so that is the major axis, and the *x* axis is the minor axis. Because of this, the vertices are the endpoints of the ellipse on the *y* axis, and the minor axis endpoints (co-vertices) are the endpoints on the *x* axis.

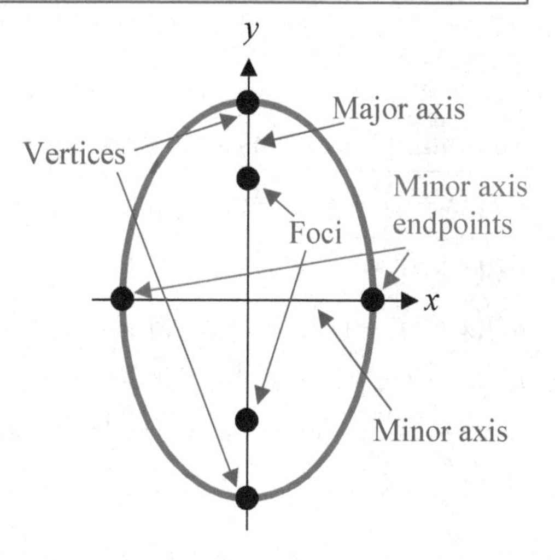

Ellipses Centered at the Origin

From the definition above we can find an equation for an ellipse. We will find it for a ellipse centered at the origin $C(0,0)$ with foci at $F_1(c,0)$ and $F_2(-c,0)$ where $c > 0$.

Suppose $Q(x, y)$ is some point on the ellipse. The distance from F_1 to Q is
$d(Q,F_1)=\sqrt{(x-c)^2+(y-0)^2}=\sqrt{(x-c)^2+y^2}$

Likewise, the distance from F_2 to Q is
$d(Q,F_2)=\sqrt{(x-(-c))^2+(y-0)^2}=\sqrt{(x+c)^2+y^2}$

From the definition of the ellipse, the sum of these distances should be constant:
$d(Q,F_1)+d(Q,F_2)=k$ so that
$\sqrt{(x-c)^2+y^2}+\sqrt{(x+c)^2+y^2}=k$

If we label one of the vertices $(a,0)$, it should satisfy the equation above since it is a point on the ellipse. This allows us to write k in terms of a.

$\sqrt{(a-c)^2+0^2}+\sqrt{(a+c)^2+0^2}=k$
$|a-c|+|a+c|=k$ Since $a >$ c, these will be positive
$(a-c)+(a+c)=k$
$2a=k$

Substituting that into our equation, we will now try to rewrite the equation in a friendlier form.

$\sqrt{(x-c)^2+y^2}+\sqrt{(x+c)^2+y^2}=2a$ Move one radical
$\sqrt{(x-c)^2+y^2}=2a-\sqrt{(x+c)^2+y^2}$ Square both sides
$\left(\sqrt{(x-c)^2+y^2}\right)^2=\left(2a-\sqrt{(x+c)^2+y^2}\right)^2$ Expand
$(x-c)^2+y^2=4a^2-4a\sqrt{(x+c)^2+y^2}+(x+c)^2+y^2$ Expand more
$x^2-2xc+c^2+y^2=4a^2-4a\sqrt{(x+c)^2+y^2}+x^2+2xc+c^2+y^2$

Combining like terms and isolating the radical leaves

$4a\sqrt{(x+c)^2+y^2}=4a^2+4xc$ Divide by 4
$a\sqrt{(x+c)^2+y^2}=a^2+xc$ Square both sides again
$a^2\left((x+c)^2+y^2\right)=a^4+2a^2xc+x^2c^2$ Expand
$a^2\left(x^2+2xc+c^2+y^2\right)=a^4+2a^2xc+x^2c^2$ Distribute
$a^2x^2+2a^2xc+a^2c^2+a^2y^2=a^4+2a^2xc+x^2c^2$ Combine like terms

$$a^2x^2 - x^2c^2 + a^2y^2 = a^4 - a^2c^2$$ Factor common terms
$$(a^2 - c^2)x^2 + a^2y^2 = a^2(a^2 - c^2)$$

Let $b^2 = a^2 - c^2$. Since $a > c$, we know $b > 0$. Substituting b^2 for $a^2 - c^2$ leaves
$$b^2x^2 + a^2y^2 = a^2b^2$$ Divide both sides by a^2b^2
$$\frac{x^2}{a^2} + \frac{y^2}{b^2} = 1$$

This is the standard equation for an ellipse. We typically swap *a* and *b* when the major axis of the ellipse is vertical.

Equation of an Ellipse Centered at the Origin in Standard Form

The standard form of an equation of an ellipse centered at the origin $C(0,0)$ depends on whether the major axis is horizontal or vertical. The table below gives the standard equation, vertices, minor axis endpoints, foci, and graph for each.

Major Axis	Horizontal	Vertical
Standard Equation	$\frac{x^2}{a^2} + \frac{y^2}{b^2} = 1$	$\frac{x^2}{b^2} + \frac{y^2}{a^2} = 1$
Vertices	$(-a, 0)$ and $(a, 0)$	$(0, -a)$ and $(0, a)$
Minor Axis Endpoints	$(0, -b)$ and $(0, b)$	$(-b, 0)$ and $(b, 0)$
Foci	$(-c, 0)$ and $(c, 0)$ where $b^2 = a^2 - c^2$	$(0, -c)$ and $(0, c)$ where $b^2 = a^2 - c^2$
Graph	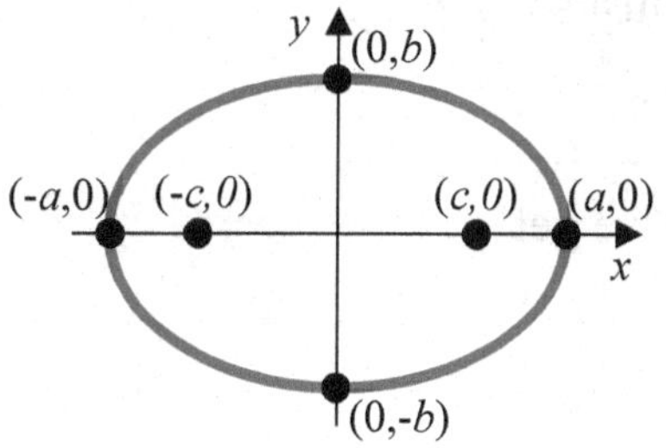	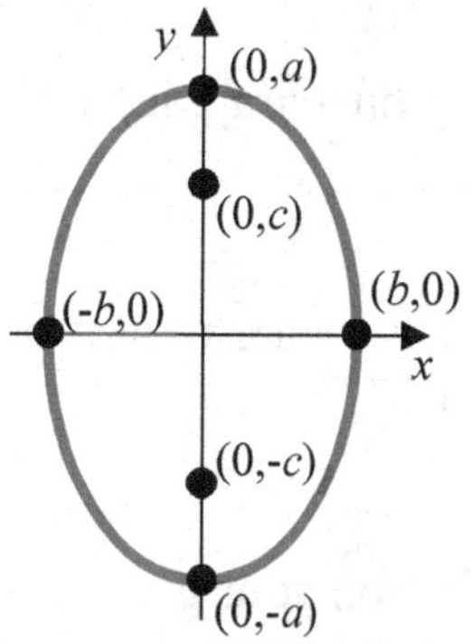

Example 1

Put the equation of the ellipse $9x^2 + y^2 = 9$ in standard form. Find the vertices, minor axis endpoints, length of the major axis, and length of the minor axis. Sketch the graph, then check using a graphing utility.

The standard equation has a 1 on the right side, so this equation can be put in standard form by dividing by 9:

$$\frac{x^2}{1} + \frac{y^2}{9} = 1$$

Since the y-denominator is greater than the x-denominator, the ellipse has a vertical major axis. Comparing to the general standard form equation $\frac{x^2}{b^2} + \frac{y^2}{a^2} = 1$, we see the value of $a = \sqrt{9} = 3$ and the value of $b = \sqrt{1} = 1$.

The vertices lie on the y-axis at $(0, \pm a) = (0, \pm 3)$.
The minor axis endpoints lie on the x-axis at $(\pm b, 0) = (\pm 1, 0)$.
The length of the major axis is $2(a) = 2(3) = 6$.
The length of the minor axis is $2(b) = 2(1) = 2$.

To sketch the graph we plot the vertices and the minor axis endpoints. Then we sketch the ellipse, rounding at the vertices and the minor axis endpoints.

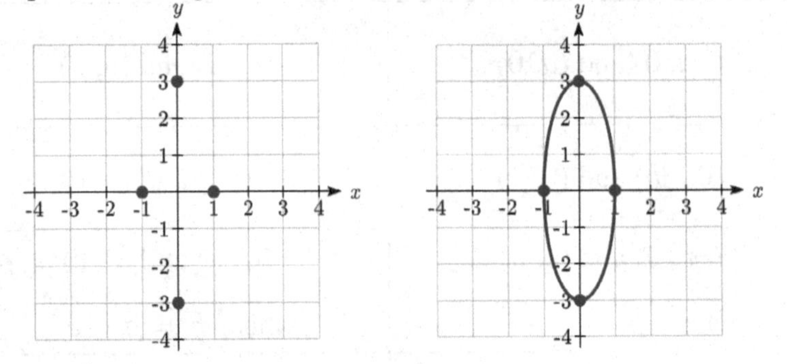

To check on a graphing utility, we must solve the equation for y. Isolating y^2 gives us

$$y^2 = 9(1 - x^2)$$

Taking the square root of both sides we get

$$y = \pm 3\sqrt{1 - x^2}$$

Under **Y=** on your graphing utility enter the two halves of the ellipse as $y = 3\sqrt{1 - x^2}$ and $y = -3\sqrt{1 - x^2}$. Set the window to a comparable scale to the sketch with xmin = -5, xmax = 5, ymin= -5, and ymax = 5.
Here's an example output on a TI-84 calculator:

Plot1 Plot2 Plot3 \Y₁■3√(1-X²) \Y₂■-3√(1-X²)	WINDOW Xmin=-5 Xmax=5 Xscl=1 Ymin=-5 Ymax=5 Yscl=1 Xres=1	

Sometimes we are given the equation. Sometimes we need to find the equation from a graph or other information.

Example 2

Find the standard form of the equation for an ellipse centered at (0,0) with horizontal major axis length 28 and minor axis length 16.

Since the center is at (0,0) and the major axis is horizontal, the ellipse equation has the standard form $\frac{x^2}{a^2}+\frac{y^2}{b^2}=1$. The major axis has length $2a = 28$ or $a = 14$. The minor axis has length $2b = 16$ or $b = 8$. Substituting gives $\frac{x^2}{16^2}+\frac{y^2}{8^2}=1$ or $\frac{x^2}{256}+\frac{y^2}{64.}=1$.

Try it Now

1. Find the standard form of the equation for an ellipse with horizontal major axis length 20 and minor axis length 6.

Example 3

Find the standard form of the equation for the ellipse graphed here.

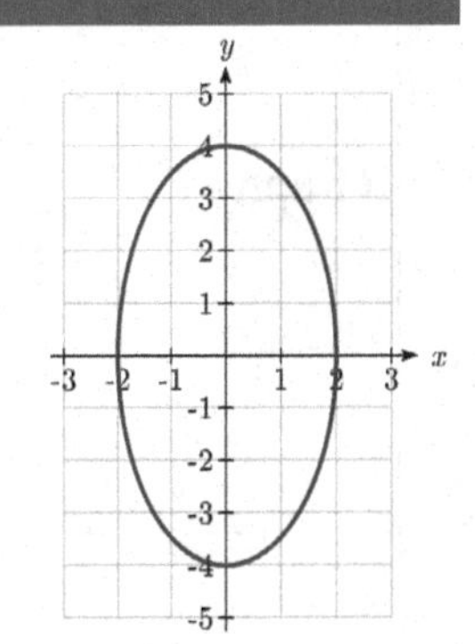

The center is at (0,0) and the major axis is vertical, so the standard form of the equation will be $\frac{x^2}{b^2}+\frac{y^2}{a^2}=1$.

From the graph we can see the vertices are (0,4) and (0,-4), giving $a = 4$.
The minor-axis endpoints are (2,0) and (-2,0), giving $b = 2$.

The equation will be $\frac{x^2}{2^2}+\frac{y^2}{4^2}=1$ or $\frac{x^2}{4}+\frac{y^2}{16}=1$.

Ellipses Not Centered at the Origin

Not all ellipses are centered at the origin. The graph of such an ellipse is a shift of the graph centered at the origin, so the standard equation for one centered at (h, k) is slightly different. We can shift the graph right h units and up k units by replacing x with $x - h$ and y with $y - k$, similar to what we did when we learned transformations.

Equation of an Ellipse Centered at (*h*, *k*) in Standard Form

The standard form of an equation of an ellipse centered at the point $C(h,k)$ depends on whether the major axis is horizontal or vertical. The table below gives the standard equation, vertices, minor axis endpoints, foci, and graph for each.

Major Axis	Horizontal	Vertical
Standard Equation	$\frac{(x-h)^2}{a^2}+\frac{(y-k)^2}{b^2}=1$	$\frac{(x-h)^2}{b^2}+\frac{(y-k)^2}{a^2}=1$
Vertices	$(h \pm a, k)$	$(h, k \pm a)$
Minor Axis Endpoints	$(h, k \pm b)$	$(h \pm b, k)$
Foci	$(h \pm c, k)$ where $b^2 = a^2 - c^2$	$(h, k \pm c)$ where $b^2 = a^2 - c^2$
Graph	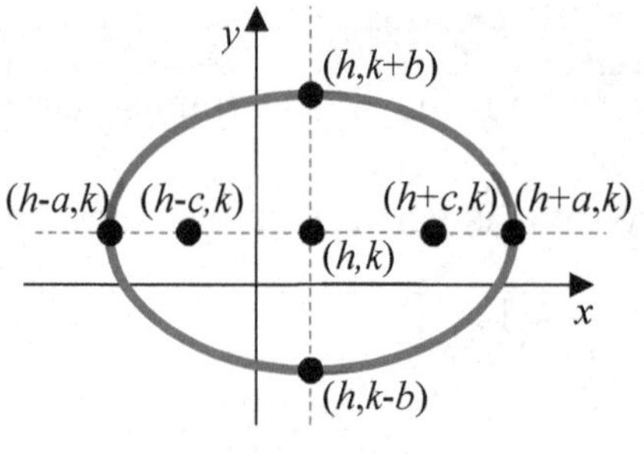 y (h,k+b) (h-a,k) (h-c,k) (h+c,k) (h+a,k) (h,k) x (h,k-b)	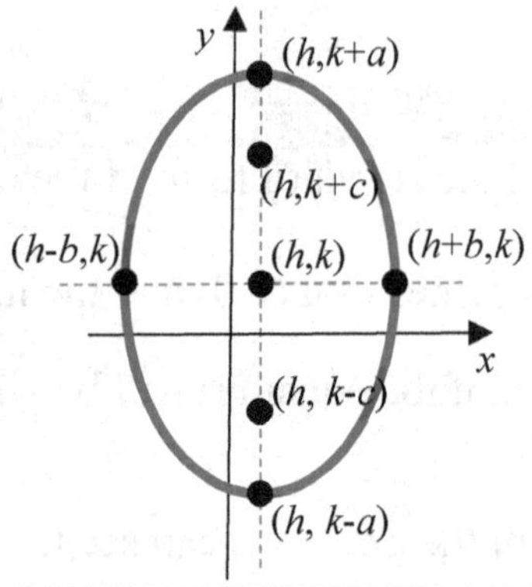 y (h,k+a) (h,k+c) (h-b,k) (h,k) (h+b,k) x (h, k-c) (h, k-a)

Example 4

Put the equation of the ellipse $x^2+2x+4y^2-24y=-33$ in standard form. Find the vertices, minor axis endpoints, length of the major axis, and length of the minor axis. Sketch the graph.

To rewrite this in standard form, we will need to complete the square, twice.

Looking at the x terms, x^2+2x, we like to have something of the form $(x+n)^2$. Notice that if we were to expand this, we'd get $x^2+2nx+n^2$, so in order for the coefficient on x to match, we'll need $(x+1)^2=x^2+2x+1$. However, we don't have a +1 on the left side of the equation to allow this factoring. To accommodate this, we will add 1 to both sides of the equation, which then allows us to factor the left side as a perfect square:

$$x^2+2x+1+4y^2-24y=-33+1$$
$$(x+1)^2+4y^2-24y=-32$$

Repeating the same approach with the y terms, first we'll factor out the 4.

$$4y^2-24y=4(y^2-6y)$$

Now we want to be able to write $4(y^2-6y)$ as $4(y+n)^2=4(y^2+2ny+n^2)$.
For the coefficient of y to match, n will have to -3, giving

$$4(y-3)^2=4(y^2-6y+9)=4y^2-24y+36.$$

To allow this factoring, we can add 36 to both sides of the equation.

$$(x+1)^2+4y^2-24y+36=-32+36$$
$$(x+1)^2+4(y^2-6y+9)=4$$
$$(x+1)^2+4(y-3)^2=4$$

Dividing by 4 gives the standard form of the equation for the ellipse

$$\frac{(x+1)^2}{4}+\frac{(y-3)^2}{1}=1$$

Since the x-denominator is greater than the y-denominator, the ellipse has a horizontal major axis. From the general standard equation $\frac{(x-h)^2}{a^2}+\frac{(h-k)^2}{b^2}=1$ we see the value of $a=\sqrt{4}=2$ and the value of $b=\sqrt{1}=1$.

The center is at (h, k) = (-1, 3).
The vertices are at $(h\pm a, k)$ or (-3, 3) and (1,3).
The minor axis endpoints are at $(h, k\pm b)$ or (-1, 2) and (-1,4).

The length of the major axis is $2(a) = 2(2) = 4$.
The length of the minor axis is $2(b) = 2(1) = 2$.

To sketch the graph we plot the vertices and the minor axis endpoints. Then we sketch the ellipse, rounding at the vertices and the minor axis endpoints.

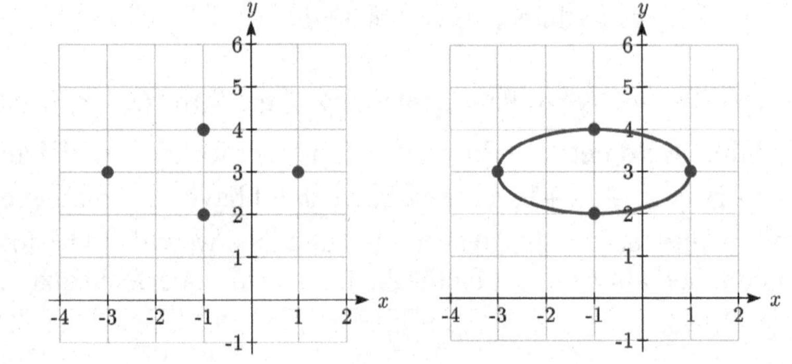

Example 5

Find the standard form of the equation for an ellipse centered at (-2,1), a vertex at (-2,4) and passing through the point (0,1).

The center at (-2,1) and vertex at (-2,4) means the major axis is vertical since the *x*-values are the same. The ellipse equation has the standard form $\frac{(x-h)^2}{b^2} + \frac{(y-k)^2}{a^2} = 1$.

The value of a = 4-1=3. Substituting a = 3, h = -2, and k = 1 gives

$\frac{(x+2)^2}{b^2} + \frac{(y-1)^2}{3^2} = 1$. Substituting for x and y using the point (0,1) gives

$\frac{(0+2)^2}{b^2} + \frac{(1-1)^2}{3^2} = 1$.

Solving for b gives b=2.

The equation of the ellipse in standard form is $\frac{(x+2)^2}{2^2} + \frac{(y-1)^2}{3^2} = 1$ or

$\frac{(x+2)^2}{4} + \frac{(y-1)^2}{9} = 1$.

Try it Now

2. Find the center, vertices, minor axis endpoints, length of the major axis, and length of the minor axis for the ellipse $(x-4)^2 + \frac{(y+2)^2}{4} = 1$.

Bridges with Semielliptical Arches

Arches have been used to build bridges for centuries, like in the Skerton Bridge in England which uses five semielliptical arches for support[3]. Semielliptical arches can have engineering benefits such as allowing for longer spans between supports.

Example 6

A bridge over a river is supported by a single semielliptical arch. The river is 50 feet wide. At the center, the arch rises 20 feet above the river. The roadway is 4 feet above the center of the arch. What is the vertical distance between the roadway and the arch 15 feet from the center?

Put the center of the ellipse at (0,0) and make the span of the river the major axis.

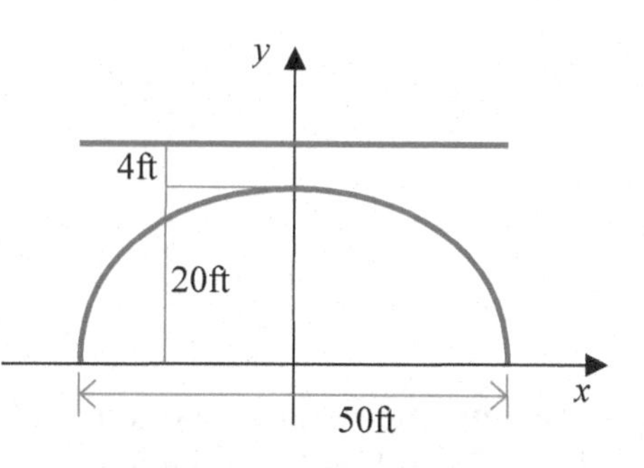

Since the major axis is horizontal, the equation has the form $\frac{x^2}{a^2}+\frac{y^2}{b^2}=1$.

The value of $a=\frac{1}{2}(50)=25$ and the value of $b = 20$, giving $\frac{x^2}{25^2}+\frac{y^2}{15^2}=1$.

Substituting $x = 15$ gives $\frac{15^2}{25^2}+\frac{y^2}{20^2}=1$. Solving for y, $y=20\sqrt{1-\frac{225}{625}}=16$.

The roadway is $20 + 4 = 24$ feet above the river. The vertical distance between the roadway and the arch 15 feet from the center is $24 - 16 = 8$ feet.

[3] Maxine Armstrong (https://commons.wikimedia.org/wiki/File:Skerton_Bridge,_Lancaster,_England.JPG), "Skerton Bridge, Lancaster, England", CC BY-SA

Ellipse Foci

The location of the foci can play a key role in ellipse application problems. Standing on a focus in a whispering gallery allows you to hear someone whispering at the other focus. To find the foci, we need to find the length from the center to the foci, *c*, using the equation $b^2 = a^2 - c^2$. It looks similar to, but is not the same as, the Pythagorean Theorem.

Example 7

The National Statuary Hall whispering chamber is an elliptical room 46 feet wide and 96 feet long. To hear each other whispering, two people need to stand at the foci of the ellipse. Where should they stand?

We could represent the hall with a horizontal ellipse centered at the origin. The major axis length would be 96 feet, so $a = \frac{1}{2}(96) = 48$, and the minor axis length would be 46 feet, so $b = \frac{1}{2}(46) = 23$. To find the foci, we can use the equation $b^2 = a^2 - c^2$.

$$23^2 = 48^2 - c^2$$
$$c^2 = 48^2 - 23^2$$
$$c = \sqrt{1775} \approx \pm 42 \text{ ft.}$$

To hear each other whisper, two people would need to stand 2(42) = 84 feet apart along the major axis, each about 48 – 42 = 6 feet from the wall.

Example 8

Find the foci of the ellipse $\frac{(x-2)^2}{4} + \frac{(y+3)^2}{29} = 1$.

The ellipse is vertical with an equation of the form $\frac{(x-h)^2}{b^2} + \frac{(y-k)^2}{a^2} = 1$.
The center is at $(h, k) = (2, -3)$. The foci are at $(h, k \pm c)$.

To find length *c* we use $b^2 = a^2 - c^2$.
Substituting gives $4 = 29 - c^2$ or $c = \sqrt{25} = 5$.

The ellipse has foci $(2, -3 \pm 5)$, or $(2, -8)$ and $(2, 2)$.

Example 9

Find the standard form of the equation for an ellipse with foci (-1,4) and (3,4) and major axis length 10.

Since the foci differ in the x -coordinates, the ellipse is horizontal with an equation of the form $\frac{(x-h)^2}{a^2}+\frac{(h-k)^2}{b^2}=1$.

The center is at the midpoint of the foci $\left(\frac{x_1+x_2}{2},\frac{y_1+y_2}{2}\right)=\left(\frac{(-1)+3}{2},\frac{4+4}{2}\right)=(1,4)$.

The value of a is half the major axis length: $a=\frac{1}{2}(10)=5$.

The value of c is half the distance between the foci: $c=\frac{1}{2}(3-(-1))=\frac{1}{2}(4)=2$.

To find length b we use $b^2=a^2-c^2$. Substituting a and c gives $b^2=5^2-2^2=21$.

The equation of the ellipse in standard form is $\frac{(x-1)^2}{5^2}+\frac{(y-4)^2}{21}=1$ or $\frac{(x-1)^2}{25}+\frac{(y-4)^2}{21}=1$.

Try it Now

3. Find the standard form of the equation for an ellipse with focus (2,4), vertex (2,6), and center (2,1).

Planetary Orbits

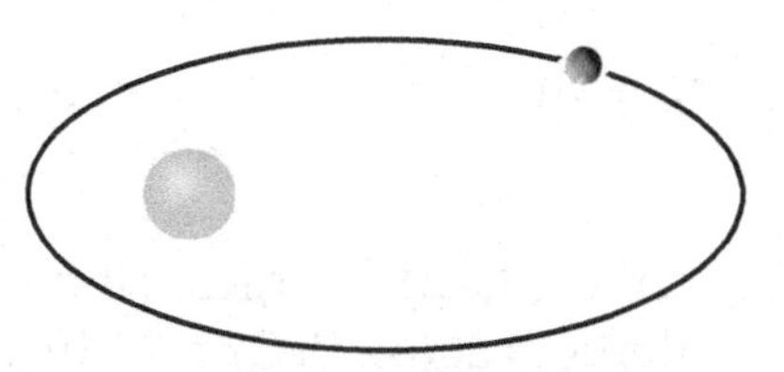

It was long thought that planetary orbits around the sun were circular. Around 1600, Johannes Kepler discovered they were actually elliptical[4]. His first law of planetary motion says that planets travel around the sun in an elliptical orbit with the sun as one of the foci. The length of the major axis can be found by measuring the planet's *aphelion*, its greatest distance from the sun, and *perihelion*, its shortest distance from the sun, and summing them together.

[4] Technically, they're approximately elliptical. The orbits of the planets are not exactly elliptical because of interactions with each other and other celestial bodies.

Example 10

Mercury's aphelion is 35.98 million miles and its perihelion is 28.58 million miles. Write an equation for Mercury's orbit.

Let the center of the ellipse be (0,0) and its major axis be horizontal so the equation will have form $\frac{x^2}{a^2}+\frac{y^2}{b^2}=1$.

The length of the major axis is $2a = 35.98 + 28.58 = 64.56$ giving $a = 32.28$ and $a^2 = 1041.9984$.

Since the perihelion is the distance from the focus to one vertex, we can find the distance between the foci by subtracting twice the perihelion from the major axis length: $2c = 64.56 - 2(28.58) = 7.4$ giving $c = 3.7$.
Substitution of a and c into $b^2 = a^2 - c^2$ yields $b^2 = 32.28^2 - 3.7^2 = 1028.3084$.

The equation is $\frac{x^2}{1041.9984}+\frac{y^2}{1028.3084}=1$.

Important Topics of This Section
Ellipse Definition
Ellipse Equations in Standard Form
Ellipse Foci
Applications of Ellipses

Try it Now Answers

1. $2a = 20$, so $a = 10$. $2b = 6$, so $b = 3$. $\frac{x^2}{100}+\frac{y^2}{9}=1$

2. Center (4, -2). Vertical ellipse with $a = 2$, $b = 1$.
   Vertices at (4, -2±2) = (4,0) and (4,-4),
   minor axis endpoints at (4±1, -2) = (3,-2) and (5,-2),
   major axis length 4, minor axis length 2

3. Vertex, center, and focus have the same x-value, so it's a vertical ellipse.
   Using the vertex and center, $a = 6 - 1 = 5$
   Using the center and focus, $c = 4 - 1 = 3$
   $b^2 = 5^2 - 3^2$. $b = 4$.
   $\frac{(x-2)^2}{16}+\frac{(y-1)^2}{25}=1$

Section 9.1 Exercises

In problems 1–4, match each graph with one of the equations A–D.

A. $\frac{x^2}{4}+\frac{y^2}{9}=1$ B. $\frac{x^2}{9}+\frac{y^2}{4}=1$ C. $\frac{x^2}{9}+y^2=1$ D. $x^2+\frac{y^2}{9}=1$

1. 2. 3. 4.

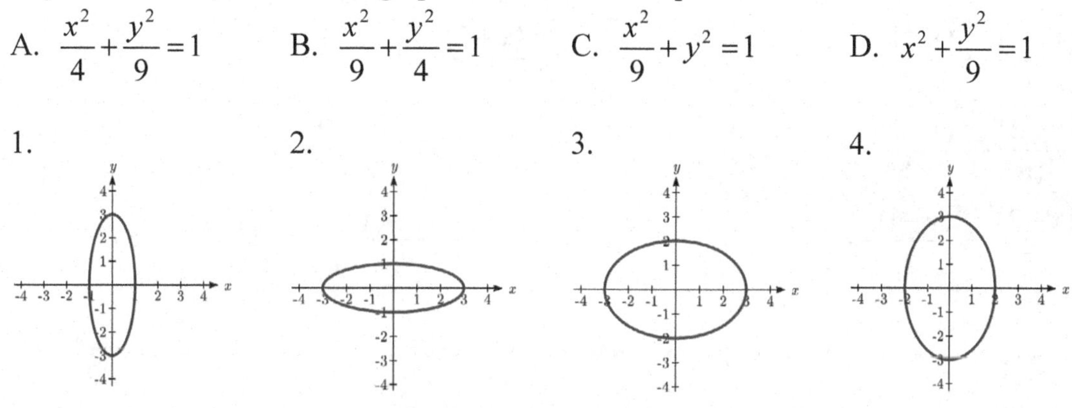

In problems 5–14, find the vertices, the minor axis endpoints, length of the major axis, and length of the minor axis. Sketch the graph. Check using a graphing utility.

5. $\frac{x^2}{4}+\frac{y^2}{25}=1$ 6. $\frac{x^2}{16}+\frac{y^2}{4}=1$ 7. $\frac{x^2}{4}+y^2=1$ 8. $x^2+\frac{y^2}{25}=1$

9. $x^2+25y^2=25$ 10. $16x^2+y^2=16$ 11. $16x^2+9y^2=144$

12. $16x^2+25y^2=400$ 13. $9x^2+y^2=18$ 14. $x^2+4y^2=12$

In problems 15–16, write an equation for the graph.

15. 16.

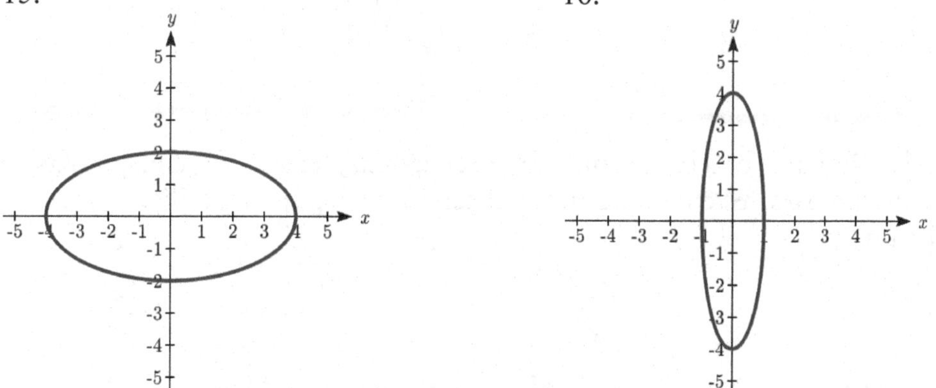

In problems 17–20, find the standard form of the equation for an ellipse satisfying the given conditions.

17. Center (0,0), horizontal major axis length 64, minor axis length 14

18. Center (0,0), vertical major axis length 36, minor axis length 18

19. Center (0,0), vertex (0,3), $b=2$

20. Center (0,0), vertex (4,0), $b=3$

In problems 21–28, match each graph to equations A-H.

A. $\dfrac{(x-2)^2}{4}+\dfrac{(y-1)^2}{9}=1$ E. $\dfrac{(x+2)^2}{4}+\dfrac{(y+1)^2}{9}=1$

B. $\dfrac{(x-2)^2}{4}+\dfrac{(y-1)^2}{16}=1$ F. $\dfrac{(x+2)^2}{4}+\dfrac{(y+1)^2}{16}=1$

C. $\dfrac{(x-2)^2}{16}+\dfrac{(y-1)^2}{4}=1$ G. $\dfrac{(x+2)^2}{16}+\dfrac{(y+1)^2}{4}=1$

D. $\dfrac{(x-2)^2}{9}+\dfrac{(y-1)^2}{4}=1$ H. $\dfrac{(x+2)^2}{9}+\dfrac{(y+1)^2}{4}=1$

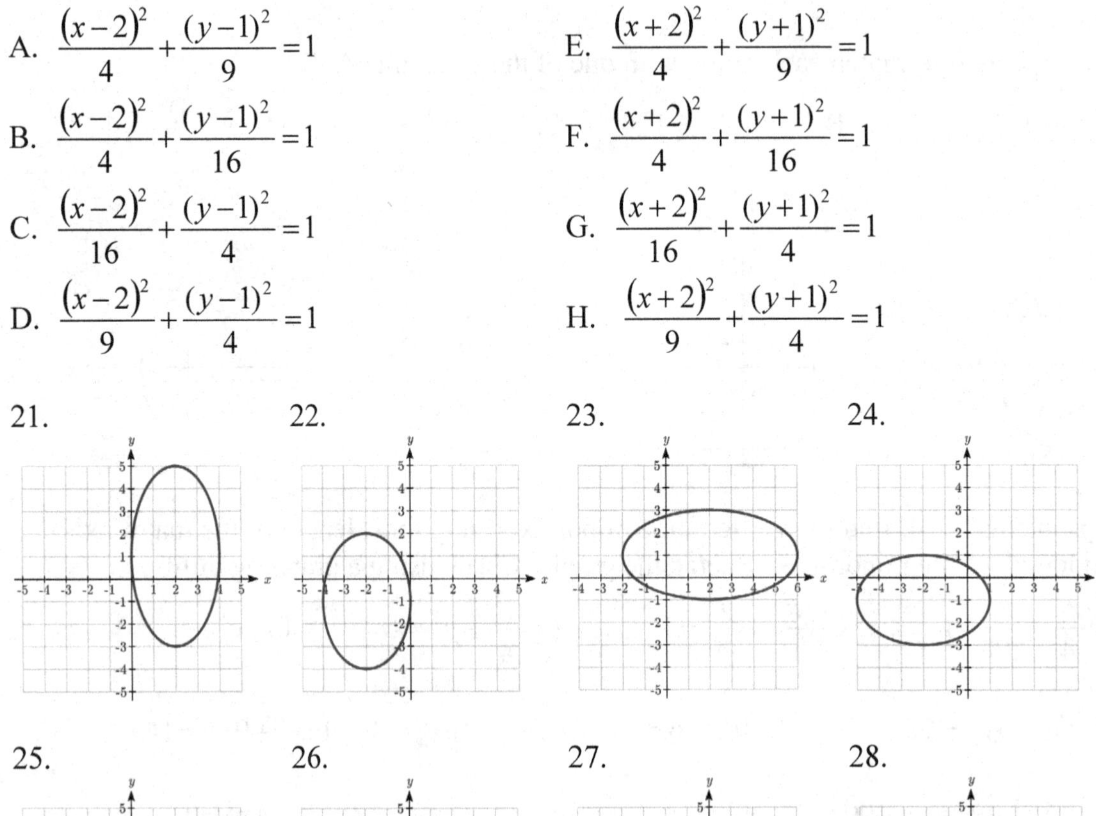

21. 22. 23. 24.

25. 26. 27. 28.

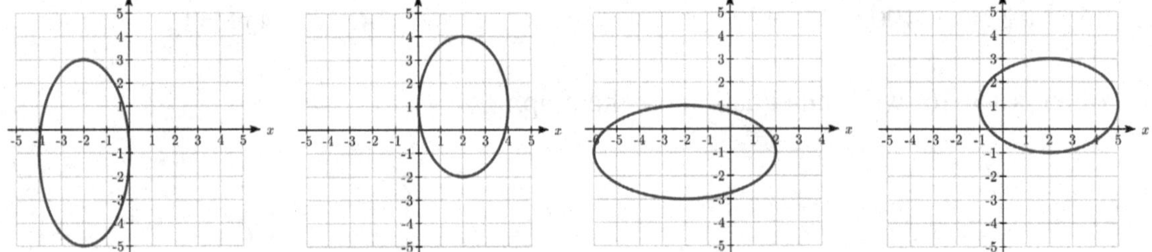

In problems 29–38, find the vertices, the minor axis endpoints, length of the major axis, and length of the minor axis. Sketch the graph. Check using a graphing utility.

29. $\dfrac{(x-1)^2}{25}+\dfrac{(y+2)^2}{4}=1$ 30. $\dfrac{(x+5)^2}{16}+\dfrac{(y-3)^2}{36}=1$

31. $(x+2)^2+\dfrac{(y-3)^2}{25}=1$ 32. $\dfrac{(x-1)^2}{25}+(y-6)^2=1$

33. $4x^2+8x+4+y^2=16$ 34. $x^2+4y^2+16y+16=36$

35. $x^2+2x+4y^2+16y=-1$ 36. $4x^2+16x+y^2-8y=4$

37. $9x^2-36x+4y^2+8y=104$ 38. $4x^2+8x+9y^2+36y=-4$

In problems 39–40, write an equation for the graph.

39. 40.

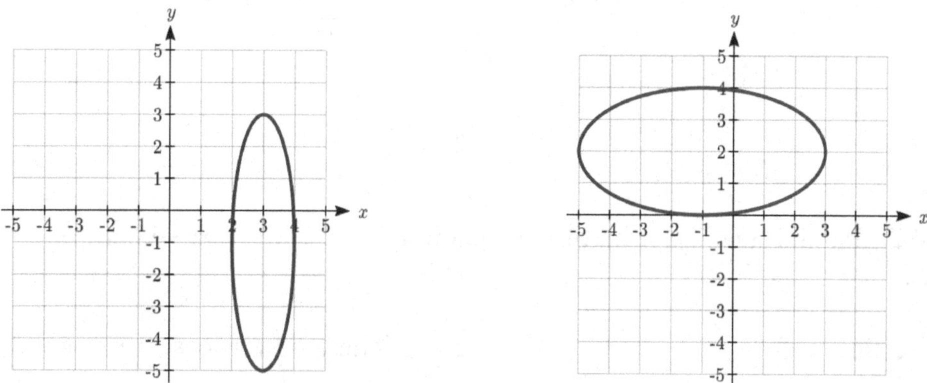

In problems 41–42, find the standard form of the equation for an ellipse satisfying the given conditions.

41. Center (-4,3), vertex(-4,8), point on the graph (0,3)

42. Center (1,-2), vertex(-5,-2), point on the graph (1,0)

43. **Window** A window in the shape of a semiellipse is 12 feet wide and 4 feet high. What is the height of the window above the base 5 feet from the center ?

44. **Window** A window in the shape of a semiellipse is 16 feet wide and 7 feet high. What is the height of the window above the base 4 feet from the center?

45. **Bridge** A bridge over a river is supported by a semielliptical arch. The river is 150 feet wide. At the center, the arch rises 60 feet above the river. The roadway is 5 feet above the center of the arch. What is the vertical distance between the roadway and the arch 45 feet from the center?

46. **Bridge** A bridge over a river is supported by a semielliptical arch. The river is 1250 feet wide. At the center, the arch rises 175 feet above the river. The roadway is 3 feet above the center of the arch. What is the vertical distance between the roadway and the arch 600 feet from the center?

47. **Racetrack** An elliptical racetrack is 100 feet long and 90 feet wide. What is the width of the racetrack 20 feet from a vertex on the major axis?

48. **Racetrack** An elliptical racetrack is 250 feet long and 150 feet wide. What is the width of the racetrack 25 feet from a vertex on the major axis?

In problems 49-52, find the foci.

49. $\frac{x^2}{19}+\frac{y^2}{3}=1$

50. $\frac{x^2}{2}+\frac{y^2}{38}=1$

51. $(x+6)^2+\frac{(y-1)^2}{26}+=1$

52. $\frac{(x-3)^2}{10}+(y+5)^2=1$

In problems 53-72, find the standard form of the equation for an ellipse satisfying the given conditions.

53. Major axis vertices (±3,0), c=2

54. Major axis vertices (0,±7), c=4

55. Foci (0,±5) and major axis length 12

56. Foci (±3,0) and major axis length 8

57. Foci (±5,0), vertices (±7,0)

58. Foci (0,±2), vertices (0,±3)

59. Foci (0,±4) and x-intercepts (±2,0)

60. Foci (±3,0) and y-intercepts (0,±1)

61. Center (0,0), major axis length 8, foci on x-axis, passes through point $\left(2,\sqrt{6}\right)$

62. Center (0,0), major axis length 12, foci on y-axis, passes through point $\left(\sqrt{10},4\right)$

63. Center (-2,1), vertex (-2,5), focus (-2,3)

64. Center (-1,-3), vertex (-7,-3), focus (-4,-3)

65. Foci (8,2) and (-2,2), major axis length 12

66. Foci (-1,5) and (-1,-3), major axis length 14

67. Vertices (3,4) and (3,-6), c= 2

68. Vertices (2,2) and (-4,2), c= 2

69. Center (1,3), focus (0,3), passes through point (1,5)

70. Center (-1,-2), focus (1,-2), passes through point (2,-2)

71. Focus (-15,-1), vertices (-19,-1) and (15,-1)

72. Focus (-3,2), vertices (-3,4) and (-3,-8)

73. **Whispering Gallery** If an elliptical whispering gallery is 80 feet long and 25 feet wide, how far from the center of room should someone stand on the major axis of the ellipse to experience the whispering effect? Round to two decimal places.

74. **Billiards** Some billiards tables are elliptical and have the foci marked on the table. If such a one is 8 feet long and 6 feet wide, how far are the foci from the center of the ellipse? Round to two decimal places.

75. **Planetary Orbits** The orbits of planets around the sun are approximately elliptical with the sun as a focus. The *aphelion* is a planet's greatest distance from the sun and the *perihelion* is its shortest. The length of the major axis is the sum of the aphelion and the perihelion. Earth's aphelion is 94.51 million miles and its perihelion is 91.40 million miles. Write an equation for Earth's orbit.

76. **Satellite Orbits** The orbit of a satellite around Earth is elliptical with Earth's center as a focus. The satellite's maximum height above the Earth is 170 miles and its minimum height above the Earth is 90 miles. Write an equation for the satellite's orbit. Assume Earth is spherical and has a radius of 3960 miles.

77. **Eccentricity** e of an ellipse is the ratio $\frac{c}{a}$ where c is the distance of a focus from the center and a is the distance of a vertex from the center. Write an equation for an ellipse with eccentricity 0.8 and foci at (-4,0) and (4,0).

78. **Confocal** ellipses have the same foci. Show that, for $k > 0$, all ellipses of the form $\frac{x^2}{6+k}+\frac{y^2}{k}=1$ are confocal.

79. The **latus rectum** of an ellipse is a line segment with endpoints on the ellipse that passes through a focus and is perpendicular to the major axis. Show that $\frac{2b^2}{a}$ is the length of the latus rectum of $\frac{x^2}{a^2}+\frac{y^2}{b^2}=1$ where $a > b$.

Section 9.2 Hyperbolas

In the last section, we learned that planets have approximately elliptical orbits around the sun. When an object like a comet is moving quickly, it is able to escape the gravitational pull of the sun and follows a path with the shape of a **hyperbola**. Hyperbolas are curves that can help us find the location of a ship, describe the shape of cooling towers, or calibrate seismological equipment.

The hyperbola is another type of conic section created by intersecting a plane with a double cone, as shown below[5].

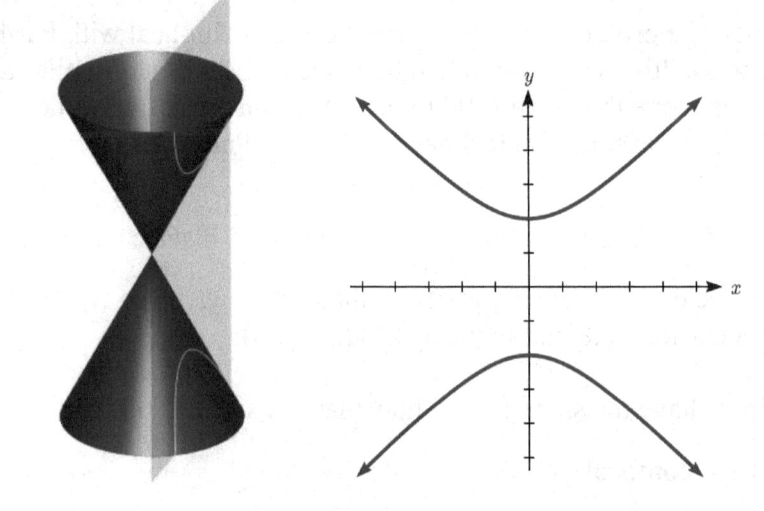

The word "hyperbola" derives from a Greek word meaning "excess." The English word "hyperbole" means exaggeration. We can think of a hyperbola as an excessive or exaggerated ellipse, one turned inside out.

We defined an ellipse as the set of all points where the sum of the distances from that point to two fixed points is a constant. A hyperbola is the set of all points where the absolute value of the difference of the distances from the point to two fixed points is a constant.

---

[5] Pbroks13 (https://commons.wikimedia.org/wiki/File:Conic_sections_with_plane.svg), "Conic sections with plane", cropped to show only a hyperbola by L Michaels, CC BY 3.0

Hyperbola Definition

A **hyperbola** is the set of all points $Q(x, y)$ for which the absolute value of the difference of the distances to two fixed points $F_1(x_1, y_1)$ and $F_2(x_2, y_2)$ called the **foci** (plural for focus) is a constant *k*: $|d(Q, F_1) - d(Q, F_2)| = k$.

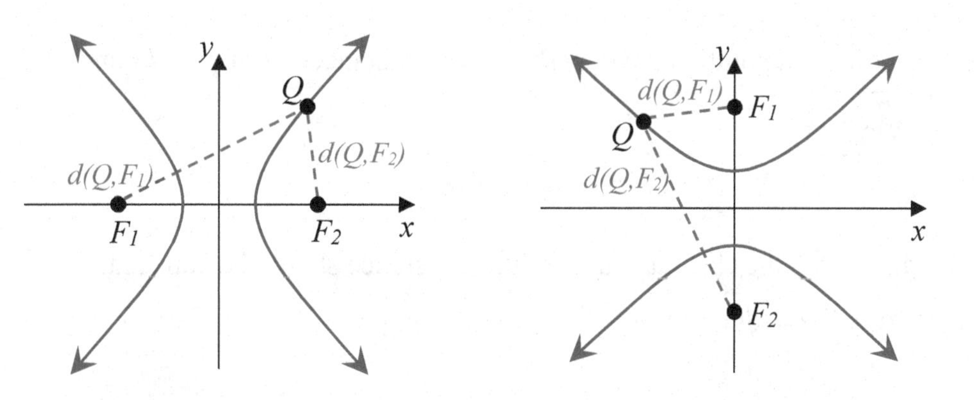

The **transverse axis** is the line passing through the foci.
Vertices are the points on the hyperbola which intersect the transverse axis.
The **transverse axis length** is the length of the line segment between the vertices.
The **center** is the midpoint between the vertices (or the midpoint between the foci).
The other axis of symmetry through the center is the **conjugate axis**.
The two disjoint pieces of the curve are called **branches**.
A hyperbola has two **asymptotes**.

Which axis is the transverse axis will depend on the orientation of the hyperbola. As a helpful tool for graphing hyperbolas, it is common to draw a **central rectangle** as a guide. This is a rectangle drawn around the center with sides parallel to the coordinate axes that pass through each vertex and co-vertex. The asymptotes will follow the diagonals of this rectangle.

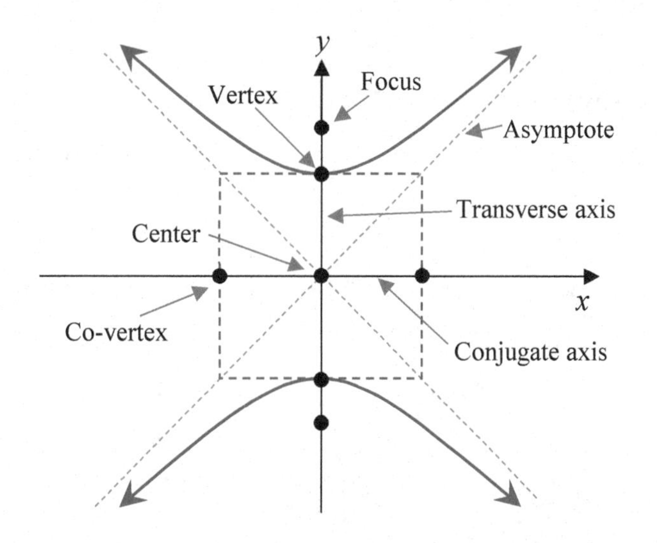

Hyperbolas Centered at the Origin

From the definition above we can find an equation of a hyperbola. We will find it for a hyperbola centered at the origin $C(0,0)$ opening horizontally with foci at $F_1(c,0)$ and $F_2(-c,0)$ where $c > 0$.

Suppose $Q(x, y)$ is a point on the hyperbola. The distances from Q to F_1 and Q to F_2 are:

$$d(Q,F_1) = \sqrt{(x-c)^2 + (y-0)^2} = \sqrt{(x-c)^2 + y^2}$$

$$d(Q,F_2) = \sqrt{(x-(-c))^2 + (y-0)^2} = \sqrt{(x+c)^2 + y^2}\,.$$

From the definition, the absolute value of the difference should be constant:

$$|d(Q,F_1) - d(Q,F_2)| = \left|\sqrt{(x-c)^2 + y^2} - \sqrt{(x+c)^2 + y^2}\right| = k$$

Substituting in one of the vertices $(a,0)$, we can determine *k* in terms of *a*:

$$\left|\sqrt{(a-c)^2 + 0^2} - \sqrt{(a+c)^2 + 0^2}\right| = k$$

$$\big||a-c| - |a+c|\big| = k \qquad \text{Since } c > a,\ |a-c| = c-a$$

$$|(c-a) - (a+c)| = k$$

$$k = |-2a| = |2a|$$

Using $k = 2a$ and removing the absolute values,

$$\sqrt{(x-c)^2 + y^2} - \sqrt{(x+c)^2 + y^2} = \pm 2a \qquad \text{Move one radical}$$

$$\sqrt{(x-c)^2 + y^2} = \pm 2a + \sqrt{(x+c)^2 + y^2} \qquad \text{Square both sides}$$

$$(x-c)^2 + y^2 = 4a^2 \pm 4a\sqrt{(x+c)^2 + y^2} + (x+c)^2 + y^2 \text{ Expand}$$

$$x^2 - 2xc + c^2 + y^2 = 4a^2 \pm 4a\sqrt{(x+c)^2 + y^2} + x^2 + 2xc + c^2 + y^2$$

Combining like terms leaves

$$-4xc = 4a^2 \pm 4a\sqrt{(x+c)^2 + y^2} \qquad \text{Divide by 4}$$

$$-xc = a^2 \pm a\sqrt{(x+c)^2 + y^2} \qquad \text{Isolate the radical}$$

$$\pm a\sqrt{(x+c)^2 + y^2} = -a^2 - xc \qquad \text{Square both sides again}$$

$$a^2\left((x+c)^2 + y^2\right) = a^4 + 2a^2xc + x^2c^2 \qquad \text{Expand and distribute}$$

$$a^2x^2 + 2a^2xc + a^2c^2 + a^2y^2 = a^4 + 2a^2xc + x^2c^2 \qquad \text{Combine like terms}$$

$$a^2y^2 + a^2c^2 - a^4 = x^2c^2 - a^2x^2 \qquad \text{Factor common terms}$$

$$a^2y^2 + a^2(c^2 - a^2) = (c^2 - a^2)x^2$$

Let $b^2 = c^2 - a^2$. Since $c > a$, $b > 0$. Substituting b^2 for $c^2 - a^2$ leaves

$a^2y^2 + a^2b^2 = b^2x^2$ Divide both sides by a^2b^2

$\frac{y^2}{b^2} + 1 = \frac{x^2}{a^2}$ Rewrite

$\frac{x^2}{a^2} - \frac{y^2}{b^2} = 1$

We can see from the graphs of the hyperbolas that the branches appear to approach asymptotes as x gets large in the negative or positive direction. The equations of the horizontal hyperbola asymptotes can be derived from its standard equation.

$\frac{x^2}{a^2} - \frac{y^2}{b^2} = 1$ Solve for y

$y^2 = b^2\left(\frac{x^2}{a^2} - 1\right)$ Rewrite 1 as $\frac{x^2}{a^2}\frac{a^2}{x^2}$

$y^2 = b^2\left(\frac{x^2}{a^2} - \frac{x^2}{a^2}\frac{a^2}{x^2}\right)$ Factor out $\frac{x^2}{a^2}$

$y^2 = b^2\frac{x^2}{a^2}\left(1 - \frac{a^2}{x^2}\right)$ Take the square root

$y = \pm\frac{b}{a}x\sqrt{1 - \frac{a^2}{x^2}}$

As $x \to \pm\infty$ the quantity $\frac{a^2}{x^2} \to 0$ and $\sqrt{1 - \frac{a^2}{x^2}} \to 1$, so the asymptotes are $y = \pm\frac{b}{a}x$.

Similarly, for vertical hyperbolas the asymptotes are $y = \pm\frac{a}{b}x$.

The standard form of an equation of a hyperbola centered at the origin $C(0,0)$ depends on whether it opens horizontally or vertically. The following table gives the standard equation, vertices, foci, asymptotes, construction rectangle vertices, and graph for each.

Equation of a Hyperbola Centered at the Origin in Standard Form

Opens	Horizontally	Vertically
Standard Equation	$\frac{x^2}{a^2}-\frac{y^2}{b^2}=1$	$\frac{y^2}{a^2}-\frac{x^2}{b^2}=1$
Vertices	(-*a*, 0) and (*a*, 0)	(0, -*a*) and (0, *a*)
Foci	(-*c*, 0) and (*c*, 0) where $b^2=c^2-a^2$	(0, -*c*) and (0, *c*) Where $b^2=c^2-a^2$
Asymptotes	$y=\pm\frac{b}{a}x$	$y=\pm\frac{a}{b}x$
Construction Rectangle Vertices	(*a*, *b*), (-*a*, *b*), (*a*,-*b*), (-*a*, -*b*)	(*b*, *a*), (-*b*, *a*), (*b*, -*a*), (-*b*, -*a*)
Graph		

Example 1

Put the equation of the hyperbola $y^2-4x^2=4$ in standard form. Find the vertices, length of the transverse axis, and the equations of the asymptotes. Sketch the graph. Check using a graphing utility.

The equation can be put in standard form $\frac{y^2}{4}-\frac{x^2}{1}=1$ by dividing by 4.

Comparing to the general standard equation $\frac{y^2}{a^2}-\frac{x^2}{b^2}=1$ we see that $a=\sqrt{4}=2$ and $b=\sqrt{1}=1$.

Since the *x* term is subtracted, the hyperbola opens vertically and the vertices lie on the *y*-axis at $(0,\pm a)=(0,\pm 2)$.

The length of the transverse axis is $2(a) = 2(2) = 4$.

Equations of the asymptotes are $y = \pm\frac{a}{b}x$ or $= \pm 2x$.

To sketch the graph we plot the vertices of the construction rectangle at $(\pm b, \pm a)$ or (-1,-2), (-1,2), (1,-2), and (1,2). The asymptotes are drawn through the diagonals of the rectangle and the vertices plotted. Then we sketch in the hyperbola, rounded at the vertices and approaching the asymptotes.

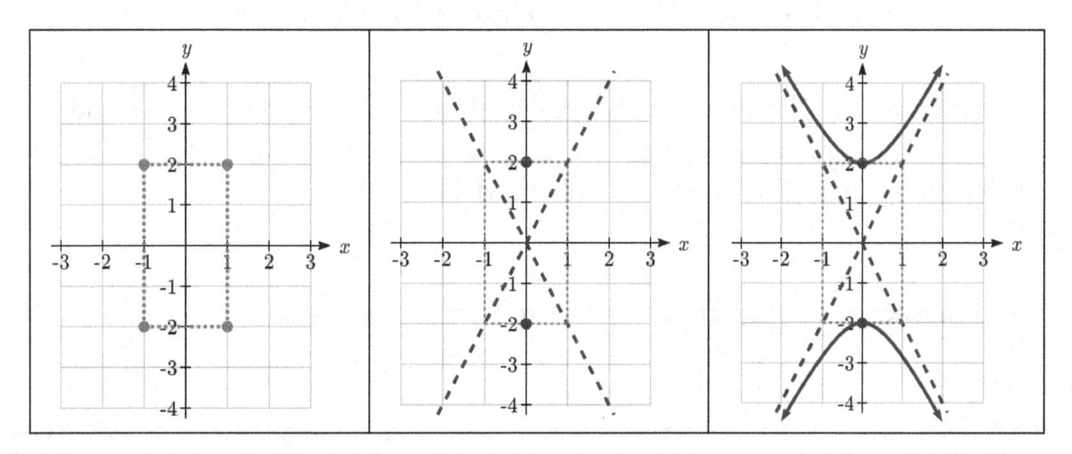

To check on a graphing utility, we must solve the equation for y. Isolating y^2 gives us $y^2 = 4(1 + x^2)$.

Taking the square root of both sides we find $y = \pm 2\sqrt{1 + x^2}$.

Under **Y=** enter the two halves of the hyperbola and the two asymptotes as $y = 2\sqrt{1 + x^2}$, $y = -2\sqrt{1 + x^2}$, $y = 2x$, and $y = -2x$. Set the window to a comparable scale to the sketch with xmin = -4, xmax = 4, ymin= -3, and ymax = 3.

Plot1 Plot2 Plot3
\Y1=2√(1+X²)
\Y2=-2√(1+X²)
\Y3=2X
\Y4=-2X

WINDOW
Xmin=-4
Xmax=4
Xscl=1
Ymin=-3
Ymax=3
Yscl=1
Xres=1

Sometimes we are given the equation. Sometimes we need to find the equation from a graph or other information.

Example 2

Find the standard form of the equation for a hyperbola with vertices at (-6,0) and (6,0) and asymptote $y = \frac{4}{3}x$.

Since the vertices lie on the x-axis with a midpoint at the origin, the hyperbola is horizontal with an equation of the form $\frac{x^2}{a^2} - \frac{y^2}{b^2} = 1$. The value of a is the distance from the center to a vertex. The distance from (6,0) to (0,0) is 6, so $a = 6$.

The asymptotes follow the form $y = \pm\frac{b}{a}x$. From $y = \frac{3}{4}x$ we see $\frac{3}{4} = \frac{b}{a}$ and substituting $a = 6$ give us $\frac{3}{4} = \frac{b}{6}$. Solving yields $b = 8$.

The equation of the hyperbola in standard form is $\frac{x^2}{6^2} - \frac{y^2}{8^2} = 1$ or $\frac{x^2}{36} - \frac{y^2}{64} = 1$.

Try it Now

1. Find the standard form of the equation for a hyperbola with vertices at (0,-8) and (0,8) and asymptote $y = 2x$

Example 3

Find the standard form of the equation for a hyperbola with vertices at (0, 9) and (0,-9) and passing through the point (8,15).

Since the vertices lie on the y-axis with a midpoint at the origin, the hyperbola is vertical with an equation of the form $\frac{y^2}{a^2} - \frac{x^2}{b^2} = 1$. The value of a is the distance from the center to a vertex. The distance from (0,9) to (0,0) is 9, so $a = 9$.

Substituting $a = 9$ and the point (8,15) gives $\frac{15^2}{9^2} - \frac{8^2}{b^2} = 1$. Solving for b yeilds

$b = \sqrt{\frac{9^2(8^2)}{15^2 - 9^2}} = 6$.

The standard equation for the hyperbola is $\frac{y^2}{9^2} - \frac{x^2}{6^2} = 1$ or $\frac{y^2}{81} - \frac{x^2}{36} = 1$.

Hyperbolas Not Centered at the Origin

Not all hyperbolas are centered at the origin. The standard equation for one centered at (h, k) is slightly different.

Equation of a Hyperbola Centered at (h, k) in Standard Form
The standard form of an equation of a hyperbola centered at $C(h,k)$ depends on whether it opens horizontally or vertically. The table below gives the standard equation, vertices, foci, asymptotes, construction rectangle vertices, and graph for each.

Opens	Horizontally	Vertically
Standard Equation	$\frac{(x-h)^2}{a^2}-\frac{(y-k)^2}{b^2}=1$	$\frac{(y-k)^2}{a^2}-\frac{(x-h)^2}{b^2}=1$
Vertices	$(h \pm a, k)$	$(h, k \pm a)$
Foci	$(h \pm c, k)$ where $b^2 = c^2 - a^2$	$(h, k \pm c)$ where $b^2 = c^2 - a^2$
Asymptotes	$y-k=\pm\frac{b}{a}(x-h)$	$y-k=\pm\frac{a}{b}(x-h)$
Construction Rectangle Vertices	$(h \pm a, k \pm b)$	$(h \pm b, k \pm a)$
Graph	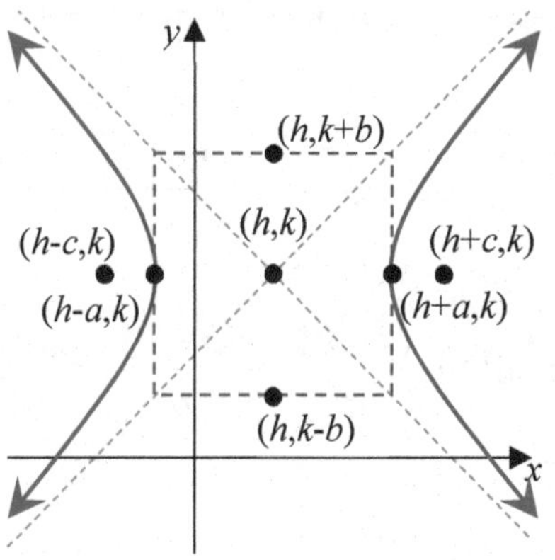 y, x, (h,k+b), (h,k), (h-c,k), (h-a,k), (h+c,k), (h+a,k), (h,k-b)	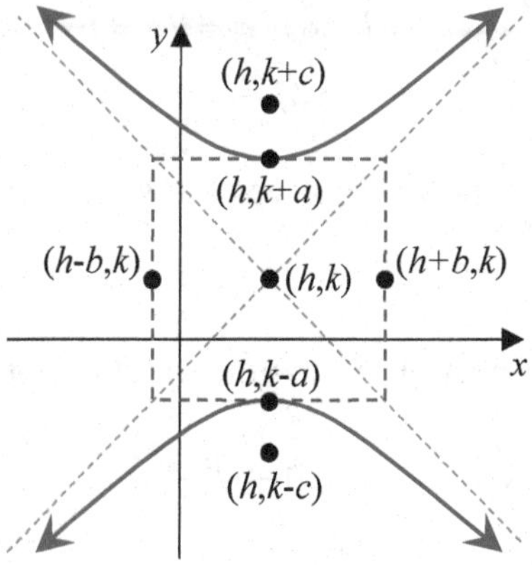 y, x, (h,k+c), (h,k+a), (h-b,k), (h,k), (h+b,k), (h,k-a), (h,k-c)

Example 4

Write an equation for the hyperbola in the graph shown.

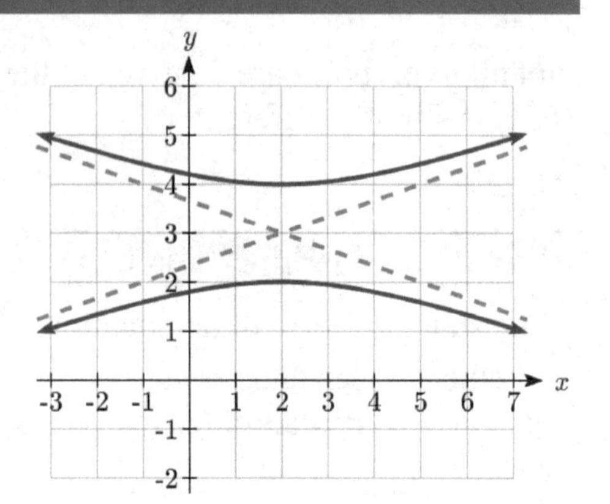

The center is at (2,3), where the asymptotes cross. It opens vertically, so the equation will look like
$\frac{(y-3)^2}{a^2}-\frac{(x-2)^2}{b^2}=1$.

The vertices are at (2,2) and (2,4). The distance from the center to a vertex is $a=4-3=1$.

If we were to draw in the construction rectangle, it would extend from $x = -1$ to $x = 5$. The distance from the center to the right side of the rectangle gives $b=5-2=3$.

The standard equation of this hyperbola is $\frac{(y-3)^2}{1^2}-\frac{(x-2)^2}{3^2}=1$, or

$(y-3)^2-\frac{(x-2)^2}{9}=1$.

Example 5

Put the equation of the hyperbola $9x^2+18x-4y^2+16y=43$ in standard form. Find the center, vertices, length of the transverse axis, and the equations of the asymptotes. Sketch the graph, then check on a graphing utility.

To rewrite the equation, we complete the square for both variables to get
$9(x^2+2x+1)-4(y^2-4y+4)=43+9-16$
$9(x+1)^2-4(y-2)^2=36$

Dividing by 36 gives the standard form of the equation, $\frac{(x+1)^2}{4}-\frac{(y-2)^2}{9}=1$

Comparing to the general standard equation $\frac{(x-h)^2}{a^2}-\frac{(h-k)^2}{b^2}=1$ we see that $a=\sqrt{4}=2$ and $b=\sqrt{9}=3$.

Since the y term is subtracted, the hyperbola opens horizontally.
The center is at $(h, k) = (-1, 2)$.
The vertices are at $(h\pm a, k)$ or (-3, 2) and (1,2).
The length of the transverse axis is $2(a)=2(2)=4$.

Equations of the asymptotes are $y-k=\pm\frac{b}{a}(x-h)$ or $y-2=\pm\frac{3}{2}(x+1)$.

To sketch the graph we plot the corners of the construction rectangle at ($h \pm a$, $k \pm b$) or (1, 5), (1, -1), (-3,5), and (-3,-1). The asymptotes are drawn through the diagonals of the rectangle and the vertices plotted. Then we sketch in the hyperbola rounded at the vertices and approaching the asymptotes.

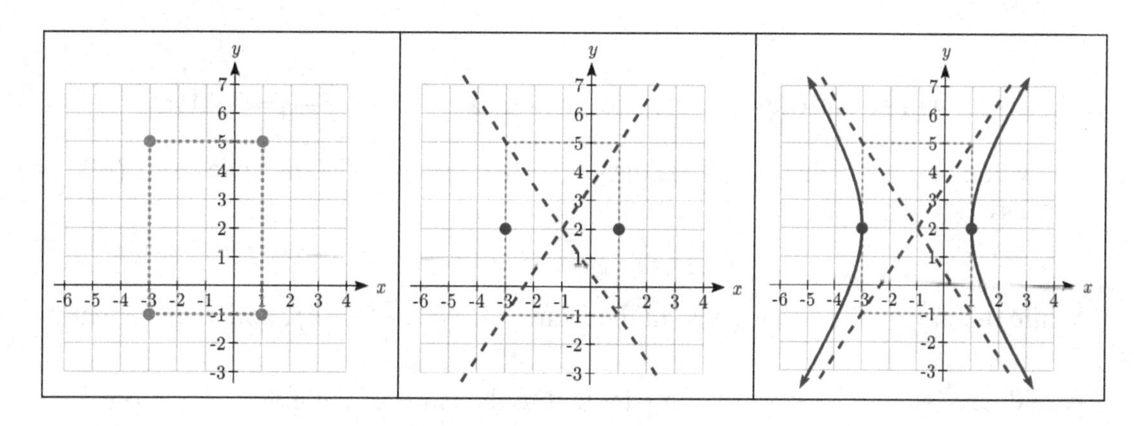

To check on a graphing utility, we must solve the equation for y.

$$y = 2 \pm \sqrt{9\left(\frac{(x+1)^2}{4} - 1\right)}.$$

Under **Y=** enter the two halves of the hyperbola and the two asymptotes as $y = 2 + \sqrt{9\left(\frac{(x+1)^2}{4} - 1\right)}$, $y = 2 - \sqrt{9\left(\frac{(x+1)^2}{4} - 1\right)}$, $y = \frac{3}{2}(x+1) + 2$, and $y = -\frac{3}{2}(x+1) + 2$. Set the window to a comparable scale to the sketch, then graph.

Note that the gaps you see on the calculator are not really there; they're a limitation of the technology.

```
Plot1 Plot2 Plot3
\Y1=2+√(9((X+1)²/4-1))
\Y2=2-√(9((X+1)²/4-1))
\Y3=3/2(X+1)+2
\Y4=-3/2(X+1)+2
```

```
WINDOW
 Xmin=-6
 Xmax=4
 Xscl=1
 Ymin=-3
 Ymax=7
 Yscl=1
 Xres=1
```

Example 6

Find the standard form of the equation for a hyperbola with vertices at $(-2,-5)$ and $(-2,7)$, and asymptote $y = \frac{3}{2}x + 4$.

Since the vertices differ in the y -coordinates, the hyperbola opens vertically with an equation of the form $\frac{(y-k)^2}{a^2}-\frac{(x-h)^2}{b^2}=1$ and asymptote equations of the form $y-k=\pm\frac{a}{b}(x-h)$.

The center will be halfway between the vertices, at $\left(-2,\frac{-5+7}{2}\right)=(-2,1)$.
The value of a is the distance from the center to a vertex. The distance from $(-2,1)$ to $(-2,-5)$ is 6, so $a=6$.

While our asymptote is not given in the form $y-k=\pm\frac{a}{b}(x-h)$, notice this equation would have slope $\frac{a}{b}$. We can compare that to the slope of the given asymptote equation to find b. Setting $\frac{3}{2}=\frac{a}{b}$ and substituting $a=6$ gives us $b=4$.

The equation of the hyperbola in standard form is $\frac{(y-1)^2}{6^2}-\frac{(x+2)^2}{4^2}=1$ or $\frac{(y-1)^2}{36}-\frac{(x+2)^2}{16}=1$.

Try it Now
2. Find the center, vertices, length of the transverse axis, and equations of the asymptotes for the hyperbola $\frac{(x+5)^2}{9}-\frac{(y-2)^2}{36}=1$.

Hyperbola Foci

The location of the foci can play a key role in hyperbola application problems. To find them, we need to find the length from the center to the foci, c, using the equation $b^2=c^2-a^2$. It looks similar to, but is not the same as, the Pythagorean Theorem.

Compare this with the equation to find length c for ellipses, which is $b^2=a^2-c^2$. If you remember that for the foci to be inside the ellipse they have to come before the vertices $(c<a)$, it's clear why we would calculate a^2 minus c^2. To be inside a hyperbola, the foci have to go beyond the vertices $(c>a)$, so we can see for hyperbolas we need c^2 minus a^2, the opposite.

Example 7

Find the foci of the hyperbola $\frac{(y+1)^2}{4}-\frac{(x-3)^2}{5}=1$.

The hyperbola is vertical with an equation of the form $\frac{(y-k)^2}{a^2}-\frac{(x-h)^2}{b^2}=1$.
The center is at $(h, k) = (3, -1)$. The foci are at $(h, k \pm c)$.

To find length c we use $b^2=c^2-a^2$. Substituting gives $5=c^2-4$ or $c=\sqrt{9}=3$.

The hyperbola has foci (3, -4) and (3, 2).

Example 8

Find the standard form of the equation for a hyperbola with foci (5, -8) and (-3, -8) and vertices (4, -8) and (-2, -8).

Since the vertices differ in the x -coordinates, the hyperbola opens horizontally with an equation of the form $\frac{(x-h)^2}{a^2}-\frac{(y-k)^2}{b^2}=1$.

The center is at the midpoint of the vertices
$\left(\frac{x_1+x_2}{2},\frac{y_1+y_2}{2}\right)=\left(\frac{4+(-2)}{2},\frac{-8+(-8)}{2}\right)=(1,-8)$.

The value of a is the horizontal length from the center to a vertex, or $a=4-1=3$.
The value of c is the horizontal length from the center to a focus, or $=5-1=4$.
To find length b we use $b^2=c^2-a^2$. Substituting gives $b^2=16-9=7$.
The equation of the hyperbola in standard form is $\frac{(x-1)^2}{3^2}-\frac{(y-(-8))^2}{7}=1$ or
$\frac{(x-1)^2}{9}-\frac{(y+8)^2}{7}=1$.

Try it Now

3. Find the standard form of the equation for a hyperbola with focus (1,9), vertex (1,8), center (1,4).

LORAN

Before GPS, the Long Range Navigation (LORAN) system was used to determine a ship's location. Two radio stations A and B simultaneously sent out a signal to a ship. The difference in time it took to receive the signal was computed as a distance locating the ship on the hyperbola with the A and B radio stations as the foci. A second pair of radio stations C and D sent simultaneous signals to the ship and computed its location on the hyperbola with C and D as the foci. The point P where the two hyperbolas intersected gave the location of the ship.

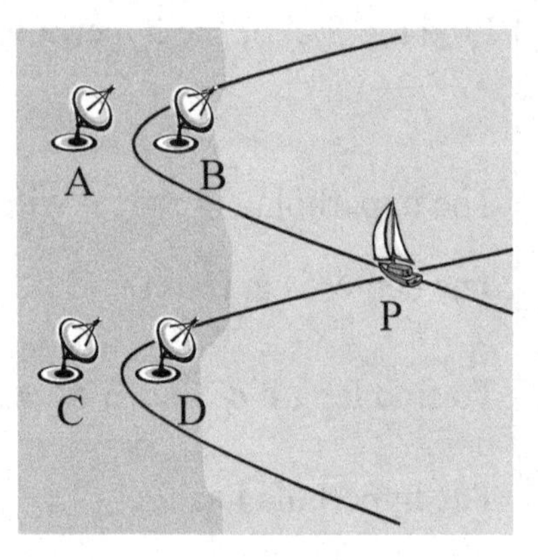

Example 9

Stations A and B are 150 kilometers apart and send a simultaneous radio signal to the ship. The signal from B arrives 0.0003 seconds before the signal from A. If the signal travels 300,000 kilometers per second, find the equation of the hyperbola on which the ship is positioned.

Stations A and B are at the foci, so the distance from the center to one focus is half the distance between them, giving $c = \frac{1}{2}(150) = 75$ km.

By letting the center of the hyperbola be at (0,0) and placing the foci at (±75,0), the equation $\frac{x^2}{a^2} - \frac{y^2}{b^2} = 1$ for a hyperbola centered at the origin can be used.

The difference of the distances of the ship from the two stations is $k = 300{,}000\frac{\text{km}}{\text{s}} \cdot (0.0003\text{s}) = 90\,\text{km}$. From our derivation of the hyperbola equation we determined $k = 2a$, so $a = \frac{1}{2}(90) = 45$.

Substituting a and c into $b^2 = c^2 - a^2$ yields $b^2 = 75^2 - 45^2 = 3600$.

The equation of the hyperbola in standard form is $\frac{x^2}{45^2} - \frac{y^2}{3600} = 1$ or $\frac{x^2}{2025} - \frac{y^2}{3600} = 1$.

To determine the position of a ship using LORAN, we would need an equation for the second hyperbola and would solve for the intersection. We will explore how to do that in the next section.

Important Topics of This Section
Hyperbola Definition
Hyperbola Equations in Standard Form
Hyperbola Foci
Applications of Hyperbolas
Intersections of Hyperbolas and Other Curves

Try it Now Answers

1. The vertices are on the y axis so this is a vertical hyperbola.
   The center is at the origin.
   $a = 8$
   Using the asymptote slope, $\frac{8}{b} = 2$, so $b = 4$.
   $\frac{y^2}{64} - \frac{x^2}{16} = 1$

2. Center (-5, 2). This is a horizontal hyperbola. $a = 3$. $b = 6$.
   transverse axis length 6,
   Vertices will be at (-5±3,2) = (-2,2) and (-8,2),
   Asymptote slope will be $\frac{6}{3} = 2$. Asymptotes: $y - 2 = \pm 2(x + 5)$

3. Focus, vertex, and center have the same x value so this is a vertical hyperbola.
   Using the vertex and center, $a = 9 - 4 = 5$
   Using the focus and center, $c = 8 - 4 = 4$
   $b^2 = 5^2 - 4^2$. $b = 3$.
   $\frac{(y-4)^2}{16} - \frac{(x-1)^2}{9} = 1$

Section 9.2 Exercises

In problems 1–4, match each graph to equations A–D.

A. $\frac{x^2}{4}-\frac{y^2}{9}=1$ B. $\frac{x^2}{9}-\frac{y^2}{4}=1$ C. $y^2-\frac{x^2}{9}=1$ D. $\frac{y^2}{9}-x^2=1$

1. 2. 3. 4.

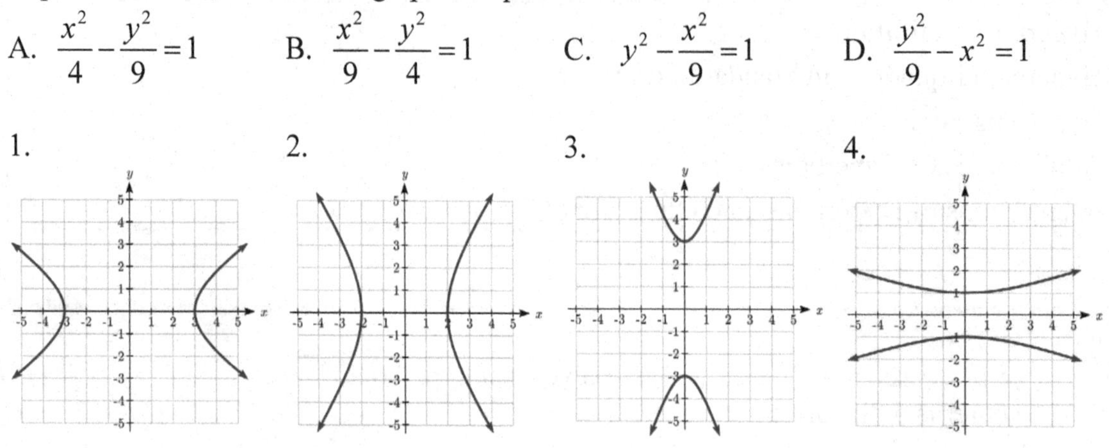

In problems 5–14, find the vertices, length of the transverse axis, and equations of the asymptotes. Sketch the graph. Check using a graphing utility.

5. $\frac{x^2}{4}-\frac{y^2}{25}=1$ 6. $\frac{y^2}{16}-\frac{x^2}{9}=1$ 7. $y^2-\frac{x^2}{4}=1$ 8. $x^2-\frac{y^2}{25}=1$

9. $x^2-9y^2=9$ 10. $y^2-4x^2=4$ 11. $9y^2-16x^2=144$

12. $16x^2-25y^2=400$ 13. $9x^2-y^2=18$ 14. $4y^2-x^2=12$

In problems 15–16, write an equation for the graph.

15.

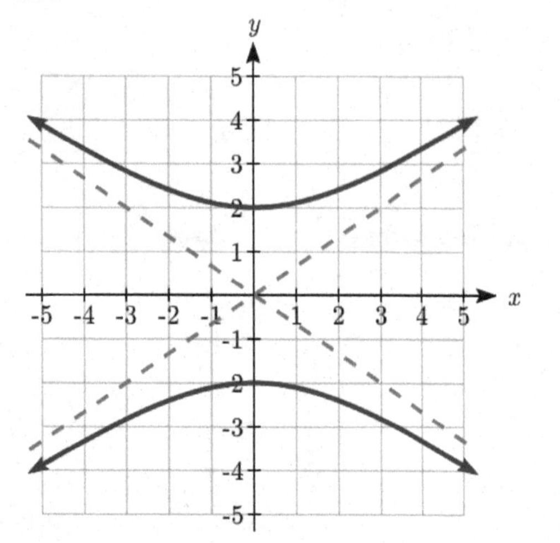

16.

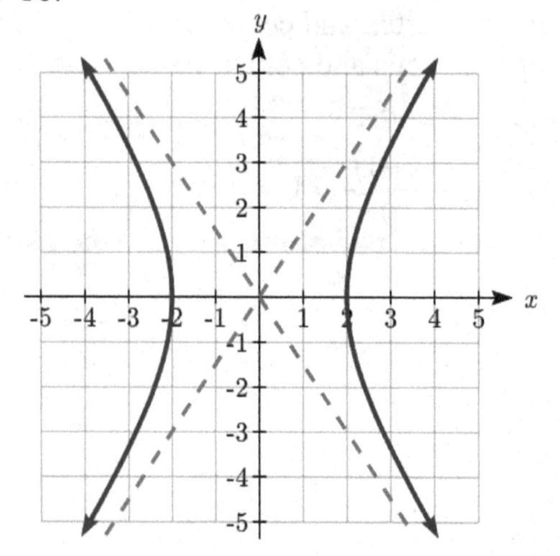

In problems 17–22, find the standard form of the equation for a hyperbola satisfying the given conditions.

17. Vertices at (0,4) and (0, -4); asymptote $y = \frac{1}{2}x$

18. Vertices at (-6,0) and (6,0); asymptote $y = 3x$

19. Vertices at (-3,0) and (3,0); passes through (5,8)

20. Vertices at (0, 4) and (0, -4); passes through (6, 5)

21. Asymptote y = x; passes through (5, 3)

22. Asymptote y = x; passes through (12, 13)

In problems 23–30, match each graph to equations A–H.

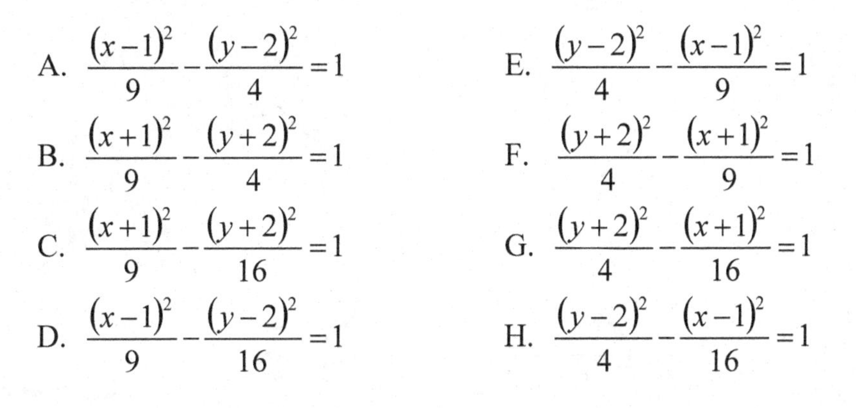

A. $\frac{(x-1)^2}{9} - \frac{(y-2)^2}{4} = 1$

B. $\frac{(x+1)^2}{9} - \frac{(y+2)^2}{4} = 1$

C. $\frac{(x+1)^2}{9} - \frac{(y+2)^2}{16} = 1$

D. $\frac{(x-1)^2}{9} - \frac{(y-2)^2}{16} = 1$

E. $\frac{(y-2)^2}{4} - \frac{(x-1)^2}{9} = 1$

F. $\frac{(y+2)^2}{4} - \frac{(x+1)^2}{9} = 1$

G. $\frac{(y+2)^2}{4} - \frac{(x+1)^2}{16} = 1$

H. $\frac{(y-2)^2}{4} - \frac{(x-1)^2}{16} = 1$

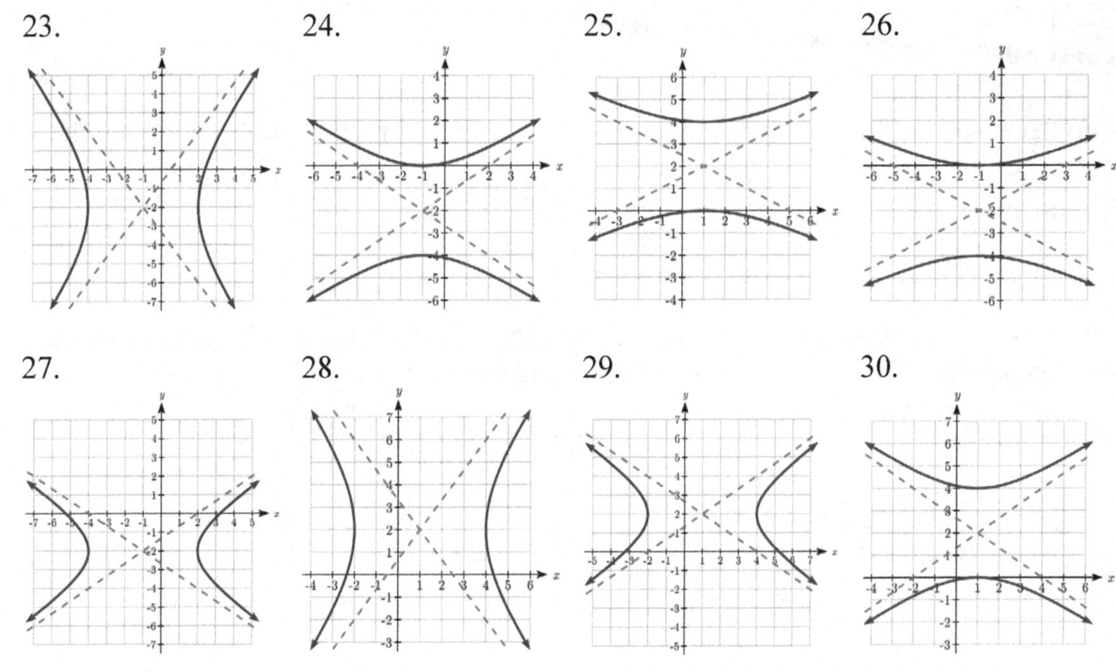

In problems 31–40, find the center, vertices, length of the transverse axis, and equations of the asymptotes. Sketch the graph. Check using a graphing utility.

31. $\dfrac{(x-1)^2}{25}-\dfrac{(y+2)^2}{4}=1$

32. $\dfrac{(y-3)^2}{16}-\dfrac{(x+5)^2}{36}=1$

33. $\dfrac{(y-1)^2}{9}-(x+2)^2=1$

34. $\dfrac{(x-1)^2}{25}-(y-6)^2=1$

35. $4x^2-8x-y^2=12$

36. $4y^2+16y-9x^2=20$

37. $4y^2-16y-x^2-2x=1$

38. $4x^2-16x-y^2+6y=29$

39. $9x^2+36x-4y^2+8y=4$

40. $9y^2+36y-16x^2-96x=-36$

In problems 41–42, write an equation for the graph.

41. 42.

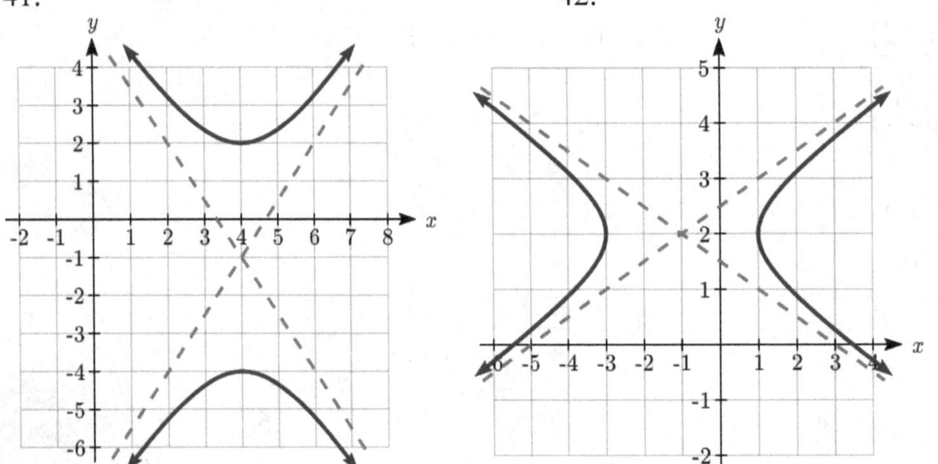

In problems 43–44, find the standard form of the equation for a hyperbola satisfying the given conditions.

43. Vertices (-1,-2) and (-1,6); asymptote $y-2=2(x+1)$

44. Vertices (-3,-3) and (5,-3); asymptote $y+3=\dfrac{1}{2}(x-1)$

In problems 45–48, find the center, vertices, length of the transverse axis, and equations of the asymptotes. Sketch the graph. Check using a graphing utility.

45. $y=\pm 4\sqrt{9x^2-1}$

46. $y=\pm\dfrac{1}{4}\sqrt{9x^2+1}$

47. $y=1\pm\dfrac{1}{2}\sqrt{9x^2+18x+10}$

48. $=-1\pm 2\sqrt{9x^2-18x+8}$

In problems 49–54, find the foci.

49. $\frac{y^2}{6}-\frac{x^2}{19}=1$

50. $x^2-\frac{y^2}{35}=1$

51. $\frac{(x-1)^2}{15}-(y-6)^2=1$

52. $\frac{(y-3)^2}{47}-\frac{(x+5)^2}{2}=1$

53. $y=1\pm\frac{4}{3}\sqrt{x^2+8x+25}$

54. $y=-3\pm\frac{12}{5}\sqrt{x^2-4x-21}$

In problems 55–66, find the standard form of the equation for a hyperbola satisfying the given conditions.

55. Foci (5,0) and (-5,0), vertices (4,0) and (4,0)

56. Foci (0,26) and (0, -26), vertices (0,10) and (0,-10)

57. Focus (0, 13), vertex (0,12), center (0,0)

58. Focus (15, 0), vertex (12, 0), center (0,0)

59. Focus (17, 0) and (-17,0), asymptotes $y=\frac{8}{15}x$ and $y=-\frac{8}{15}x$

60. Focus (0, 25) and (0, 25), asymptotes $y=\frac{24}{7}x$ and $y=-\frac{24}{7}x$

61. Focus (10, 0) and (-10, 0), transverse axis length 16

62. Focus (0, 34) and (0, -34), transverse axis length 32

63. Foci (1, 7) and (1, -3), vertices (1, 6) and (1,-2)

64. Foci (4, -2) and (-6, -2), vertices (2, -2) and (-4, -2)

65. Focus (12, 3), vertex (4, 3), center (-1, 3)

66. Focus (-3, 15), vertex (-3, 13), center (-3, -2)

67. **LORAN** Stations A and B are 100 kilometers apart and send a simultaneous radio signal to a ship. The signal from A arrives 0.0002 seconds before the signal from B. If the signal travels 300,000 kilometers per second, find an equation of the hyperbola on which the ship is positioned if the foci are located at A and B.

68. **Thunder and Lightning** Anita and Samir are standing 3050 feet apart when they see a bolt of light strike the ground. Anita hears the thunder 0.5 seconds before Samir does. Sound travels at 1100 feet per second. Find an equation of the hyperbola on which the lighting strike is positioned if Anita and Samir are located at the foci.

69. **Cooling Tower** The cooling tower for a power plant has sides in the shape of a hyperbola. The tower stands 179.6 meters tall. The diameter at the top is 72 meters. At their closest, the sides of the tower are 60 meters apart. Find an equation that models the sides of the cooling tower.

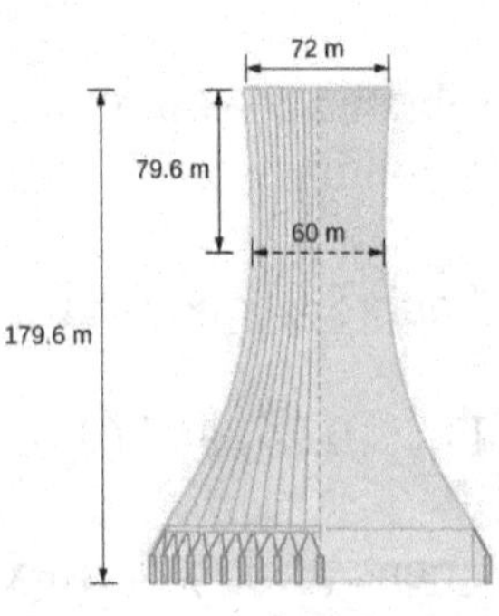

70. **Calibration** A seismologist positions two recording devices 340 feet apart at points A and B. To check the calibration, an explosive is detonated between the devices 90 feet from point A. The time the explosions register on the devices is noted and the difference calculated. A second explosion will be detonated east of point A. How far east should the second explosion be positioned so that the measured time difference is the same as for the first explosion?

71. **Target Practice** A gun at point A and a target at point B are 200 feet apart. A person at point C hears the gun fire and hit the target at exactly the same time. Find an equation of the hyperbola on which the person is standing if the foci are located at A and B. A fired bullet has a velocity of 2000 feet per second. The speed of sound is 1100 feet per second.

72. **Comet Trajectories** A comet passes through the solar system following a hyperbolic trajectory with the sun as a focus. The closest it gets to the sun is 3×10^8 miles. The figure shows the trajectory of the comet, whose path of entry is at a right angle to its path of departure. Find an equation for the comet's trajectory. Round to two decimal places.

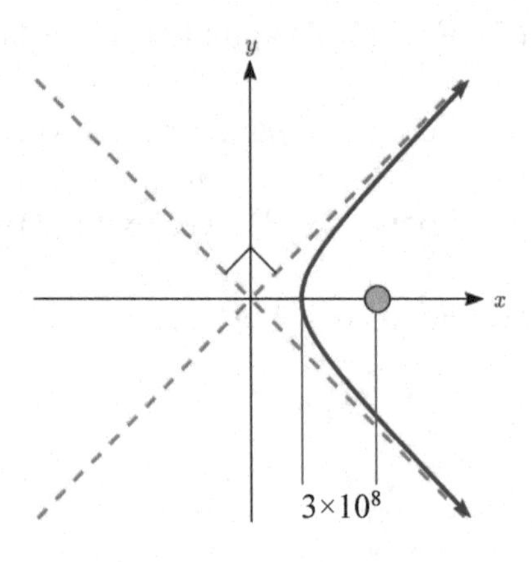

73. The **conjugate** of the hyperbola $\frac{x^2}{a^2}-\frac{y^2}{b^2}=1$ is $\frac{x^2}{a^2}-\frac{y^2}{b^2}=-1$. Show that $5y^2-x^2+25=0$ is the conjugate of $x^2-5y^2+25=0$.

74. The **eccentricity** e of a hyperbola is the ratio $\frac{c}{a}$, where c is the distance of a focus from the center and a is the distance of a vertex from the center. Find the eccentricity of $\frac{x^2}{9}-\frac{y^2}{16}=1$.

75. An **equilateral hyperbola** is one for which $a = b$. Find the eccentricity of an equilateral hyperbola.

76. The **latus rectum** of a hyperbola is a line segment with endpoints on the hyperbola that passes through a focus and is perpendicular to the transverse axis. Show that $\frac{2b^2}{a}$ is the length of the latus rectum of $\frac{x^2}{a^2}-\frac{y^2}{b^2}=1$.

77. **Confocal** hyperbolas have the same foci. Show that, for $0 < k < 6$, all hyperbolas of the form $\frac{x^2}{k}-\frac{y^2}{6-k}=1$ are confocal.

Section 9.3 Parabolas and Non-Linear Systems

To listen for signals from space, a radio telescope uses a dish in the shape of a parabola to focus and collect the signals in the receiver.

While we studied parabolas earlier when we explored quadratics, at the time we didn't discuss them as a conic section. A parabola is the shape resulting from when a plane parallel to the side of the cone intersects the cone[6].

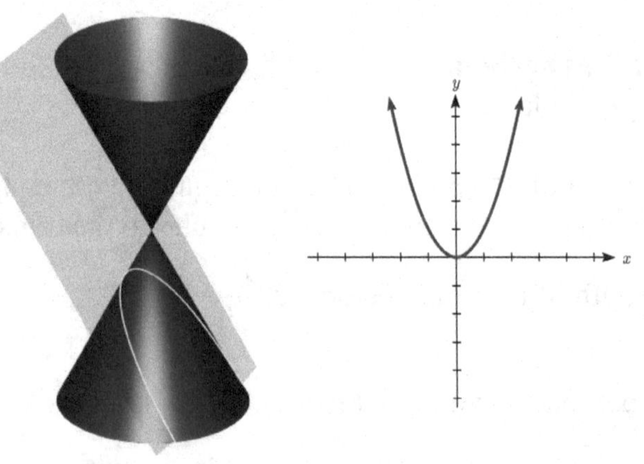

Parabola Definition and Vocabulary

A **parabola** with vertex at the origin can be defined by placing a fixed point at $F(0, p)$ called the **focus**, and drawing a line at $y = -p$, called the **directrix**. The parabola is the set of all points $Q(x, y)$ that are an equal distance between the fixed point and the directrix.

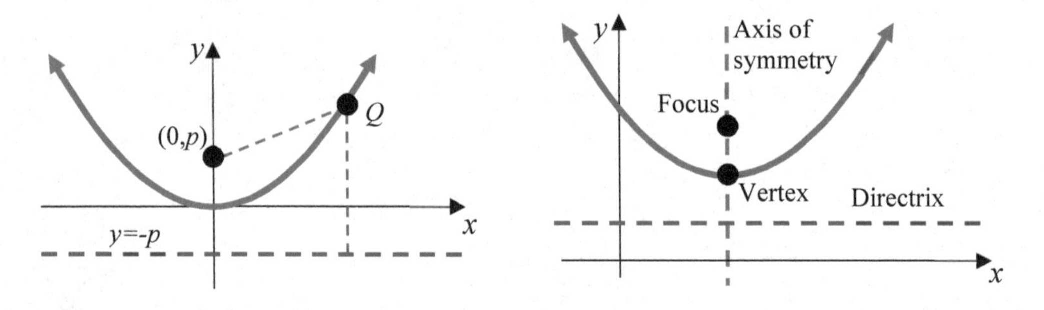

For general parabolas,

The **axis of symmetry** is the line passing through the foci, perpendicular to the directrix.

The **vertex** is the point where the parabola crosses the axis of symmetry.

The distance from the vertex to the focus, *p*, is the **focal length**.

[6] Pbroks13 (https://commons.wikimedia.org/wiki/File:Conic_sections_with_plane.svg), "Conic sections with plane", cropped to show only parabola, CC BY 3.0

Equations for Parabolas with Vertex at the Origin

From the definition above we can find an equation of a parabola. We will find it for a parabola with vertex at the origin, $C(0,0)$, opening upward with focus at $F(0, p)$ and directrix at $y = -p$.

Suppose $Q(x, y)$ is some point on the parabola. The distance from Q to the focus is
$d(Q,F) = \sqrt{(x-0)^2 + (y-p)^2} = \sqrt{x^2 + (y-p)^2}$

The distance from the point Q to the directrix is the difference of the y-values:
$d = y - (-p) = y + p$

From the definition of the parabola, these distances should be equal:

$\sqrt{x^2 + (y-p)^2} = y + p$	Square both sides
$x^2 + (y-p)^2 = (y+p)^2$	Expand
$x^2 + y^2 - 2py + p^2 = y^2 + 2py + p^2$	Combine like terms
$x^2 = 4py$	

This is the standard conic form of a parabola that opens up or down (vertical axis of symmetry), centered at the origin. Note that if we divided by 4p, we would get a more familiar equation for the parabola, $y = \dfrac{x^2}{4p}$. We can recognize this as a transformation of the parabola $y = x^2$, vertically compressed or stretched by $\dfrac{1}{4p}$.

Using a similar process, we could find an equation of a parabola with vertex at the origin opening left or right. The focus will be at (p,0) and the graph will have a horizontal axis of symmetry and a vertical directrix. The standard conic form of its equation will be $y^2 = 4px$, which we could also write as $x = \dfrac{y^2}{4p}$.

Example 1

Write the standard conic equation for a parabola with vertex at the origin and focus at (0, -2).

With focus at (0, -2), the axis of symmetry is vertical, so the standard conic equation is $x^2 = 4py$. Since the focus is (0, -2), p = -2.

The standard conic equation for the parabola is $x^2 = 4(-2)y$, or
$x^2 = -8y$

For parabolas with vertex not at the origin, we can shift these equations, leading to the equations summarized next.

Equation of a Parabola with Vertex at (*h*, *k*) in Standard Conic Form
The standard conic form of an equation of a parabola with vertex at the point (h,k) depends on whether the axis of symmetry is horizontal or vertical. The table below gives the standard equation, vertex, axis of symmetry, directrix, focus, and graph for each.

	Horizontal	Vertical
Standard Equation	$(y-k)^2 = 4p(x-h)$	$(x-h)^2 = 4p(y-k)$
Vertex	(h, k)	(h, k)
Axis of symmetry	$y = k$	$x = h$
Directrix	$x = h - p$	$y = k - p$
Focus	$(h + p, k)$	$(h, k + p)$
Graph	An example with $p < 0$ 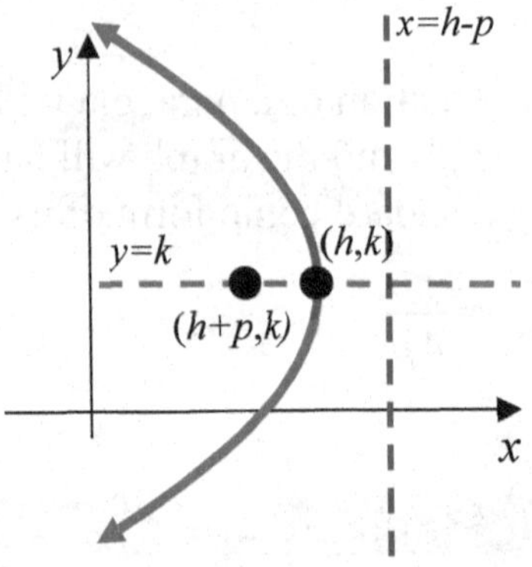 x=h-p, y, y=k, (h,k), (h+p,k), x	An example with $p > 0$ 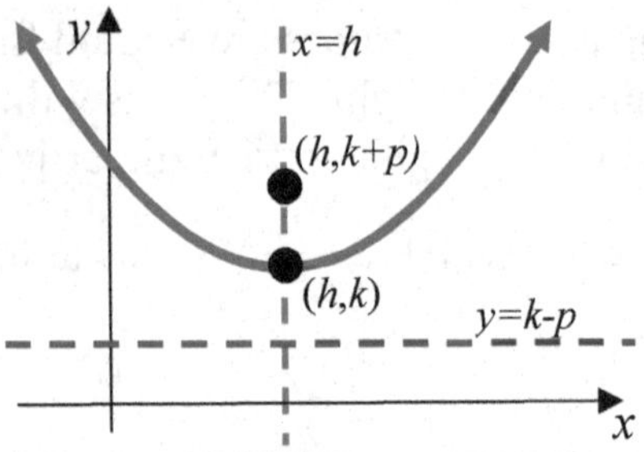 y, x=h, (h,k+p), (h,k), y=k-p, x

Since you already studied quadratics in some depth earlier, we will primarily explore the new concepts associated with parabolas, particularly the focus.

Example 2

Put the equation of the parabola $y = 8(x-1)^2 + 2$ in standard conic form. Find the vertex, focus, and axis of symmetry.

From your earlier work with quadratics, you may already be able to identify the vertex as (1,2), but we'll go ahead and put the parabola in the standard conic form. To do so, we need to isolate the squared factor.

$y = 8(x-1)^2 + 2$ Subtract 2 from both sides

$y - 2 = 8(x-1)^2$ Divide by 8

$$\frac{(y-2)}{8} = (x-1)^2$$

This matches the general form for a vertical parabola, $(x-h)^2 = 4p(y-k)$, where $4p = \frac{1}{8}$. Solving this tells us $p = \frac{1}{32}$. The standard conic form of the equation is $(x-1)^2 = 4\left(\frac{1}{32}\right)(y-2)$.

The vertex is at (1,2). The axis of symmetry is at $x = 1$.

The directrix is at $y = 2 - \frac{1}{32} = \frac{63}{32}$.

The focus is at $\left(1, 2 + \frac{1}{32}\right) = \left(1, \frac{65}{32}\right)$.

Example 3

A parabola has its vertex at (1,5) and focus at (3,5). Find an equation for the parabola.

Since the vertex and focus lie on the line $y = 5$, that is our axis of symmetry.

The vertex (1,5) tells us $h = 1$ and $k = 5$.

Looking at the distance from the vertex to the focus, $p = 3 - 1 = 2$.

Substituting these values into the standard conic form of an equation for a horizontal parabola gives the equation

$(y-5)^2 = 4(2)(x-1)$

$(y-5)^2 = 8(x-1)$

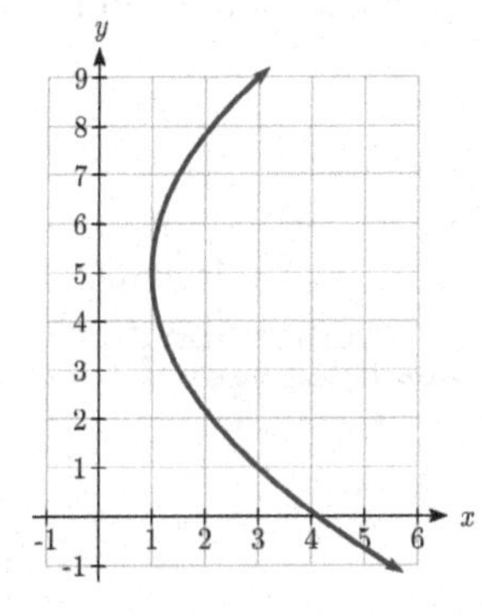

Note this could also be rewritten by solving for x, resulting in

$$x = \frac{1}{8}(y-5)^2 + 1$$

Try it Now

1. A parabola has its vertex at (-2,3) and focus at (-2,2). Find an equation for this parabola.

Applications of Parabolas

In an earlier section, we learned that ellipses have a special property that a ray eminating from one focus will be reflected back to the other focus, the property that enables the whispering chamber to work. Parabolas also have a special property, that any ray eminating from the focus will be reflected parallel to the axis of symmetry. Reflectors in flashlights take advantage of this property to focus the light from the bulb into a collimated beam. The same property can be used in reverse, taking parallel rays of sunlight or radio signals and directing them all to the focus.

Example 4

A solar cooker is a parabolic dish that reflects the sun's rays to a central point allowing you to cook food. If a solar cooker has a parabolic dish 16 inches in diameter and 4 inches tall, where should the food be placed?

We need to determine the location of the focus, since that's where the food should be placed. Positioning the base of the dish at the origin, the shape from the side looks like:

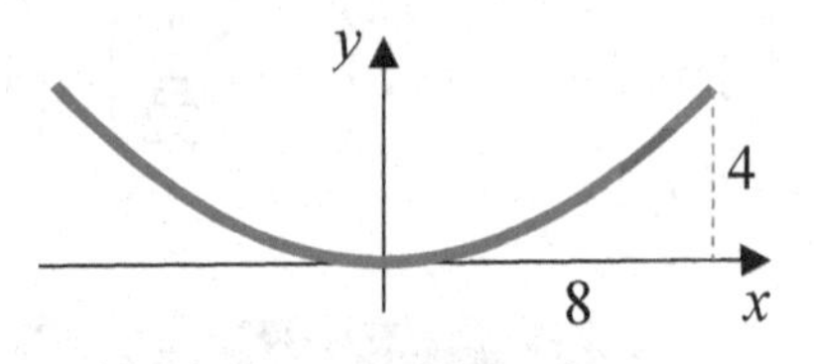

The standard conic form of an equation for the parabola would be $x^2 = 4py$. The parabola passes through (4, 8), so substituting that into the equation, we can solve for p:

$8^2 = 4(p)(4)$

$p = \frac{8^2}{16} = 4$

The focus is 4 inches above the vertex. This makes for a very convenient design, since then a grate could be placed on top of the dish to hold the food.

Try it Now

2. A radio telescope is 100 meters in diameter and 20 meters deep. Where should the receiver be placed?

Non-Linear Systems of Equations

In many applications, it is necessary to solve for the intersection of two curves. Many of the techniques you may have used before to solve systems of linear equations will work for non-linear equations as well, particularly substitution. You have already solved some examples of non-linear systems when you found the intersection of a parabola and line while studying quadratics, and when you found the intersection of a circle and line while studying circles.

Example 4

Find the points where the ellipse $\frac{x^2}{4}+\frac{y^2}{25}=1$ intersects the circle $x^2+y^2=9$.

To start, we might multiply the ellipse equation by 100 on both sides to clear the fractions, giving $25x^2+4y^2=100$.

A common approach for finding intersections is substitution. With these equations, rather than solving for *x* or *y*, it might be easier to solve for x^2 or y^2. Solving the circle equation for x^2 gives $x^2=9-y^2$. We can then substitute that expression for x^2 into the ellipse equation.

$25x^2+4y^2=100$ Substitute $x^2=9-y^2$

$25(9-y^2)+4y^2=100$ Distribute

$225-25y^2+4y^2=100$ Combine like terms

$-21y^2=-125$ Divide by -21

$y^2=\frac{125}{21}$ Use the square root to solve

$y=\pm\sqrt{\frac{125}{21}}=\pm\frac{5\sqrt{5}}{\sqrt{21}}$

We can substitute each of these *y* values back in to $x^2=9-y^2$ to find *x*

$$x^2=9-\left(\sqrt{\frac{125}{21}}\right)^2=9-\frac{125}{21}=\frac{189}{21}-\frac{125}{21}=\frac{64}{21}$$

$$x=\pm\sqrt{\frac{64}{21}}=\pm\frac{8}{\sqrt{21}}$$

There are four points of intersection: $\left(\pm\frac{8}{\sqrt{21}},\pm\frac{5\sqrt{5}}{\sqrt{21}}\right)$.

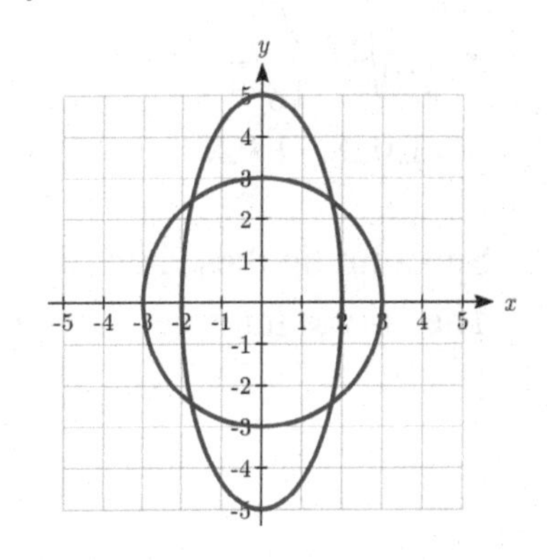

It's worth noting there is a second technique we could have used in the previous example, called elimination. If we multiplied the circle equation by -4 to get $-4x^2 - 4y^2 = -36$, we can then add it to the ellipse equation, eliminating the variable y.

$25x^2 + 4y^2 = 100$

$-4x^2 - 4y^2 = -36$ Add the left sides, and add the right sides

$21x^2 = 64$ Solve for x

$$x = \pm\sqrt{\frac{64}{21}} = \pm\frac{8}{\sqrt{21}}$$

Example 5

Find the points where the hyperbola $\frac{y^2}{4} - \frac{x^2}{9} = 1$ intersects the parabola $y = 2x^2$.

We can solve this system of equations by substituting $y = 2x^2$ into the hyperbola equation.

$\frac{(2x^2)^2}{4} - \frac{x^2}{9} = 1$ Simplify

$\frac{4x^4}{4} - \frac{x^2}{9} = 1$ Simplify, and multiply by 9

$9x^4 - x^2 = 9$ Move the 9 to the left

$9x^4 - x^2 - 9 = 0$

While this looks challenging to solve, we can think of it as a "quadratic in disguise," since $x^4 = (x^2)^2$. Letting $u = x^2$, the equation becomes

$9u^2 - u^2 - 9 = 0$ Solve using the quadratic formula

$u = \frac{-(-1) \pm \sqrt{(-1)^2 - 4(9)(-9)}}{2(9)} = \frac{1 \pm \sqrt{325}}{18}$ Solve for x

$x^2 = \frac{1 \pm \sqrt{325}}{18}$ But $1 - \sqrt{325} < 0$, so

$x = \pm\sqrt{\frac{1 + \sqrt{325}}{18}}$ This leads to two real solutions

$x \approx 1.028, -1.028$

Substituting these into $y = 2x^2$, we can find the corresponding y values.
The curves intersect at the points (1.028, 2.114) and (-1.028, 2.114).

Try it Now

3. Find the points where the line $y = 4x$ intersect the ellipse $\frac{y^2}{4} - \frac{x^2}{16} = 1$

Solving for the intersection of two hyperbolas allows us to utilize the LORAN navigation approach described in the last section.

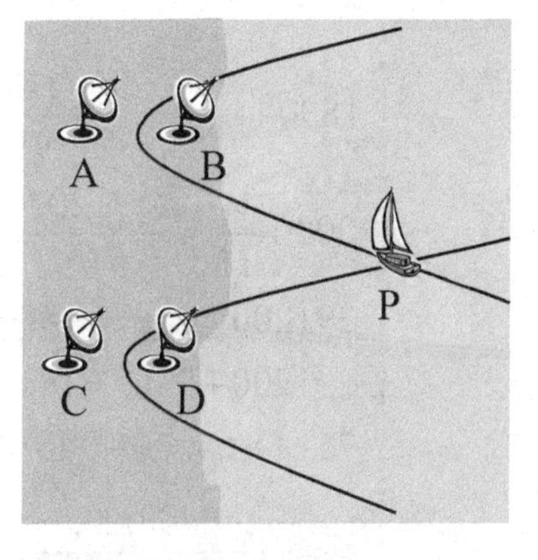

In our example, stations A and B are 150 kilometers apart and send a simultaneous radio signal to the ship. The signal from B arrives 0.0003 seconds before the signal from A. We found the equation of the hyperbola in standard form would be

$$\frac{x^2}{2025} - \frac{y^2}{3600} = 1$$

Example 10

Continuing the situation from the last section, suppose stations C and D are located 200 km due south of stations A and B and 100 km apart. The signal from D arrives 0.0001 seconds before the signal from C, leading to the equation $\frac{x^2}{225} - \frac{(y+200)^2}{2275} = 1$. Find the position of the ship.

To solve for the position of the boat, we need to find where the hyperbolas intersect. This means solving the system of equations. To do this, we could start by solving both equations for x^2. With the first equation from the previous example,

$$\frac{x^2}{2025} - \frac{y^2}{3600} = 1$$ Move the y term to the right

$$\frac{x^2}{2025} = 1 + \frac{y^2}{3600}$$ Multiply both sides by 2025

$$x^2 = 2025 + \frac{2025y^2}{3600}$$ Simplify

$$x^2 = 2025 + \frac{9y^2}{16}$$

With the second equation, we repeat the same process

$$\frac{x^2}{225} - \frac{(y+200)^2}{2275} = 1$$ Move the y term to the right and multiply by 225

$$x^2 = 225 + \frac{225(y+200)^2}{2275}$$ Simplify

$$x^2 = 225 + \frac{9(y+200)^2}{91}$$

Now set these two expressions for x^2 equal to each other and solve.

$$2025 + \frac{9y^2}{16} = 225 + \frac{9(y+200)^2}{91}$$ Subtract 225 from both sides

$$1800 + \frac{9y^2}{16} = \frac{9(y+200)^2}{91}$$ Divide by 9

$$200 + \frac{y^2}{16} = \frac{(y+200)^2}{91}$$ Multiply both sides by $16 \cdot 91 = 1456$

$$291200 + 91y^2 = 16(y+200)^2$$ Expand and distribute

$$291200 + 91y^2 = 16y^2 + 6400y + 640000$$ Combine like terms on one side

$$75y^2 - 6400y - 348800 = 0$$ Solve using the quadratic formula

$$y = \frac{-(-6400) \pm \sqrt{(-6400)^2 - 4(75)(-348800)}}{2(75)} \approx 123.11 \text{ km or } -37.78 \text{ km}$$

We can find the associated *x* values by substituting these *y*-values into either hyperbola equation. When $y \approx 123.11$,

$$x^2 \approx 2025 + \frac{9(123.11)^2}{16}$$

$$x \approx \pm 102.71$$

When $y \approx$ -37.78km,

$$x^2 \approx 2025 + \frac{9(-37.78)^2}{16}$$

$$x \approx \pm 53.18$$

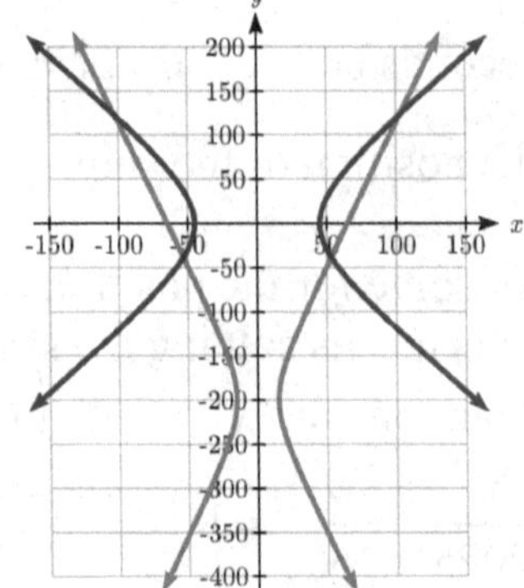

This provides 4 possible locations for the ship. Two can be immediately discarded, as they're on land. Navigators would use other navigational techniques to decide between the two remaining locations.

Important Topics of This Section
Parabola Definition
Parabola Equations in Standard Form
Applications of Parabolas
Solving Non-Linear Systems of Equations

Try it Now Answers

1. Axis of symmetry is vertical, and the focus is below the vertex.
   $p = 2 - 3 = -1$.
   $(x-(-2))^2 = 4(-1)(y-3)$, or $(x+2)^2 = -4(y-3)$.

2. The standard conic form of the equation is $x^2 = 4py$.
   Using (50,20), we can find that $50^2 = 4p(20)$, so $p = 31.25$ meters.
   The receiver should be placed 31.25 meters above the vertex.

3. Substituting $y = 4x$ gives $\frac{(4x)^2}{4} - \frac{x^2}{16} = 1$. Simplify
   $\frac{16x^2}{4} - \frac{x^2}{16} = 1$. Multiply by 16 to get
   $64x^2 - x^2 = 16$
   $x = \pm\sqrt{\frac{16}{63}} = \pm 0.504$
   Substituting those into $y = 4x$ gives the corresponding y values.
   The curves intersect at (0.504, 2.016) and (-0.504, -2.016).

Section 9.3 Exercises

In problems 1–4, match each graph with one of the equations A–D.

A. $y^2 = 4x$ B. $x^2 = 4y$ C. $x^2 = 8y$ D. $y^2 + 4x = 0$

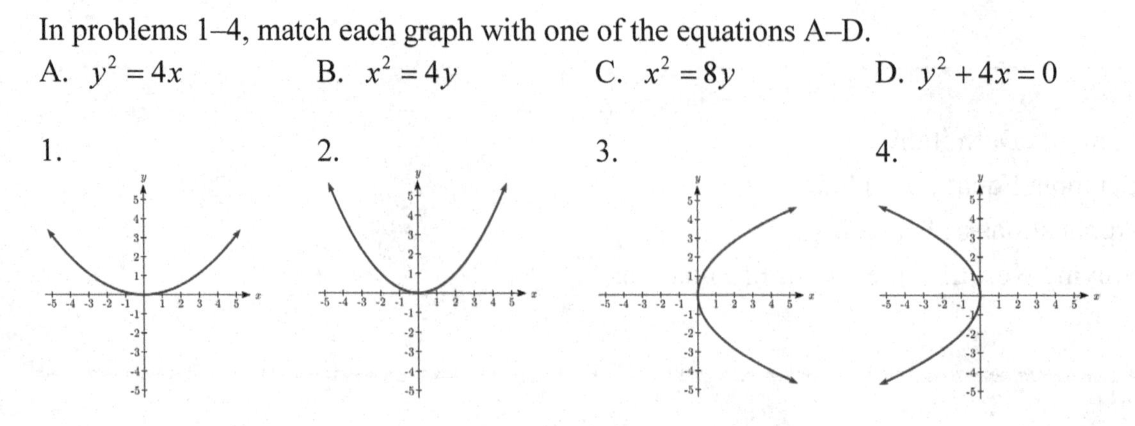

In problems 5–14, find the vertex, axis of symmetry, directrix, and focus of the parabola.

5. $y^2 = 16x$ 6. $x^2 = 12y$ 7. $y = 2x^2$ 8. $x = -\dfrac{y^2}{8}$

9. $x + 4y^2 = 0$ 10. $8y + x^2 = 0$ 11. $(x-2)^2 = 8(y+1)$

12. $(y+3)^2 = 4(x-2)$ 13. $y = \dfrac{1}{4}(x+1)^2 + 4$ 14. $x = -\dfrac{1}{12}(y+1)^2 + 1$

In problems 15–16, write an equation for the graph.

15. 16.

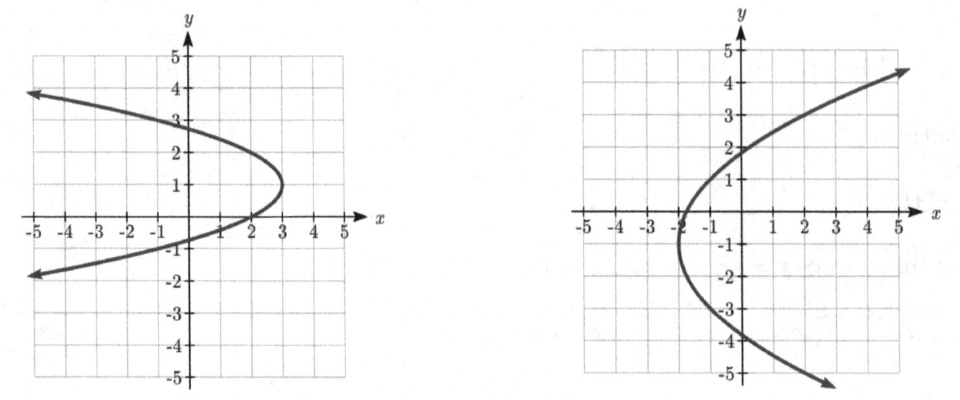

In problems 17-20, find the standard form of the equation for a parabola satisfying the given conditions.

17. Vertex at (2,3), opening to the right, focal length 3

18. Vertex at (-1,2), opening down, focal length 1

19. Vertex at (0,3), focus at (0,4)

20. Vertex at (1,3), focus at (0,3)

21. The mirror in an automobile headlight has a parabolic cross-section with the light bulb at the focus. On a schematic, the equation of the parabola is given as $x^2 = 4y^2$. At what coordinates should you place the light bulb?

22. If we want to construct the mirror from the previous exercise so that the focus is located at (0,0.25), what should the equation of the parabola be?

23. A satellite dish is shaped like a paraboloid of revolution. This means that it can be formed by rotating a parabola around its axis of symmetry. The receiver is to be located at the focus. If the dish is 12 feet across at its opening and 4 feet deep at its center, where should the receiver be placed?

24. Consider the satellite dish from the previous exercise. If the dish is 8 feet across at the opening and 2 feet deep, where should we place the receiver?

25. A searchlight is shaped like a paraboloid of revolution. A light source is located 1 foot from the base along the axis of symmetry. If the opening of the searchlight is 2 feet across, find the depth.

26. If the searchlight from the previous exercise has the light source located 6 inches from the base along the axis of symmetry and the opening is 4 feet wide, find the depth.

In problems 27–34, solve each system of equations for the intersections of the two curves.

27. $\begin{aligned} y &= 2x \\ y^2 - x^2 &= 1 \end{aligned}$

28. $\begin{aligned} y &= x+1 \\ 2x^2 + y^2 &= 1 \end{aligned}$

29. $\begin{aligned} x^2 + y^2 &= 11 \\ x^2 - 4y^2 &= 1 \end{aligned}$

30. $\begin{aligned} 2x^2 + y^2 &= 4 \\ y^2 - x^2 &= 1 \end{aligned}$

31. $\begin{aligned} y &= x^2 \\ y^2 - 6x^2 &= 16 \end{aligned}$

32. $\begin{aligned} x &= y^2 \\ \frac{x^2}{4} + \frac{y^2}{9} &= 1 \end{aligned}$

33. $\begin{aligned} x^2 - y^2 &= 1 \\ 4y^2 - x^2 &= 1 \end{aligned}$

34. $\begin{aligned} x^2 &= 4(y-2) \\ x^2 &= 8(y+1) \end{aligned}$

35. A LORAN system has transmitter stations A, B, C, and D at (-125,0), (125,0), (0, 250), and (0,-250), respectively. A ship in quadrant two computes the difference of its distances from A and B as 100 miles and the difference of its distances from C and D as 180 miles. Find the *x*- and *y*-coordinates of the ship's location. Round to two decimal places.

36. A LORAN system has transmitter stations A, B, C, and D at (-100,0), (100,0), (-100, -300), and (100,-300), respectively. A ship in quadrant one computes the difference of its distances from A and B as 80 miles and the difference of its distances from C and D as 120 miles. Find the *x*- and *y*-coordinates of the ship's location. Round to two decimal places.

Section 9.4 Conics in Polar Coordinates

In the preceding sections, we defined each conic in a different way, but each involved the distance between a point on the curve and the focus. In the previous section, the parabola was defined using the focus and a line called the directrix. It turns out that all conic sections (circles, ellipses, hyperbolas, and parabolas) can be defined using a single relationship.

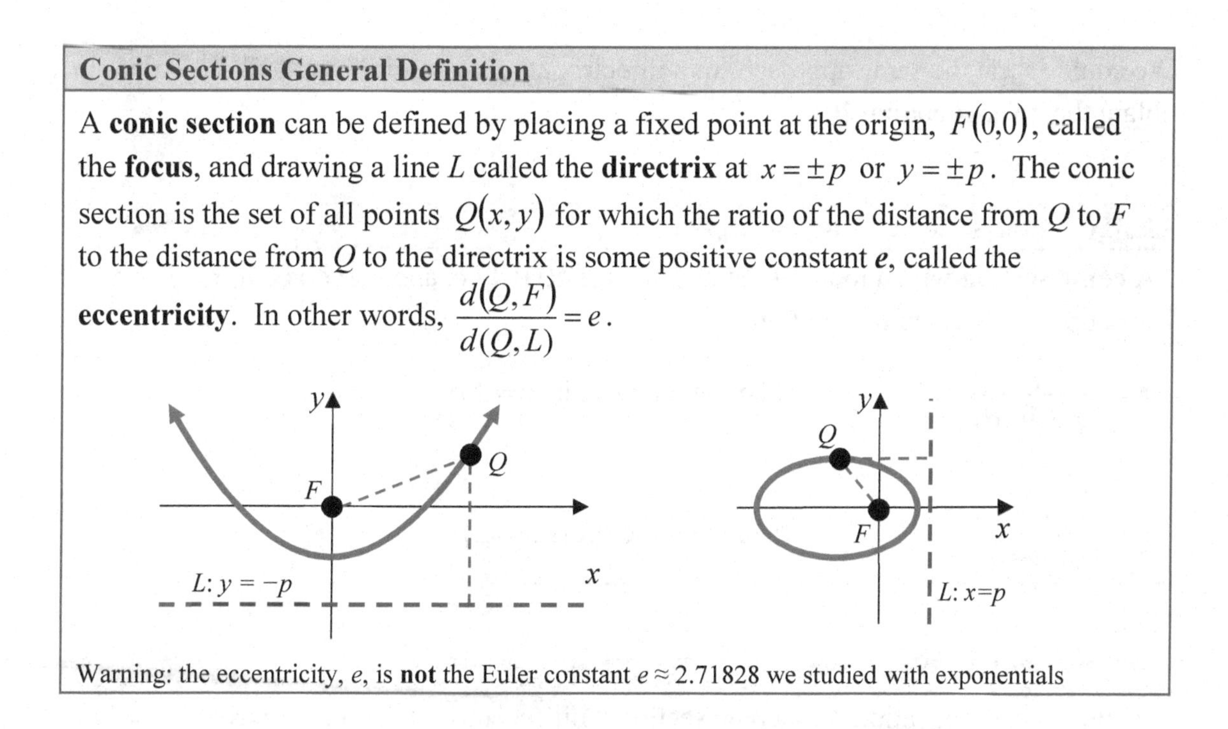

Conic Sections General Definition

A **conic section** can be defined by placing a fixed point at the origin, $F(0,0)$, called the **focus**, and drawing a line L called the **directrix** at $x = \pm p$ or $y = \pm p$. The conic section is the set of all points $Q(x, y)$ for which the ratio of the distance from Q to F to the distance from Q to the directrix is some positive constant ***e***, called the **eccentricity**. In other words, $\dfrac{d(Q,F)}{d(Q,L)} = e$.

Warning: the eccentricity, *e*, is **not** the Euler constant $e \approx 2.71828$ we studied with exponentials

The Polar Form of a Conic

To create a general equation for a conic section using the definition above, we will use polar coordinates. Represent $Q(x, y)$ in polar coordinates so $(x, y) = (r\cos(\theta), r\sin(\theta))$. For now, we'll focus on the case of a horizontal directrix at $y = -p$, as in the picture above on the left.

The distance from the focus to the point Q in polar is just r.
The distance from the point Q to the directrix $y = -p$ is $r\sin(\theta) - (-p) = p + r\sin(\theta)$

The ratio of these should be the constant eccentricity e, so

$$\frac{d(Q,F)}{d(Q,L)} = e$$

Substituting in the expressions for the distances,

$$\frac{r}{p + r\sin(\theta)} = e$$

To have a standard polar equation, we need to solve for r. Start by clearing the fraction.

$r = e(p + r\sin(\theta))$	Distribute
$r = ep + er\sin(\theta)$	Move terms with r to the left
$r - er\sin(\theta) = ep$	Factor the r
$r(1 - e\sin(\theta)) = ep$	Divide

$$r = \frac{ep}{1 - e\sin(\theta)}$$

We could repeat the same approach for a directrix at $y = p$ and for vertical directrices to obtain the polar equations below.

Polar Equation for a Conic Section

A **conic section** with a focus at the origin, **eccentricity** e, and **directrix** at $x = \pm p$ or $y = \pm p$ will have polar equation:

$$r = \frac{ep}{1 \pm e\sin(\theta)}$$ when the directrix is $y = \pm p$

$$r = \frac{ep}{1 \pm e\cos(\theta)}$$ when the directrix is $x = \pm p$

Example 1

Write the polar equation for a conic section with eccentricity 3 and directrix at $x = 2$.

We are given $e = 3$ and $p = 2$. Since the directrix is vertical and at a positive x value, we use the equation involving cos with the positive sign.

$$r = \frac{(3)(2)}{1 + 3\cos(\theta)} = \frac{6}{1 + 3\cos(\theta)}$$

Graphing that using technology reveals it's an equation for a hyperbola.

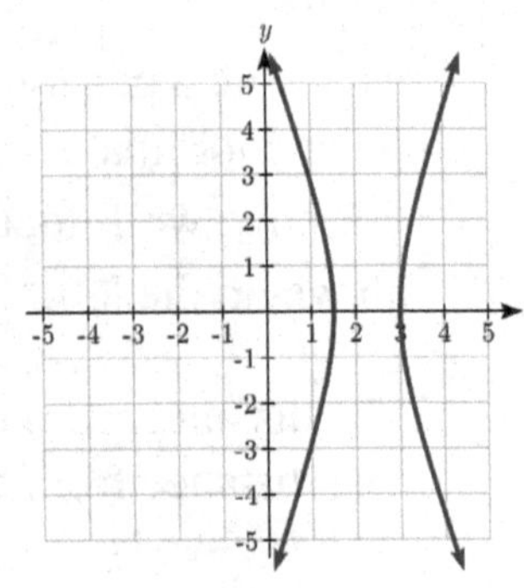

Try it Now

1. Write a polar equation for a conic with eccentricity 1 and directrix at $y = -3$.

Relating the Polar Equation to the Shape

It was probably not obvious to you that the polar equation in the last example would give the graph of a hyperbola. To explore the relationship between the polar equation and the shape, we will try to convert the polar equation into a Cartesian one. For simplicity, we will consider the case where the directrix is $x = 1$.

$r = \dfrac{e}{1 + e\cos(\theta)}$	Multiply by the denominator
$r(1 + e\cos(\theta)) = e$	Rewrite $\cos(\theta) = \dfrac{x}{r}$
$r\left(1 + e\dfrac{x}{r}\right) = e$	Distribute
$r + ex = e$	Isolate *r*
$r = e - ex$	Square both sides
$r^2 = (e - ex)^2$	Rewrite $r^2 = x^2 + y^2$ and expand
$x^2 + y^2 = e^2 - 2e^2x + e^2x^2$	Move variable terms to the left
$x^2 + 2e^2x - e^2x^2 + y^2 = e^2$	Combine like terms
$x^2(1 - e^2) + 2e^2x + y^2 = e^2$	

We could continue, by completing the square with the *x* terms, to eventually rewrite this in the standard form as $\left(\dfrac{(1-e^2)^2}{e^2}\right)\left(x - \dfrac{e^2}{1-e^2}\right)^2 + \left(\dfrac{1-e^2}{e^2}\right)y^2 = 1$, but happily there's no need for us to do that.

In the equation $x^2(1 - e^2) + 2e^2x + y^2 = e^2$, we can see that:
When $e < 1$, the coefficients of both x^2 and y^2 are positive, resulting in ellipse.
When $e > 1$, the coefficient of x^2 is negative while the coefficient of y^2 is positive, resulting in a hyperbola.
When $e = 1$, the x^2 will drop out of the equation, resulting in a parabola.

Relation Between the Polar Equation of a Conic and its Shape
For a **conic section** with a focus at the origin, **eccentricity** *e*, and **directrix** at $x = \pm p$ or $y = \pm p$, when $0 < e < 1$, the graph is an ellipse when $e = 1$, the graph is a parabola when $e > 1$, the graph is a hyperbola

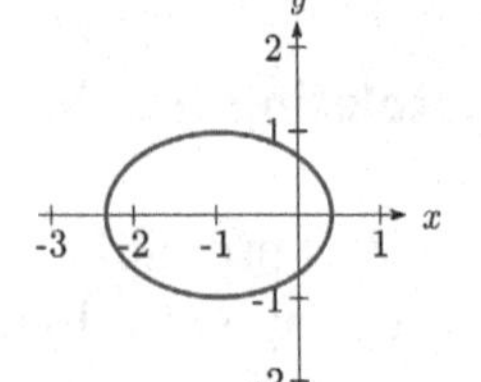

Taking a more intuitive approach, notice that if $e < 1$, the denominator $1 + e\cos(\theta)$ will always be positive and so r will always be positive. This means that the radial distance r is defined and finite for every value of θ, including $\frac{\pi}{2}$, with no breaks. The only conic with this characteristic is an ellipse.

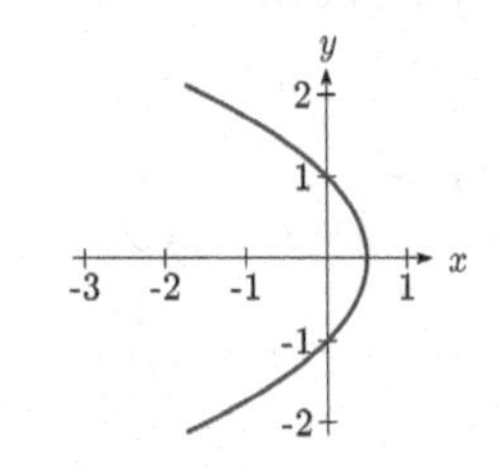

If $e = 1$, the denominator will be positive for all values of θ, except $-\pi$ where the denominator is 0 and r is undefined. This fits with a parabola, which has a point at every angle except at the angle pointing along the axis of symmetry away from the vertex.

If $e > 1$, then the denominator will be zero at two angles other than $\pm\frac{\pi}{2}$, and r will be negative for a set of θ values. This division of positive and negative radius values would result in two distinct branches of the graph, fitting with a hyperbola.

Example 2

For each of the following conics with focus at the origin, identify the shape, the directrix, and the eccentricity.

a. $r = \dfrac{8}{1 - 2\sin(\theta)}$ b. $r = \dfrac{6}{3 - 2\cos(\theta)}$ c. $r = \dfrac{8}{5 + 5\sin(\theta)}$

a. This equation is already in standard form $r = \dfrac{ep}{1 \pm e\sin(\theta)}$ for a conic with horizontal directrix at $y = -p$.

The eccentricity is the coefficient of $\sin(\theta)$, so $e = 2$.

Since $e = 2 > 1$, the shape will be a hyperbola.

Looking at the numerator, $ep = 8$, and substituting $e = 2$ gives $p = 4$. The directrix is $y = -4$.

b. This equation is not in standard form, since the constant in the denominator is not 1. To put it into standard form, we can multiply the numerator and denominator by 1/3.

$$r=\frac{6}{3-2\cos(\theta)}\cdot\frac{\frac{1}{3}}{\frac{1}{3}}=\frac{6\left(\frac{1}{3}\right)}{(3-2\cos(\theta))\left(\frac{1}{3}\right)}=\frac{2}{1-\frac{2}{3}\cos(\theta)}$$

This is the standard form for a conic with vertical directrix $x=-p$. The eccentricity is the coefficient on $\cos(\theta)$, so $e=\frac{2}{3}$.
Since $0<e<1$, the shape is an ellipse.

Looking at the numerator, $ep=2$, so $\frac{2}{3}p=2$, giving $p=3$. The directrix is $x=-3$.

c. This equation is also not in standard form. Multiplying the numerator and denominator by 1/5 will put it in standard form.

$$r=\frac{8}{5+5\sin(\theta)}\cdot\frac{\frac{1}{5}}{\frac{1}{5}}=\frac{8\left(\frac{1}{5}\right)}{(5+5\sin(\theta))\left(\frac{1}{5}\right)}=\frac{\frac{8}{5}}{1+\sin(\theta)}$$

This is the standard form for a conic with horizontal directrix at $y=p$.
The eccentricity is the coefficient on $\sin(\theta)$, so $e=1$. The shape will be a parabola.

Looking at the numerator, $ep=\frac{8}{5}$. Since $e=1$, $p=\frac{8}{5}$. The directrix is $y=\frac{8}{5}$.

Notice that since the directrix is above the focus at the origin, the parabola will open downward.

Try it Now

2. Identify the shape, the directrix, and the eccentricity of $r=\frac{9}{4+2\cos(\theta)}$

Graphing Conics from the Polar Form

Identifying additional features of a conic in polar form can be challenging, which makes graphing without technology likewise challenging. We can utilize our understanding of the conic shapes from earlier sections to aid us.

Example 3

Sketch a graph of $r = \dfrac{3}{1 - 0.5\sin(\theta)}$ and write its Cartesian equation.

This is in standard form, and we can identify that $e = 0.5$, so the shape is an ellipse. From the numerator, $ep = 3$, so $0.5p = 3$, giving $p = 6$. The directrix is $y = -6$.

To sketch a graph, we can start by evaluating the function at a few convenient θ values, and finding the corresponding Cartesian coordinates.

$\theta = 0$ $\quad r = \dfrac{3}{1 - 0.5\sin(0)} = \dfrac{3}{1} = 3$ $\quad (x, y) = (3,0)$

$\theta = \dfrac{\pi}{2}$ $\quad r = \dfrac{3}{1 - 0.5\sin\left(\dfrac{\pi}{2}\right)} = \dfrac{3}{1 - 0.5} = 6$ $\quad (x, y) = (0,6)$

$\theta = \pi$ $\quad r = \dfrac{3}{1 - 0.5\sin(\pi)} = \dfrac{3}{1} = 3$ $\quad (x, y) = (-3,0)$

$\theta = \dfrac{3\pi}{2}$ $\quad r = \dfrac{3}{1 - 0.5\sin\left(\dfrac{3\pi}{2}\right)} = \dfrac{3}{1 + 0.5} = 2$ $\quad (x, y) = (0,-2)$

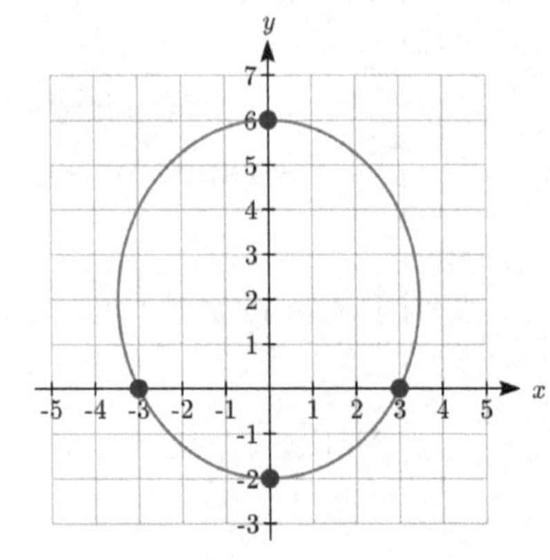

Plotting these points and remembering the origin is one of the foci gives an idea of the shape, which we could sketch in. To get a better understanding of the shape, we could use these features to find more.

The vertices are at (0, -2) and (0, 6), so the center must be halfway between, at $\left(0, \dfrac{-2+6}{2}\right)$ = (0, 2). Since the vertices are a distance a from the center, $a = 6 - 2 = 4$.

One focus is at (0, 0), a distance of 2 from the center, so $c = 2$, and the other focus must be 2 above the center, at (0, 4).

We can now solve for b: $b^2 = a^2 - c^2$, so $b^2 = 4^2 - 2^2 = 10$, hence $b = \pm\sqrt{10}$. The minor axis endpoints would be at $\left(-\sqrt{10}, 2\right)$ and $\left(\sqrt{10}, 2\right)$.

We can now use the center, a, and b to write the Cartesian equation for this curve:

$$\frac{x^2}{10} + \frac{(y-2)^2}{16} = 1$$

Try it Now

3. Sketch a graph of $r = \dfrac{6}{1+2\cos(\theta)}$ and identify the important features.

Important Topics of This Section
Polar equations for Conic Sections
Eccentricity and Directrix
Determining the shape of a polar conic section

Try it Now Answers

1. $r = \dfrac{(1)(3)}{1-\sin(\theta)}$. $r = \dfrac{3}{1-\sin(\theta)}$

2. We can convert to standard form by multiplying the top and bottom by $\dfrac{1}{4}$.

$r = \dfrac{\frac{9}{4}}{1+\frac{1}{2}\cos(\theta)}$. Eccentricity $= \dfrac{1}{2}$, so the shape is an ellipse.

The numerator is $ep = \dfrac{1}{2}p = \dfrac{9}{4}$. The directrix is $x = \dfrac{9}{2}$.

3. The eccentricity is $e = 2$, so the graph of the equation is a hyperbola. The directrix is $x = 3$. Since the directrix is a vertical line and the focus is at the origin, the hyperbola is horizontal.

$\theta = 0$ $\quad r = \dfrac{6}{1+2\cos(0)} = \dfrac{6}{1+2} = 2$ $\quad (x, y) = (2,0)$

$\theta = \dfrac{\pi}{2}$ $\quad r = \dfrac{6}{1+2\cos\left(\frac{\pi}{2}\right)} = \dfrac{6}{1} = 6$ $\quad (x, y) = (0,6)$

$\theta = \pi$ $\quad r = \dfrac{6}{1+2\cos(\pi)} = \dfrac{6}{1-2} = -6$ $\quad (x, y) = (6,0)$

$\theta = \dfrac{3\pi}{2}$ $\quad r = \dfrac{6}{1+2\cos\left(\frac{3\pi}{2}\right)} = \dfrac{6}{1} = 6$ $\quad (x, y) = (0,-6)$

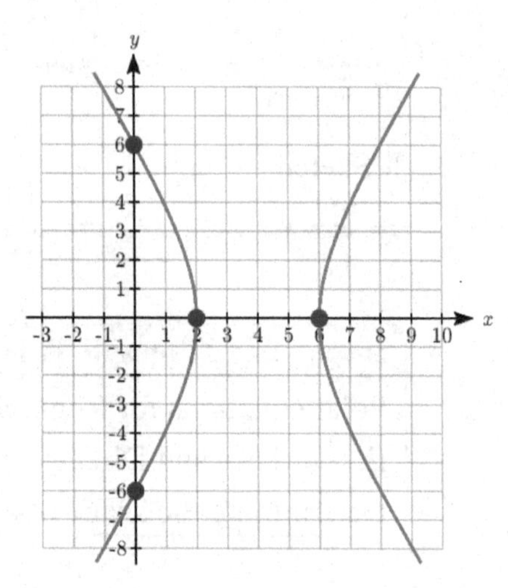

Plotting those points, we can connect the three on the left with a smooth curve to form one branch of the hyperbola, and the other branch will be a mirror image passing through the last point.

The vertices are at (2,0) and (6,0).

The center of the hyperbola would be at the midpoint of the vertices, at (4,0).
The vertices are a distance $a = 2$ from the center.
The focus at the origin is a distance $c = 4$ from the center.
Solving for b, $b^2 = 4^2 - 2^2 = 12$. $b = \pm\sqrt{12} = \pm 2\sqrt{3}$.

The asymptotes would be $y = \pm\sqrt{3}(x-4)$.

The Cartesian equation of the hyperbola would be:

$$\frac{(x-4)^2}{4} - \frac{y^2}{12} = 1$$

Section 9.4 Exercises

In problems 1–8, find the eccentricity and directrix, then identify the shape of the conic.

1. $r = \dfrac{12}{1 + 3\cos(\theta)}$

2. $r = \dfrac{4}{1 - \sin(\theta)}$

3. $r = \dfrac{2}{4 - 3\sin(\theta)}$

4. $r = \dfrac{7}{2 - \cos(\theta)}$

5. $r = \dfrac{1}{5 - 5\cos(\theta)}$

6. $r = \dfrac{6}{3 + 8\cos(\theta)}$

7. $r = \dfrac{4}{7 + 2\cos(\theta)}$

8. $r = \dfrac{16}{4 + 3\sin(\theta)}$

In problems 9–14, find a polar equation for a conic having a focus at the origin with the given characteristics.

9. Directrix $x = -4$, eccentricity $e = 5$.

10. Directrix $y = -2$, eccentricity $e = 3$.

11. Directrix $y = 3$, eccentricity $e = \dfrac{1}{3}$.

12. Directrix $x = 5$, eccentricity $e = \dfrac{3}{4}$.

13. Directrix $y = -2$, eccentricity $e = 1$.

14. Directrix $x = -3$, eccentricity $e = 1$.

In problems 15–20, sketch a graph of the conic. Use the graph to help you find important features and write a Cartesian equation for the conic.

15. $r = \dfrac{9}{1 - 2\cos(\theta)}$

16. $r = \dfrac{4}{1 + 3\sin(\theta)}$

17. $r = \dfrac{12}{3 + \sin(\theta)}$

18. $r = \dfrac{15}{3 - 2\cos(\theta)}$

19. $r = \dfrac{6}{1 + \cos(\theta)}$

20. $r = \dfrac{4}{1 - \sin(\theta)}$

21. At the beginning of the chapter, we defined an ellipse as the set of all points Q for which the sum of the distance from each focus to Q is constant. Mathematically, $d(Q,F_1)+d(Q,F_2)=k$. It is not obvious that this definition and the one provided in this section involving the directrix are related. In this exercise, we will start with the definition from this section and attempt to derive the earlier formula from it.

   a. Draw an ellipse with foci at $(c,0)$ and $(-c,0)$, vertices at $(a,0)$ and $(-a,0)$, and directrixes at $x=p$ and $x=-p$. Label the foci F_1 and F_2. Label the directrixes L_1 and L_2. Label some point (x,y) on the ellipse Q.

   b. Find formulas for $d(Q,L_1)$ and $D(Q,L_2)$ in terms of x and p.

   c. From the definition of a conic in this section, $\dfrac{d(Q,F_1)}{d(Q,L_1)}=e$. Likewise, $\dfrac{d(Q,F_2)}{d(Q,L_2)}=e$ as well. Use these ratios, with your answers from part (b) above, to find formulas for $d(Q,F_1)$ and $D(Q,F_2)$ in terms of e, x, and p.

   d. Show that the sum, $d(Q,F_1)+d(Q,F_2)$, is constant. This establishes that the definitions are connected.

   e. Let Q be a vertex. Find the distances $d(Q,F_1)$ and $D(Q,F_2)$ in terms of a and c. Then combine this with your result from part (d) to find a formula for p in terms of a and e.

   f. Let Q be a vertex. Find the distances $D(Q,L_2)$ and $D(Q,F_2)$ in terms of a, p, and c. Use the relationship $\dfrac{d(Q,F_2)}{d(Q,L_2)}=e$, along with your result from part (e), to find a formula for e in terms of a and c.

22. When we first looked at hyperbolas, we defined them as the set of all points Q for which the absolute value of the difference of the distances to two fixed points is constant. Mathematically, $|d(Q,F_1)-d(Q,F_2)|=k$. Use a similar approach to the one in the last exercise to obtain this formula from the definition given in this section. Find a formula for e in terms of a and c.

Solutions to Selected Exercises

Chapter 5

Section 5.1

1. 10

3. $(x-8)^2+(y+10)^2=8^2$

5. $(x-7)^2+(y+2)^2=293$

7. $(x-5)^2+(y-8)^2=13$

9.

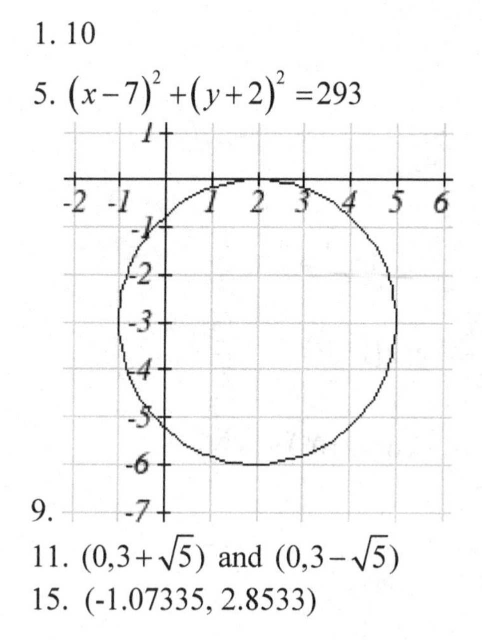

11. $(0,3+\sqrt{5})$ and $(0,3-\sqrt{5})$

13. (1.3416407865, 7.683281573)

15. (-1.07335, 2.8533)

17. 29.87 miles

Section 5.2

1.

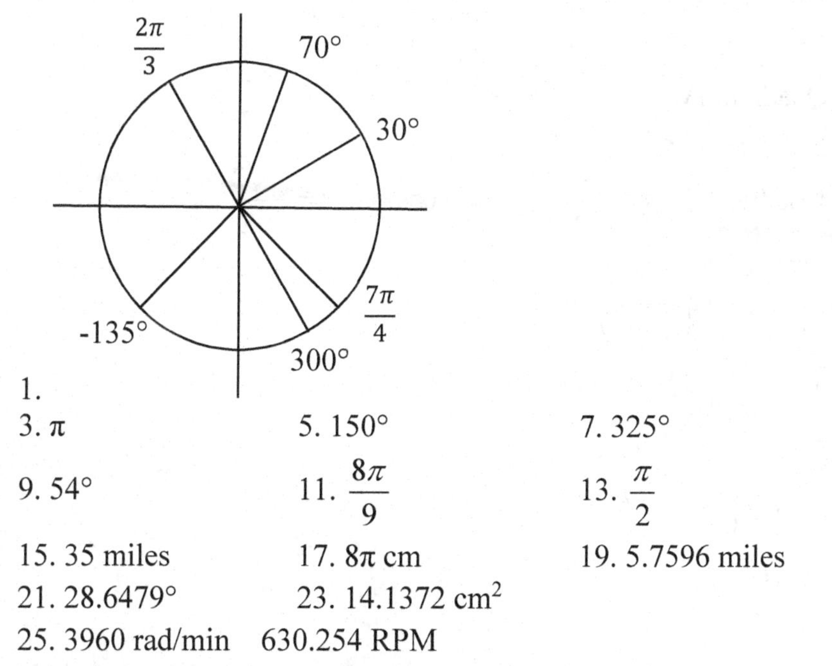

3. π

5. 150°

7. 325°

9. 54°

11. $\frac{8\pi}{9}$

13. $\frac{\pi}{2}$

15. 35 miles

17. 8π cm

19. 5.7596 miles

21. 28.6479°

23. 14.1372 cm^2

25. 3960 rad/min 630.254 RPM

27. 2.094 in/sec, π/12 rad/sec, 2.5 RPM

29. 75,398.22 mm/min = 1.257 m/sec

31. Angular speed: π/12 rad/hr. Linear speed: 1036.73 miles/hr

Section 5.3

1. a. III b. II

3. $-\frac{4}{5}$

5. $-\frac{4\sqrt{3}}{7}$

7. $-\frac{\sqrt{55}}{8}$

9. a. reference: 45°. Quadrant III. $\sin(225°) = -\frac{\sqrt{2}}{2}$. $\cos(225°) = -\frac{\sqrt{2}}{2}$

b. reference: 60°. Quadrant IV. $\sin(300°) = -\frac{\sqrt{3}}{2}$. $\cos(300°) = \frac{1}{2}$

c. reference: 45°. Quadrant II. $\sin(135°) = \frac{\sqrt{2}}{2}$. $\cos(135°) = -\frac{\sqrt{2}}{2}$

d. reference: 30°. Quadrant III. $\sin(210°) = -\frac{1}{2}$. $\cos(210°) = -\frac{\sqrt{3}}{2}$

11. a. reference: $\frac{\pi}{4}$. Quadrant III. $\sin\left(\frac{5\pi}{4}\right) = -\frac{\sqrt{2}}{2}$. $\cos\left(\frac{5\pi}{4}\right) = -\frac{\sqrt{2}}{2}$

b. reference: $\frac{\pi}{6}$. Quadrant III. $\sin\left(\frac{7\pi}{6}\right) = -\frac{1}{2}$. $\cos\left(\frac{7\pi}{6}\right) = -\frac{\sqrt{3}}{2}$

c. reference: $\frac{\pi}{3}$. Quadrant IV. $\sin\left(\frac{5\pi}{3}\right) = -\frac{\sqrt{3}}{2}$. $\cos\left(\frac{5\pi}{3}\right) = \frac{1}{2}$

d. reference: $\frac{\pi}{4}$. Quadrant II. $\sin\left(\frac{3\pi}{4}\right) = \frac{\sqrt{2}}{2}$. $\cos\left(\frac{3\pi}{4}\right) = -\frac{\sqrt{2}}{2}$

13. a. $\sin\left(-\frac{3\pi}{4}\right) = -\frac{\sqrt{2}}{2}$ $\cos\left(-\frac{3\pi}{4}\right) = -\frac{\sqrt{2}}{2}$

b. $\sin\left(\frac{23\pi}{6}\right) = -\frac{1}{2}$ $\cos\left(\frac{23\pi}{6}\right) = \frac{\sqrt{3}}{2}$

c. $\sin\left(-\frac{\pi}{2}\right) = -1$ $\cos\left(-\frac{\pi}{2}\right) = 0$

d. $\sin(5\pi) = 0$ $\cos(5\pi) = -1$

15. a. $\frac{2\pi}{3}$ b. 100° c. 40° d. $\frac{5\pi}{3}$ e. 235°

17. a. $\frac{5\pi}{3}$ b. 280° c. 220° d. $\frac{2\pi}{3}$ e. 55°

19. (-11.491, -9.642)

Section 5.4

1. $\sec(\theta)=\sqrt{2}$, $\csc(\theta)=\sqrt{2}$, $\tan(\theta)=1$, $\cot(\theta)=1$

3. $\sec(\theta)=-\frac{2\sqrt{3}}{3}$, $\csc(\theta)=2$, $\tan(\theta)=-\frac{\sqrt{3}}{3}$, $\cot(\theta)=-\sqrt{3}$

5. $\sec(\theta)=-2$, $\csc(\theta)=\frac{2\sqrt{3}}{3}$, $\tan(\theta)=-\sqrt{3}$, $\cot(\theta)=-\frac{\sqrt{3}}{3}$

7. a. $\sec(135°)=-\sqrt{2}$ b. $\csc(210°)=-2$ c. $\tan(60°)=\sqrt{3}$. d. $\cot(225°)=1$

9. $\cos(\theta)=-\frac{\sqrt{7}}{4}$, $\sec(\theta)=-\frac{4\sqrt{7}}{7}$, $\csc(\theta)=\frac{4}{3}$, $\tan(\theta)=-\frac{3\sqrt{7}}{7}$, $\cot(\theta)=-\frac{\sqrt{7}}{3}$

11. $\sin(\theta)=-\frac{2\sqrt{2}}{3}$, $\csc(\theta)=-\frac{3\sqrt{2}}{4}$, $\sec(\theta)=-3$, $\tan(\theta)=2\sqrt{2}$, $\cot(\theta)=\frac{\sqrt{2}}{4}$

13. $\sin(\theta)=\frac{12}{13}$, $\cos(\theta)=\frac{5}{13}$, $\sec(\theta)=\frac{13}{5}$, $\csc(\theta)=\frac{13}{12}$, $\cot(\theta)=\frac{5}{12}$

15. a. sin(0.15) = 0.1494 cos(0.15) = 0.9888 tan(0.15) = 0.1511
 b. sin(4) = -0.7568 cos(4) = -0.6536 tan(4) = 1.1578
 c. sin(70°) = 0.9397 cos(70°) = 0.3420 tan(70°) = 2.7475
 d. sin(283°) = -0.9744 cos(283°) = 0.2250 tan(283°) = -4.3315

17. $\sec(t)$ 19. $\tan(t)$ 21. $\tan(t)$ 23. $\cot(t)$ 25. $(\sec(t))^2$

Section 5.5

1. $\sin(A)=\frac{5\sqrt{41}}{41}, \cos(A)=\frac{4\sqrt{41}}{41}, \tan(A)=\frac{5}{4}$
 $\sec(A)=\frac{\sqrt{41}}{4}, \csc(A)=\frac{\sqrt{41}}{5}, \cot(A)=\frac{4}{5}$

3. $c=14,\ b=7\sqrt{3},\ B=60°$

5. $a=5.3171,\ c=11.3257,\ A=28°$

7. $a=9.0631,\ b=4.2262,\ B=25°$

9. 32.4987 ft

11. 836.2698 ft

13. 460.4069 ft

15. 660.35 feet

17. 28.025 ft

19. 143.0427

21. 86.6685

Chapter 6

Section 6.1

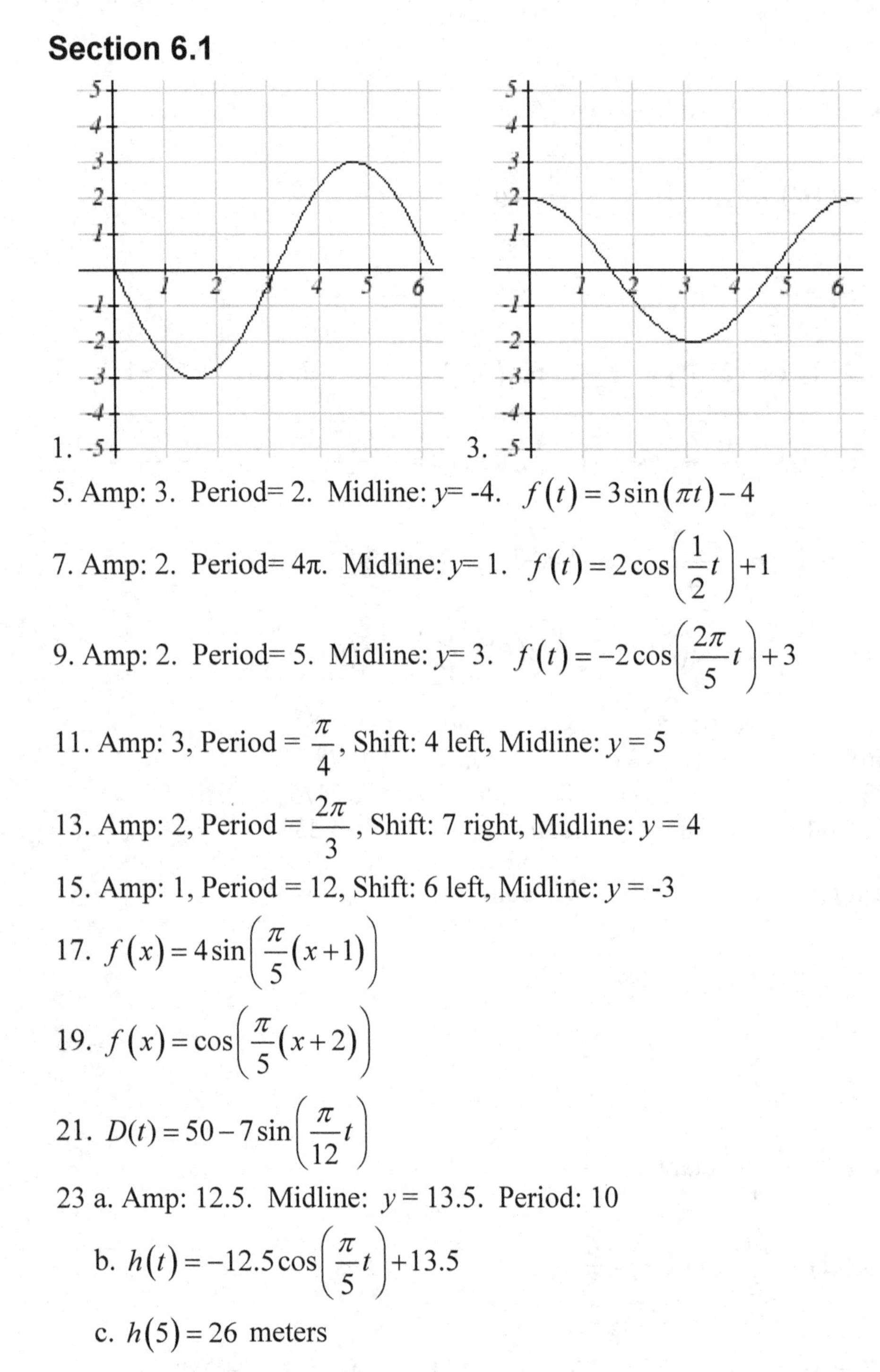

1. 3.

5. Amp: 3. Period= 2. Midline: y= -4. $f(t) = 3\sin(\pi t) - 4$

7. Amp: 2. Period= 4π. Midline: y= 1. $f(t) = 2\cos\left(\frac{1}{2}t\right) + 1$

9. Amp: 2. Period= 5. Midline: y= 3. $f(t) = -2\cos\left(\frac{2\pi}{5}t\right) + 3$

11. Amp: 3, Period = $\frac{\pi}{4}$, Shift: 4 left, Midline: $y = 5$

13. Amp: 2, Period = $\frac{2\pi}{3}$, Shift: 7 right, Midline: $y = 4$

15. Amp: 1, Period = 12, Shift: 6 left, Midline: $y = -3$

17. $f(x) = 4\sin\left(\frac{\pi}{5}(x+1)\right)$

19. $f(x) = \cos\left(\frac{\pi}{5}(x+2)\right)$

21. $D(t) = 50 - 7\sin\left(\frac{\pi}{12}t\right)$

23 a. Amp: 12.5. Midline: $y = 13.5$. Period: 10

b. $h(t) = -12.5\cos\left(\frac{\pi}{5}t\right) + 13.5$

c. $h(5) = 26$ meters

Section 6.2

1. II
3. I
5. Period: $\frac{\pi}{4}$. Horizontal shift: 8 right
7. Period: 8. Horizontal shift: 1 left
9. Period: 6. Horizontal shift: 3 left

11. 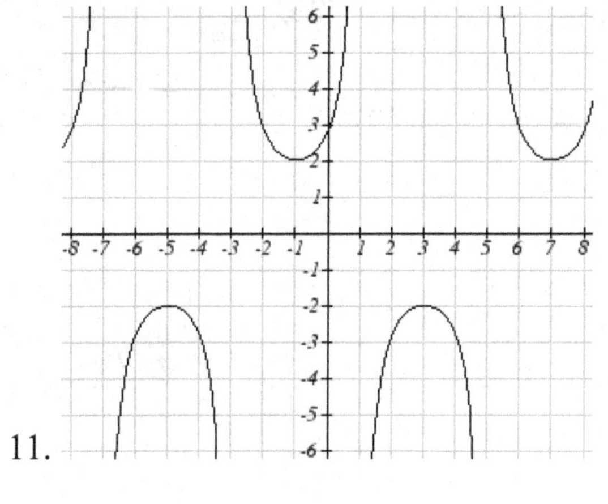

13. 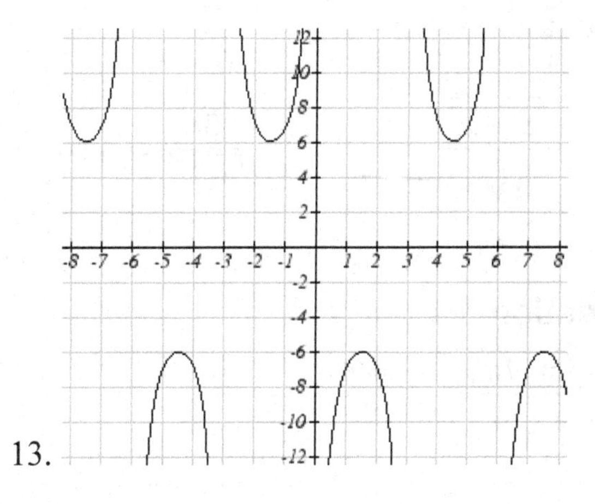

15. 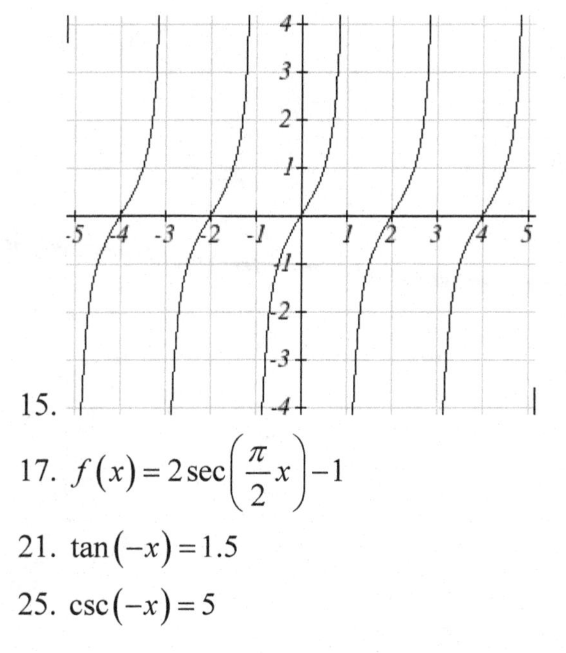

17. $f(x) = 2\sec\left(\frac{\pi}{2}x\right) - 1$

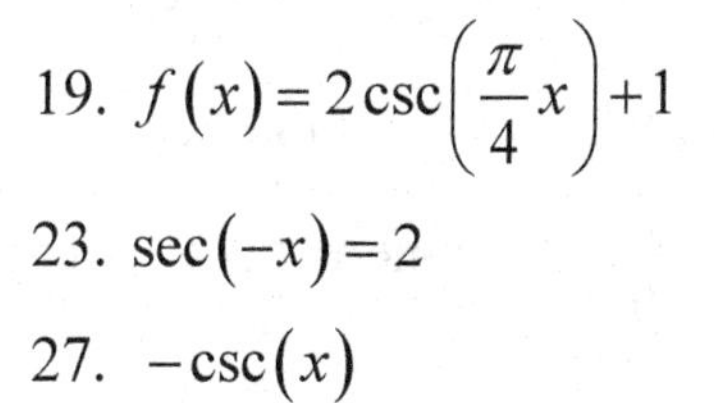

19. $f(x) = 2\csc\left(\frac{\pi}{4}x\right) + 1$

21. $\tan(-x) = 1.5$

23. $\sec(-x) = 2$

25. $\csc(-x) = 5$

27. $-\csc(x)$

Section 6.3

1. $\frac{\pi}{4}$ 3. $-\frac{\pi}{6}$ 5. $\frac{\pi}{3}$

7. $\frac{3\pi}{4}$ 9. $\frac{\pi}{4}$ 11. $-\frac{\pi}{3}$

13. 1.9823 15. -0.9273 17. 44.427°

19. $\frac{\pi}{4}$ 21. $-\frac{\pi}{6}$ 23. $\frac{2\sqrt{10}}{7}$ 25. $\frac{1}{\sqrt{17}}$

27. $\frac{\sqrt{25-x^2}}{5}$ 29. $\frac{3x}{\sqrt{9x^2+1}}$

Section 6.4

1. $\frac{5\pi}{4}, \frac{7\pi}{4}$ 3. $\frac{\pi}{3}, \frac{5\pi}{3}$ 5. $\frac{\pi}{2}$ 7. $\frac{\pi}{2}, \frac{3\pi}{2}$

9. $\frac{\pi}{4}+2\pi k, \frac{7\pi}{4}+2\pi k$, where k is an integer

11. $\frac{7\pi}{6}+2\pi k, \frac{11\pi}{6}+2\pi k$, where k is an integer

13. $\frac{\pi}{18}+\frac{2\pi}{3}k, \frac{5\pi}{18}+\frac{2\pi}{3}k$, where k is an integer

15. $\frac{5\pi}{12}+\frac{2\pi}{3}k, \frac{7\pi}{12}+\frac{2\pi}{3}k$, where k is an integer

17. $\frac{\pi}{6}+\pi k, \frac{5\pi}{6}+\pi k$, where k is an integer

19. $\frac{\pi}{4}+\frac{2\pi}{3}k, \frac{5\pi}{12}+\frac{2\pi}{3}k$, where k is an integer

21. $4+8k$, where k is an integer

23. $\frac{1}{6}+2k, \frac{5}{6}+2k$, where k is an integer

25. 0.2734, 2.8682 27. 3.7603, 5.6645 29. 2.1532, 4.1300

31. 0.7813, 5.5019 33. 0.04829, 0.47531 35. 0.7381, 1.3563

37. 0.9291, 3.0709 39. 1.3077, 4.6923

Section 6.5

1. $c = \sqrt{89}$, $A = 57.9946°$, $B = 32.0054°$

3. $b = \sqrt{176}$, $A = 27.8181°$, $B = 62.1819°$

5. $y(x) = 6\sin\left(\frac{\pi}{2}(x-1)\right) + 4$

7. $D(t) = 50 - 13\cos\left(\frac{\pi}{12}(t-5)\right)$

9. a. $P(t) = 129 - 25\cos\left(\frac{\pi}{6}t\right)$ b. $P(t) = 129 - 25\cos\left(\frac{\pi}{6}(t-3)\right)$

11. 75 degrees

13. 8

15. 2.80869431742

17. 5.035 months

Chapter 7

Section 7.1

1. $\frac{7\pi}{6}, \frac{11\pi}{6}$

3. $\frac{\pi}{3}, \frac{5\pi}{3}$

5. $\frac{2}{3} + 8k$, and $\frac{10}{3} + 8k$, where k is an integer

7. $\frac{5\pi}{12} + k\pi$ and $\frac{7\pi}{12} + k\pi$, where k is an integer

9. $0.1339 + 10k$ and $8.6614 + 10k$, where k is an integer

11. $1.1438 + \frac{2\pi}{3}k$ and $1.9978 + \frac{2\pi}{3}k$, where k is an integer

13. $\frac{\pi}{2}, \frac{3\pi}{2}, 0.644, 2.498$

15. 0.056, 1.515, 3.197, 4.647

17. $0, \pi, \frac{\pi}{3}, \frac{5\pi}{3}$

19. $\frac{\pi}{6}, \frac{5\pi}{6}, \frac{7\pi}{6}, \frac{11\pi}{6}$

21. 1.183, 1.958, 4.325, 5.100

23. $\frac{3\pi}{2}, \frac{7\pi}{6}, \frac{11\pi}{6}$

25. $\pi, \frac{\pi}{3}, \frac{5\pi}{3}$

27. 1.823, 4.460

29. 2.301, 3.983, 0.723, 5.560

31. 3.305, 6.120

33. $0, \frac{\pi}{3}, \frac{2\pi}{3}, \pi, \frac{4\pi}{3}, \frac{5\pi}{3}$

35. $0, \frac{\pi}{4}, \frac{3\pi}{4}, \pi, \frac{5\pi}{4}, \frac{7\pi}{4}$

37. $\frac{\pi}{6}, \frac{2\pi}{3}, \frac{5\pi}{6}, \frac{4\pi}{3}$

39. $0, \pi, 1.231, 5.052$

41. $\frac{\pi}{3}, \frac{5\pi}{3}$

Section 7.2

1. $\frac{\sqrt{2}+\sqrt{6}}{4}$

3. $\frac{-\sqrt{2}-\sqrt{6}}{4}$

5. $\frac{\sqrt{2}-\sqrt{6}}{4}$

7. $\frac{\sqrt{2}+\sqrt{6}}{4}$

9. $\frac{\sqrt{3}}{2}\sin(x)-\frac{1}{2}\cos(x)$

11. $-\frac{\sqrt{3}}{2}\cos(x)+\frac{1}{2}\sin(x)$

13. $\sec(t)$

15. $\tan(x)$

17. $8(\cos(5x)-\cos(27x))$

19. $\sin(8x)+\sin(2x)$

21. $2\cos(5t)\cos(t)$

23. $2\sin(5x)\cos(2x)$

25. a. $\left(\frac{2}{3}\right)\left(-\frac{1}{4}\right)+\left(-\frac{\sqrt{5}}{3}\right)\left(\frac{\sqrt{15}}{4}\right)=\frac{-2-5\sqrt{3}}{12}$

b. $\left(-\frac{\sqrt{5}}{3}\right)\left(-\frac{1}{4}\right)+\left(\frac{2}{3}\right)\left(\frac{\sqrt{15}}{4}\right)=\frac{\sqrt{5}+2\sqrt{15}}{12}$

27. $0.373+\frac{2\pi}{3}k$ and $0.674+\frac{2\pi}{3}k$, where k is an integer

29. $2\pi k$, where k is an integer

31. $\frac{\pi}{7}+\frac{4\pi}{7}k$, $\frac{3\pi}{7}+\frac{4\pi}{7}k$, $\frac{\pi}{3}+\frac{4\pi}{3}k$, and $\pi+\frac{4\pi}{3}k$, where k is an integer

33. $\frac{7\pi}{12}+\pi k$, $\frac{11\pi}{12}+\pi k$, and $\frac{\pi}{4}k$, where k is an integer

35. $2\sqrt{13}\sin(x+5.3004)$ or $2\sqrt{13}\sin(x-0.9828)$

37. $\sqrt{29}\sin(3x+0.3805)$

39. 0.3681, 3.8544

41. 0.7854, 1.8158

43. $\tan(6t)$

Section 7.3

1. a. $\frac{3\sqrt{7}}{32}$ b. $\frac{31}{32}$ c. $\frac{3\sqrt{7}}{31}$

3. $\cos(56°)$

5. $\cos(34°)$

7. $\cos(18x)$

9. $2\sin(16x)$

11. $0, \pi, 2.4189, 3.8643$

13. 0.7297, 2.4119, 3.8713, 5.5535

15. $\frac{\pi}{6}, \frac{\pi}{2}, \frac{5\pi}{6}, \frac{3\pi}{2}$

17. a. $\frac{2\pi}{9}, \frac{4\pi}{9}, \frac{8\pi}{9}, \frac{10\pi}{9}, \frac{14\pi}{9}, \frac{16\pi}{9}, 0, \frac{2\pi}{3}, \frac{4\pi}{3}$

19. $\frac{1+\cos(10x)}{2}$

21. $\frac{3}{8} - \frac{1}{2}\cos(16x) + \frac{1}{8}\cos(32x)$

23. $\frac{1}{16} - \frac{1}{16}\cos(2x) + \frac{1}{16}\cos(4x) - \frac{1}{16}\cos(2x)\cos(4x)$

25. a. $\sqrt{\frac{1}{2} + \frac{2\sqrt{3}}{7}}$ b. $\sqrt{\frac{1}{2} - \frac{2\sqrt{3}}{7}}$ c. $\frac{1}{7-4\sqrt{3}}$

Section 7.4

1. $y = 3\sin\left(\frac{\pi}{6}(x-3)\right) - 1$

3. Amplitude: 8, Period: $\frac{1}{3}$ second, Frequency: 3 Hz (cycles per second)

5. $P(t) = -19\cos\left(\frac{\pi}{6}t\right) + \frac{40}{3}t + 650$

7. $P(t) = -33\cos\left(\frac{\pi}{6}t\right) + 900(1.07)^t$

9. $D(t)=10(0.85)^t\cos(36\pi t)$

11. $D(t)=17(0.9145)^t\cos(28\pi t)$

13. a. IV b. III

15. $y=6(4)^x+5\sin\left(\frac{\pi}{2}x\right)$

17. $y=-3\sin\left(\frac{\pi}{2}\right)+2x+7$

19. $y=8\left(\frac{1}{2}\right)^x\cos\left(\frac{\pi}{2}x\right)+3$

Chapter 8

Section 8.1

1.
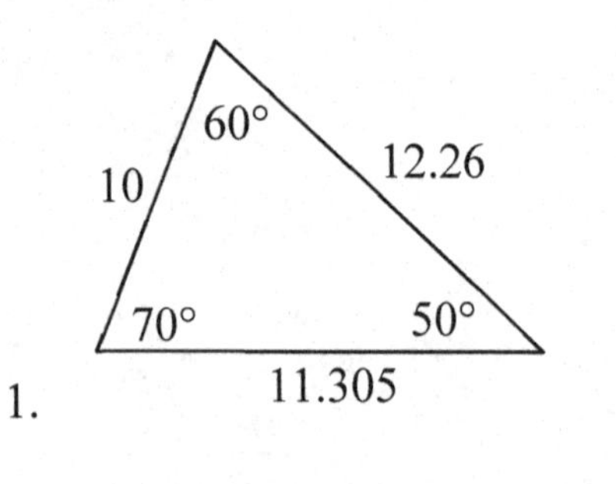

3.
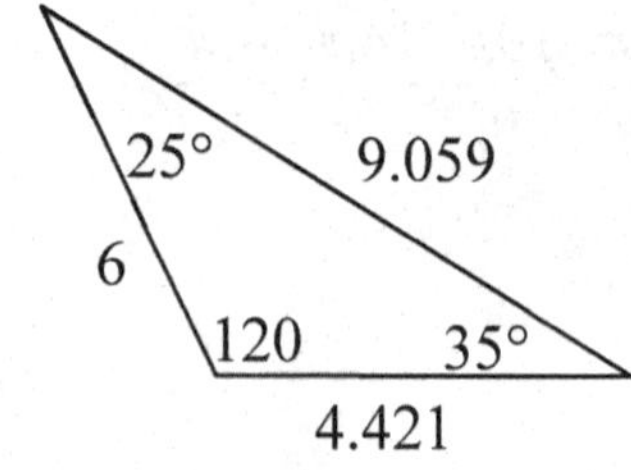
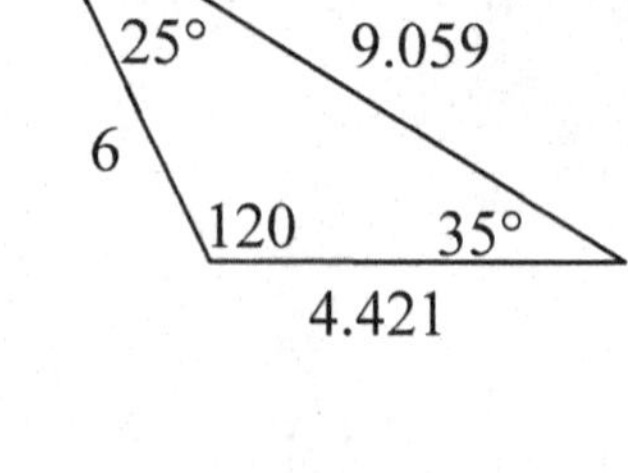

5.
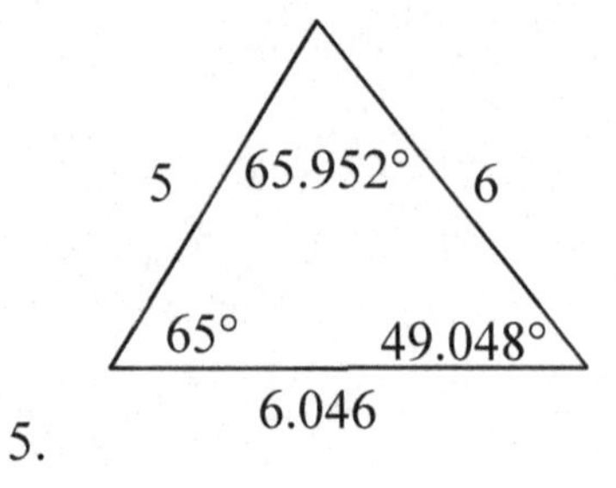

7.
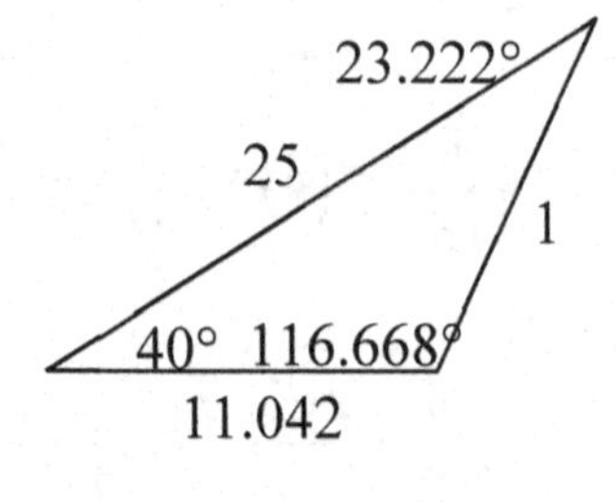

9. $\beta=68°$, $a=14.711$, $c=20.138$

11. $\beta=28.096°$, $\gamma=32.904°$, $c=16.149$

13. Not possible.

15. $\beta=64.243°$, $\gamma=72.657°$, $c=257.328$ OR $\beta=115.757°$, $\gamma=21.143°$, $c=97.238$

17.
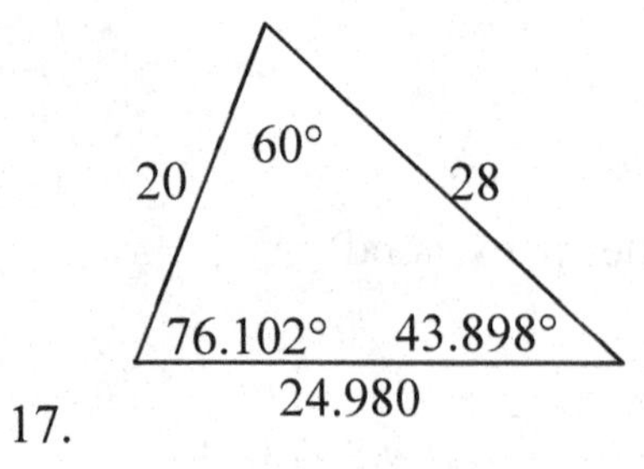

19.
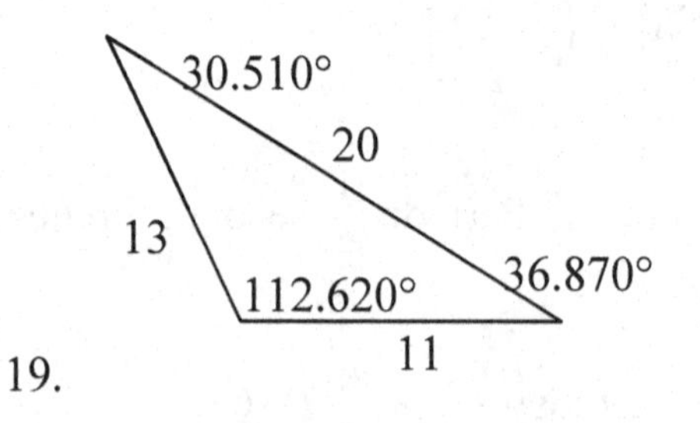

21. $c=2.066, \alpha=52.545°, \beta=86.255°$

23. $a = 11.269, \beta = 27.457°, \gamma = 32.543°$

25. 177.562

27. 978.515 ft

29. Distance to A: 565.258 ft. Distance to shore: 531.169 ft

31. 529.014 m

33. 173.877 feet

35. 4.642 km, 2.794 km

37. 757.963 ft

39. 2371.129 miles

41. 65.375 cm^2

43. 7.72

Section 8.2

1. $\left(-\frac{7\sqrt{3}}{2}, -\frac{7}{2}\right)$

3. $\left(2\sqrt{2}, -2\sqrt{2}\right)$

5. $\left(3\sqrt{2}, -3\sqrt{2}\right)$

7. $(0,3)$

9. $\left(-\frac{3\sqrt{3}}{2}, -\frac{3}{2}\right)$

11. $(-1.248, 2.728)$

13. $\left(2\sqrt{5}, 0.464\right)$

15. $\left(2\sqrt{13}, 2.159\right)$

17. $\left(\sqrt{34}, 5.253\right)$

19. $\left(\sqrt{269}, 4.057\right)$

21. $r = 3\sec(\theta)$

23. $r = \frac{\sin(\theta)}{4\cos^2(\theta)}$

25. $r = 4\sin(\theta)$

27. $r = \frac{\cos(\theta)}{\left(\cos^2(\theta) - \sin^2(\theta)\right)}$

29. $x^2 + y^2 = 3y$

31. $y + 7x = 4$

33. $x = 2$

35. $x^2 + y^2 = x + 2$

37. A

39. C

41. E

43. C

45. D

47. F

49.

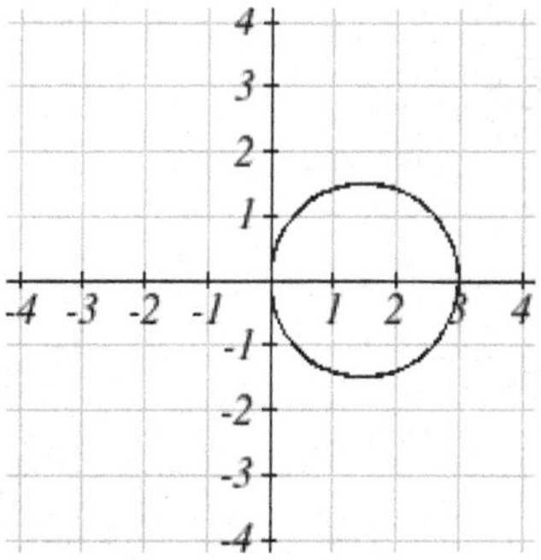

51.

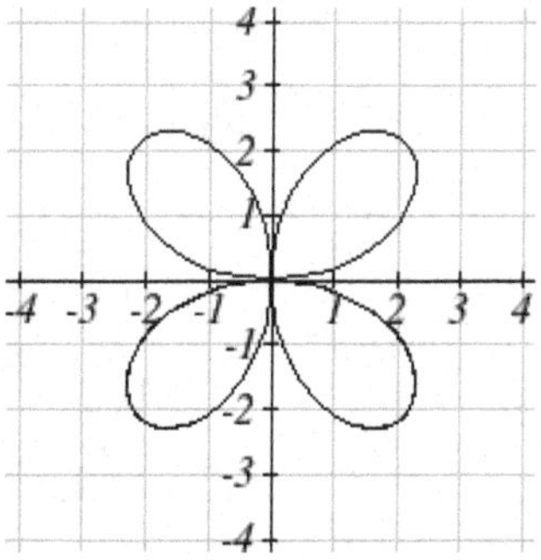

53.

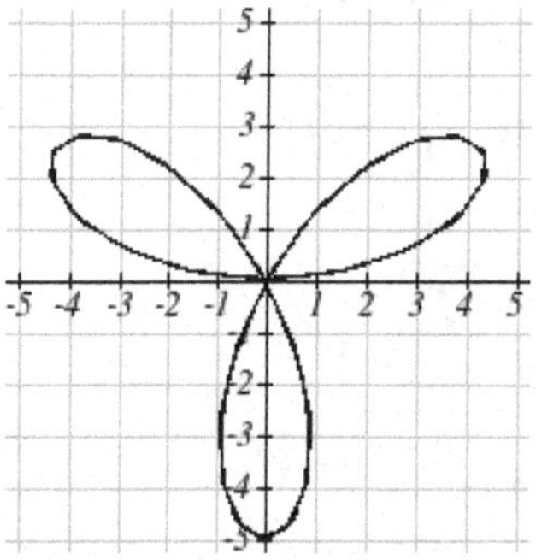

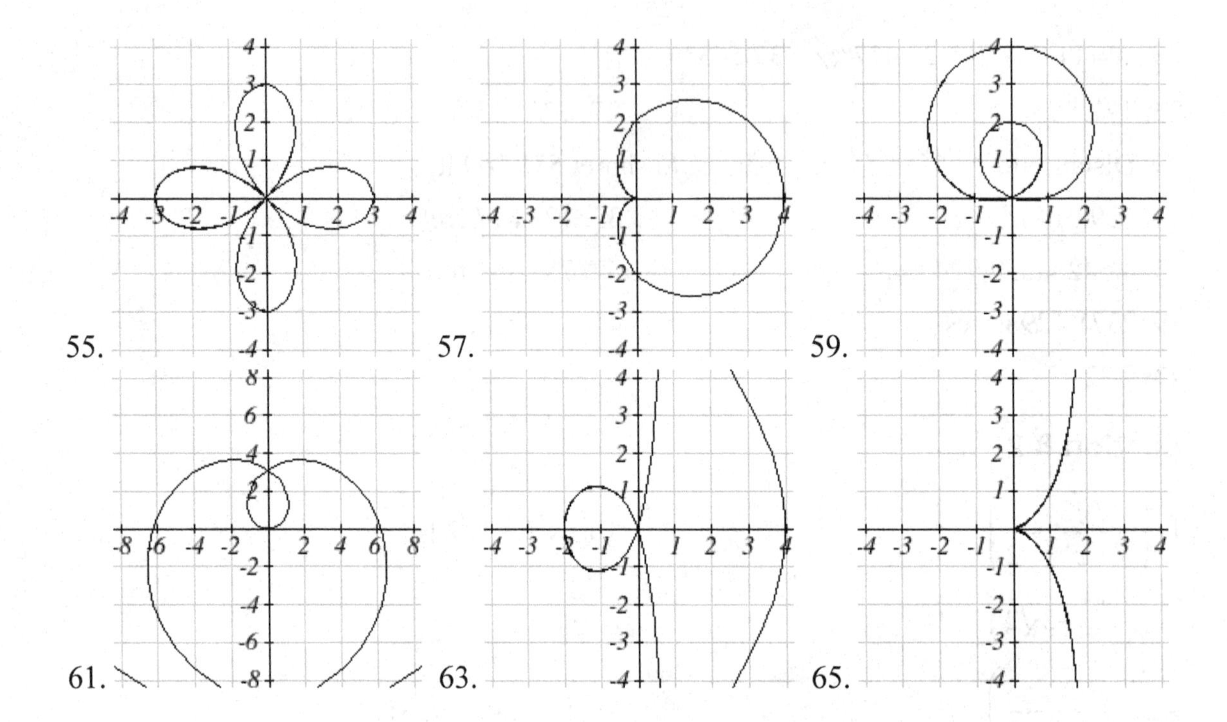

Section 8.3

1. $3i$

3. -12

5. $1+\sqrt{3}i$

7. $8-i$

9. $-11+4i$

11. $-12+8i$

13. $30-10i$

15. $11+10i$

17. 20

19. $\frac{3}{2}+2i$

21. $\frac{3}{2}+\frac{5}{2}i$

23. $-\frac{1}{25}-\frac{18}{25}i$

25. -1

27. i

29. $3\cos(2)+3\sin(2)i=-1.248+2.728i$

31. $3\sqrt{3}+3i$

33. $-\frac{3\sqrt{2}}{2}-\frac{3\sqrt{2}}{2}i$

35. $6e^{0i}$

37. $4e^{\frac{3\pi}{2}i}$

39. $2\sqrt{2}e^{\frac{\pi}{4}i}$

41. $3\sqrt{2}e^{\frac{3\pi}{4}i}$

43. $\sqrt{34}e^{0.540i}$

45. $\sqrt{10}e^{2.820i}$

47. $\sqrt{17}e^{4.467i}$

49. $\sqrt{26}e^{6.086i}$

51. $6e^{\frac{5\pi}{12}i}$

53. $2e^{\frac{7\pi}{12}i}$

55. $1024e^{\frac{5\pi}{2}i}$

57. $4e^{\frac{\pi}{3}i}$

59. 4096

61. $0.788 + 1.903i$

63. $1.771 + 0.322i$

65. $\sqrt[5]{2} \approx 1.149, 0.355 + 1.092i, -0.929 + 0.675i, -0.929 - 0.675i, 0.355 - 1.092i$

67. $1, \frac{1}{2} + \frac{\sqrt{3}}{2}i, -\frac{1}{2} + \frac{\sqrt{3}}{2}i, -1, -\frac{1}{2} - \frac{\sqrt{3}}{2}i, \frac{1}{2} - \frac{\sqrt{3}}{2}i$

Section 8.4

1. $-4, 2$

3. The vectors do not need to start at the same point

5. $3\vec{v} - \vec{u}$

7. $3\sqrt{2}, 3\sqrt{2}$

9. $-6.128, -5.142$

11. Magnitude: 4, Direction: 90°

13. Magnitude: 7.810, Direction: 39.806°

15. Magnitude: 2.236, Direction: 153.435°

17. Magnitude: 5.385, Direction: 291.801°

19. Magnitude: 7.211, Direction: 236.310°

21. $\vec{u} + \vec{v} = \langle 3, 2 \rangle$, $\vec{u} - \vec{v} = \langle 1, -8 \rangle$, $2\vec{u} - 3\vec{v} = \langle 1, -21 \rangle$

23. 4.635 miles, 17.764 deg N of E

25. 17 miles. 10.318 miles

27. $\overrightarrow{F_{net}} = -4, -11$

29. Distance: 2.868. Direction: 86.474° North of West, or 3.526° West of North

31. 4.924 degrees. 659 km/hr

33. 4.424 degrees

35. (0.081, 8.602)

37. 21.801 degrees, relative to the car's forward direction

Section 8.5

1. $6 \cdot 10 \cdot \cos(75°) = 15.529$ 3. $(0)(-3)+(4)(0)=0$ 5. $(-2)(-10)+(1)(13)=33$

7. $\cos^{-1}\left(\frac{0}{\sqrt{4}\sqrt{3}}\right) = 90°$ 9. $\cos^{-1}\left(\frac{(2)(1)+(4)(-3)}{\sqrt{2^2+4^2}\sqrt{1^2+(-3)^2}}\right) = 135°$

11. $\cos^{-1}\left(\frac{(4)(8)+(2)(4)}{\sqrt{4^2+8^2}\sqrt{2^2+4^2}}\right) = 0°$ 13. $(2)(k)+(7)(4)=0,\ k = -14$

15. $\frac{(8)(1)+(-4)(-3)}{\sqrt{1^2+(-3)^2}} = 6.325$ 17. $\left(\frac{(-6)(1)+(10)(-3)}{\sqrt{1^2+(-3)^2}^2}\right)\langle 1,-3\rangle = \langle -3.6, 10.8\rangle$

19. The vectors are $\langle 2,3\rangle$ and $\langle -5,-2\rangle$. The acute angle between the vectors is 34.509°

21. 14.142 pounds 23. $\langle 10\cos(10°), 10\sin(10°)\rangle \cdot \langle 0,-20\rangle$, so 34.7296 ft-lbs

25. $40 \cdot 120 \cdot \cos(25°) = 4350.277$ ft-lbs

Section 8.6

1. C 3. E 5. F

7.

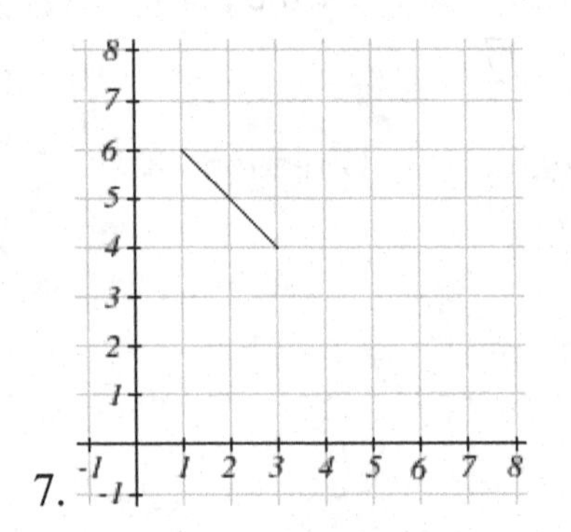

9. x(t)

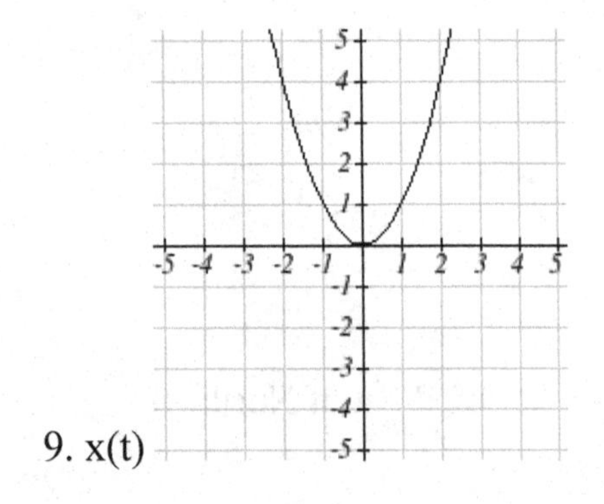

y(t)

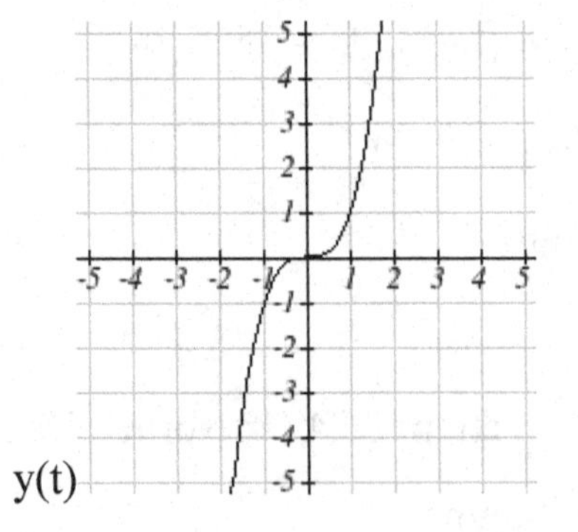

11. 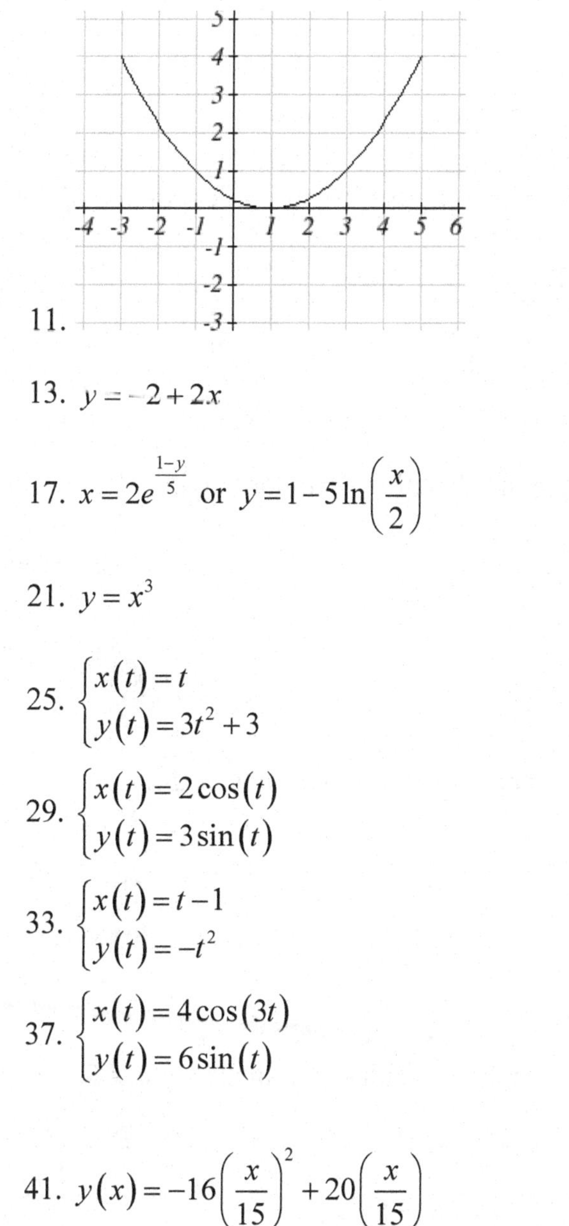

13. $y = -2 + 2x$

15. $y = 3\sqrt{\dfrac{x-1}{2}}$

17. $x = 2e^{\frac{1-y}{5}}$ or $y = 1 - 5\ln\left(\dfrac{x}{2}\right)$

19. $x = \left(\dfrac{y}{2}\right)^3 - \dfrac{y}{2}$

21. $y = x^3$

23. $\left(\dfrac{x}{4}\right)^2 + \left(\dfrac{y}{5}\right)^2 = 1$

25. $\begin{cases} x(t) = t \\ y(t) = 3t^2 + 3 \end{cases}$

27. $\begin{cases} x(t) = 3\log(t) + t \\ y(t) = t \end{cases}$.

29. $\begin{cases} x(t) = 2\cos(t) \\ y(t) = 3\sin(t) \end{cases}$

31. $\begin{cases} x(t) = t^3 \\ y(t) = t + 2 \end{cases}$

33. $\begin{cases} x(t) = t - 1 \\ y(t) = -t^2 \end{cases}$

35. $\begin{cases} x(t) = -1 + 3t \\ y(t) = 5 - 2t \end{cases}$

37. $\begin{cases} x(t) = 4\cos(3t) \\ y(t) = 6\sin(t) \end{cases}$

39. $\begin{cases} x(t) = 4\cos(2t) \\ y(t) = 3\sin(3t) \end{cases}$

41. $y(x) = -16\left(\dfrac{x}{15}\right)^2 + 20\left(\dfrac{x}{15}\right)$

43. $\begin{cases} x(t) = 20\sin\left(\dfrac{2\pi}{5}t\right) + 8\sin(\pi t) \\ y(t) = 35 - 20\cos\left(\dfrac{2\pi}{5}t\right) - 8\cos(\pi t) \end{cases}$

Chapter 9

Section 9.1

1. D 3. B

5. Vertices $(0,\pm 5)$, minor axis endpoints $(\pm 2,0)$, major length = 10, minor length = 4

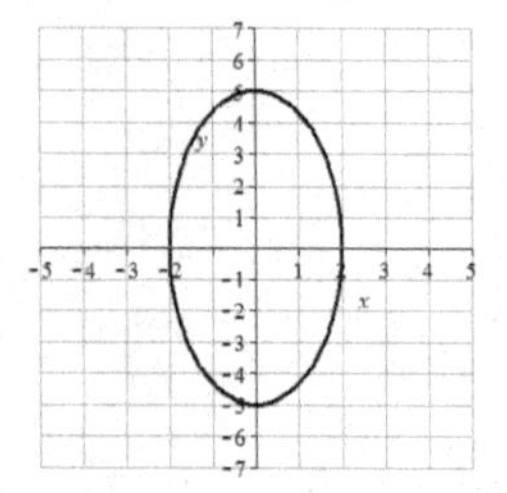

7. Vertices $(\pm 2,0)$, minor axis endpoints $(0,\pm 1)$, major length = 4, minor length = 2

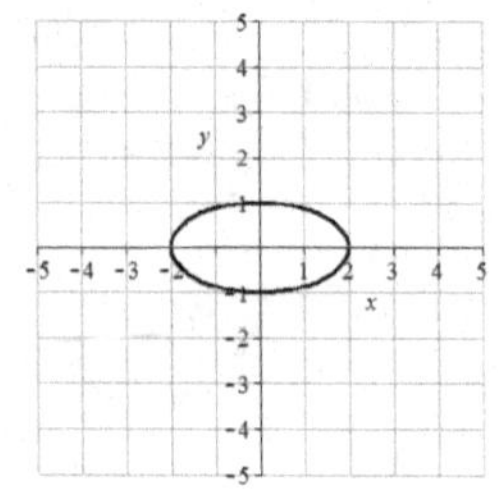

9. Vertices $(\pm 5,0)$, minor axis endpoints $(0,\pm 1)$, major length = 10, minor length = 2

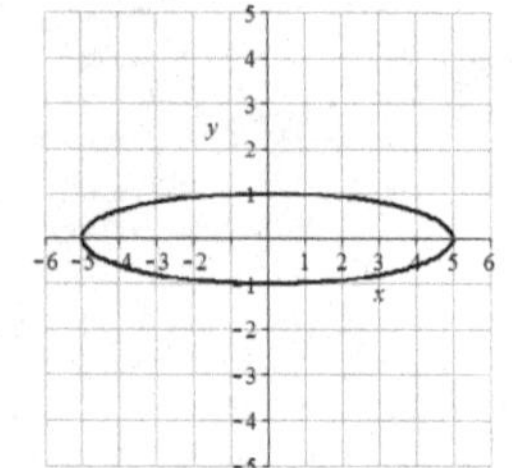

11. Vertices $(0,\pm 4)$, minor axis endpoints $(\pm 3,0)$, major length = 8, minor length = 6

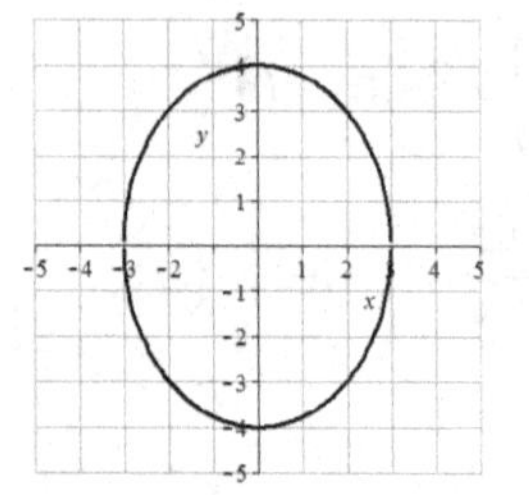

13. Vertices $\left(0,\pm 3\sqrt{2}\right)$, minor axis endpoints $\left(\pm\sqrt{2},0\right)$, major length = $6\sqrt{2}$, minor length = $2\sqrt{2}$

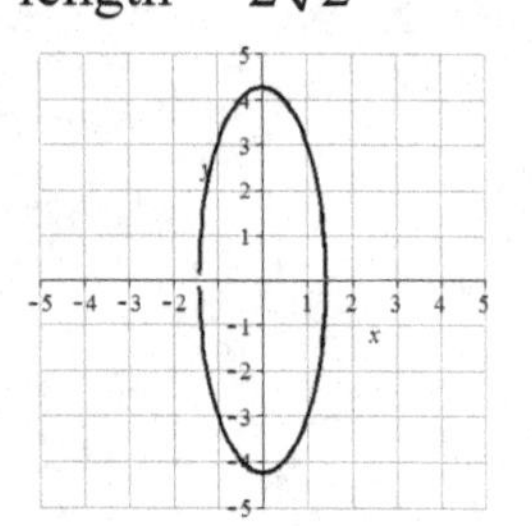

15. $\dfrac{x^2}{16}+\dfrac{y^2}{4}=1$ 17. $\dfrac{x^2}{1024}+\dfrac{y^2}{49}=1$ 19. $\dfrac{x^2}{4}+\dfrac{y^2}{9}=1$

21. B 23. C 25. F 27. G

29. Center (1,-2), vertices (6,-2) and (-4,-2), minor axis endpoints (1,0) and (1,-4), major length= 10, minor length = 4

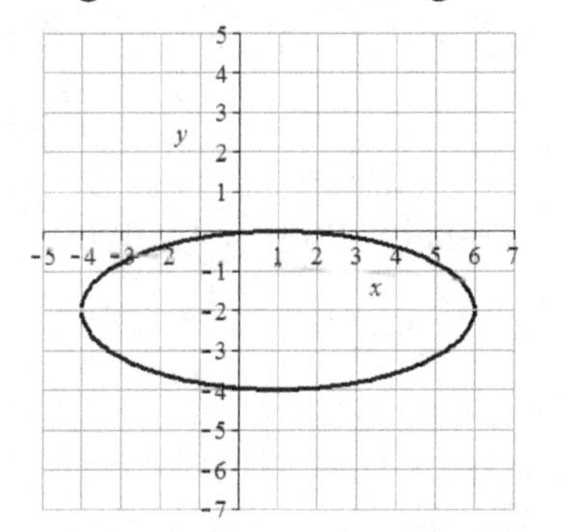

31. Center (-2,3), vertices (-2,8) and (-2,-2), minor axis endpoints (-1,3) and (-3,3), major length = 10, minor length = 2

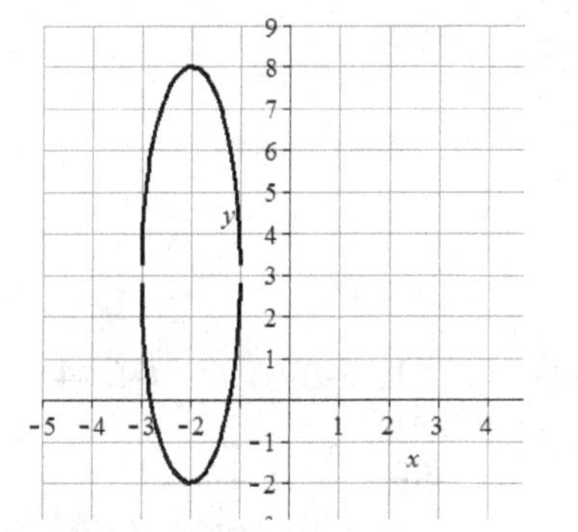

33. Center (-1,0), vertices (-1,4) and (-1,-4), minor axis endpoints (-1,0) and (3,0), major length = 8, minor length = 4

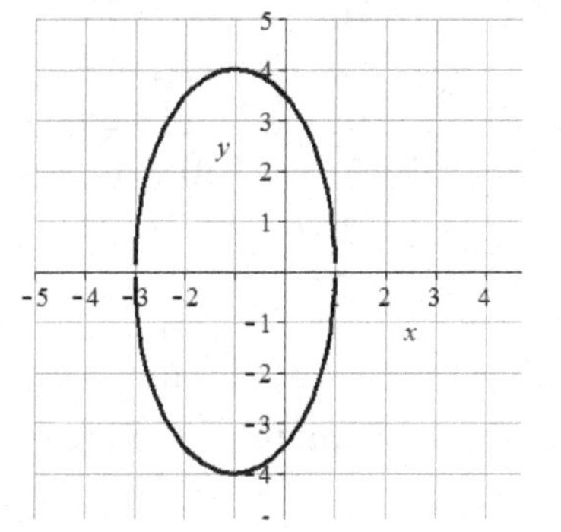

35. Center (-1,-2), vertices (3,-2) and (-5,-2), minor axis endpoints (-1,0) and (-1,-4), major length = 8, minor length = 4

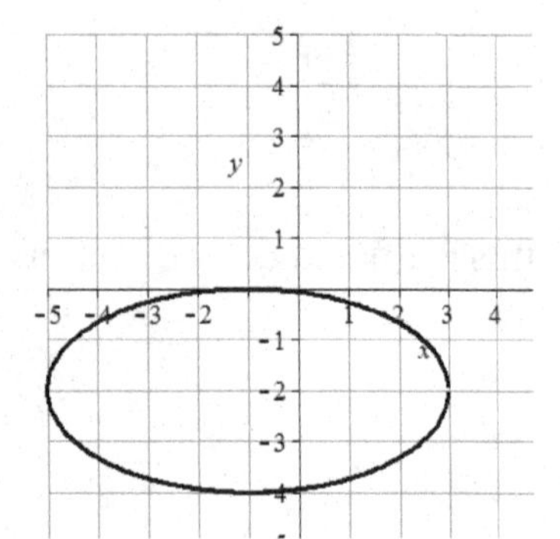

37. Center (2,-1), vertices (2,5) and (2,-7), minor axis endpoints (6,-1) and (-2,-1), major length = 12, minor length = 8

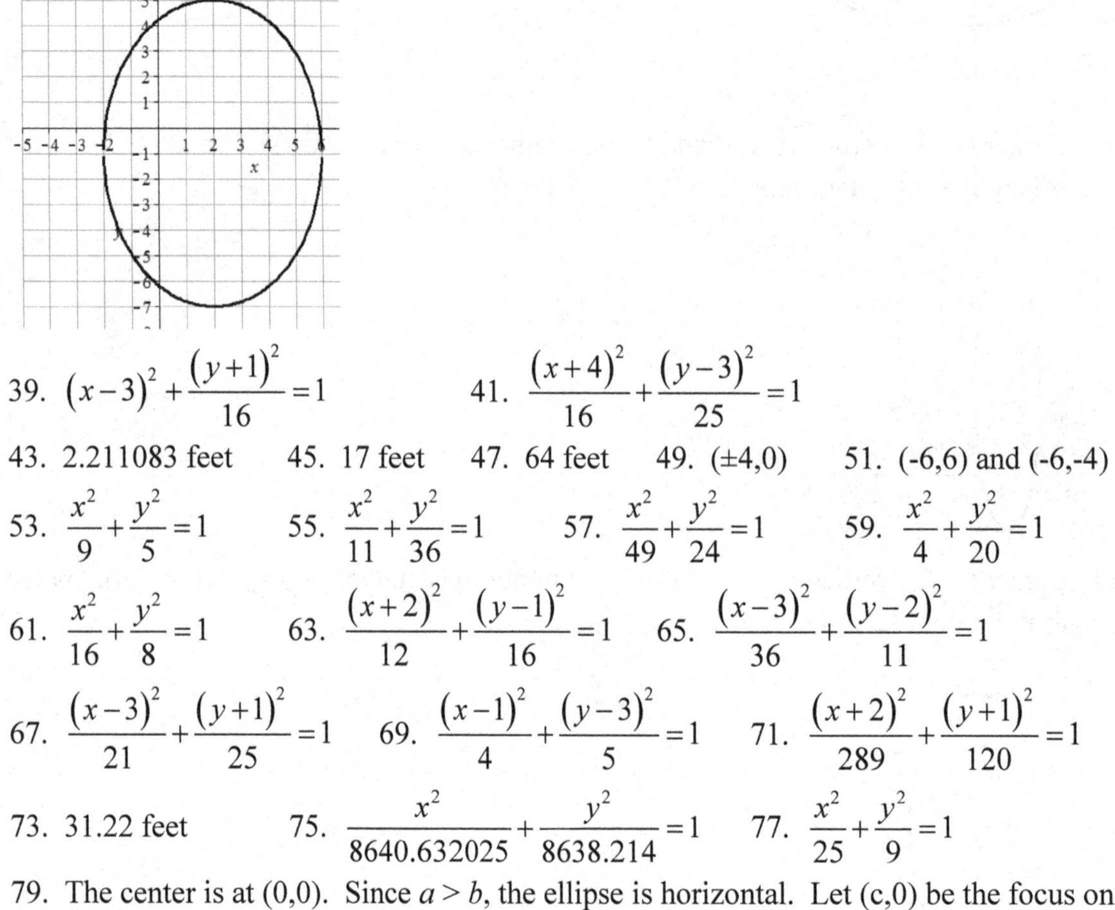

39. $(x-3)^2+\dfrac{(y+1)^2}{16}=1$ 41. $\dfrac{(x+4)^2}{16}+\dfrac{(y-3)^2}{25}=1$

43. 2.211083 feet 45. 17 feet 47. 64 feet 49. $(\pm 4,0)$ 51. (-6,6) and (-6,-4)

53. $\dfrac{x^2}{9}+\dfrac{y^2}{5}=1$ 55. $\dfrac{x^2}{11}+\dfrac{y^2}{36}=1$ 57. $\dfrac{x^2}{49}+\dfrac{y^2}{24}=1$ 59. $\dfrac{x^2}{4}+\dfrac{y^2}{20}=1$

61. $\dfrac{x^2}{16}+\dfrac{y^2}{8}=1$ 63. $\dfrac{(x+2)^2}{12}+\dfrac{(y-1)^2}{16}=1$ 65. $\dfrac{(x-3)^2}{36}+\dfrac{(y-2)^2}{11}=1$

67. $\dfrac{(x-3)^2}{21}+\dfrac{(y+1)^2}{25}=1$ 69. $\dfrac{(x-1)^2}{4}+\dfrac{(y-3)^2}{5}=1$ 71. $\dfrac{(x+2)^2}{289}+\dfrac{(y+1)^2}{120}=1$

73. 31.22 feet 75. $\dfrac{x^2}{8640.632025}+\dfrac{y^2}{8638.214}=1$ 77. $\dfrac{x^2}{25}+\dfrac{y^2}{9}=1$

79. The center is at (0,0). Since $a > b$, the ellipse is horizontal. Let (c,0) be the focus on the positive x-axis. Let (c, h) be the endpoint in Quadrant 1 of the latus rectum passing through (c,0).

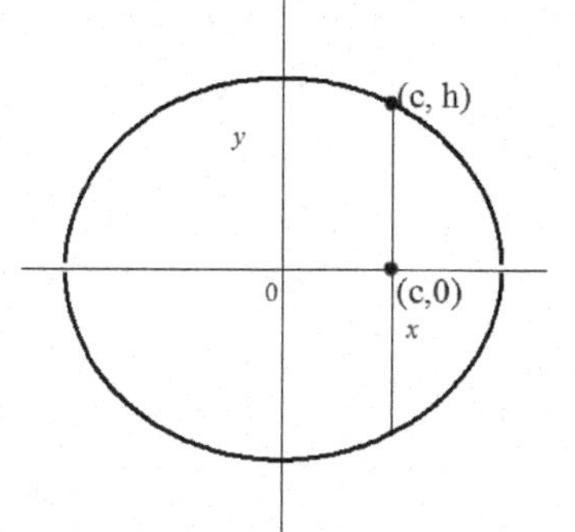

The distance between the focus and latus rectum endpoint can be found by substituting (c,0) and (c,h) into the distance formula $h=\sqrt{(x_1-x_2)^2+(y_1-y_2)^2}$ which yields $h=\sqrt{(c-c)^2+(h-0)^2}=h$. So h is half the latus rectum distance. Substituting (c,h) into the ellipse equation to find h gives $\frac{c^2}{a^2}+\frac{h^2}{b^2}=1$. Solve for h yields $h^2=b^2\left(1-\frac{c^2}{a^2}\right)=b^2\left(\frac{a^2}{a^2}-\frac{c^2}{a^2}\right)=b^2\left(\frac{a^2-c^2}{a^2}\right)=b^2\left(\frac{b^2}{a^2}\right)=\frac{b^4}{a^2}$. so $h=\sqrt{\frac{b^4}{a^2}}=\frac{b^2}{a}$. The distance of the latus rectum is $2h=\frac{2b^2}{a}$.

Section 9.2

1. B 3. D

5. Vertices (±2,0), transverse length = 4, asymptotes y = ±5/2x,

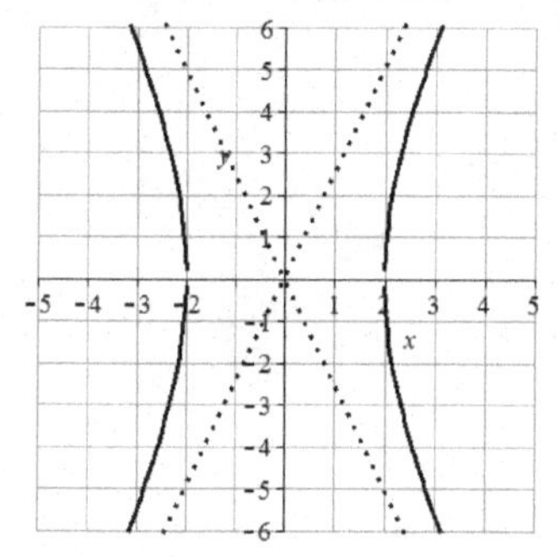

7. Vertices (0, ±1), transverse length = 2, asymptotes y = ±1/2x,

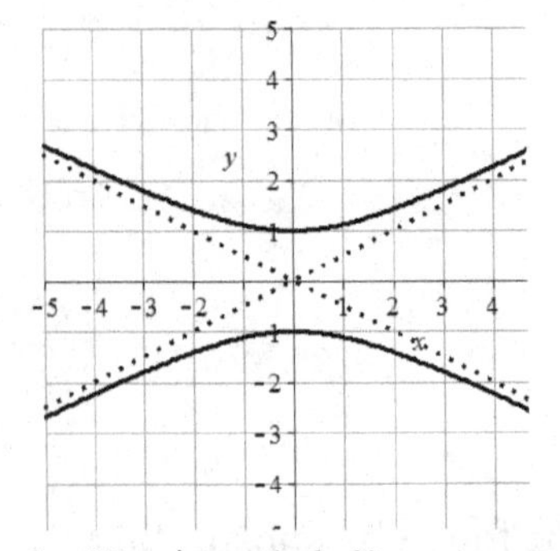

9. Vertices (±3,0), transverse length = 6, asymptotes y =±1/3x,

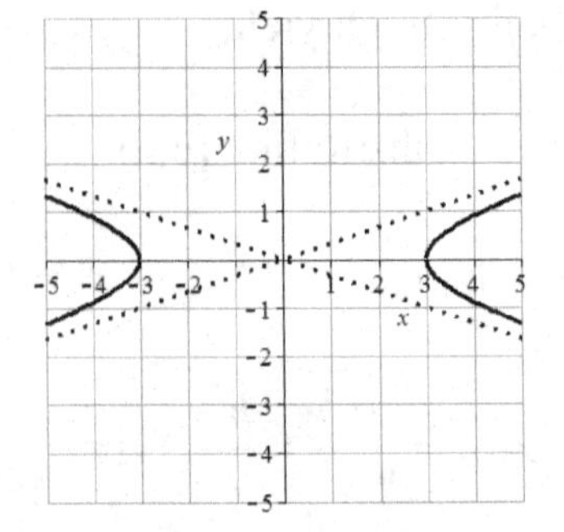

11. Vertices (0, ±4), transverse length = 8, asymptotes y =±4/3x

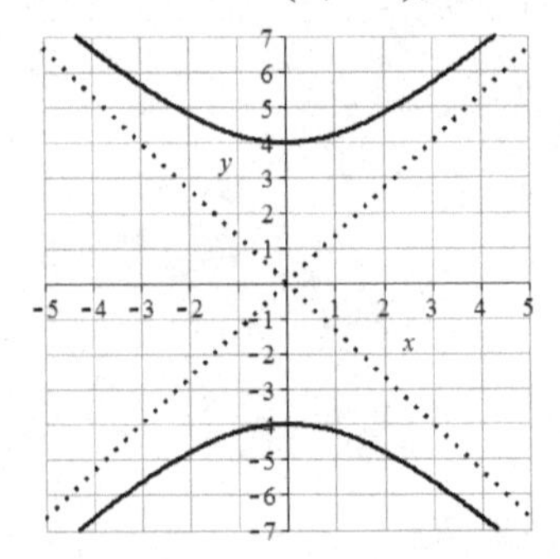

13. Vertices ($\pm\sqrt{2}$,0), transverse length = $2\sqrt{2}$, asymptotes y =±3x,

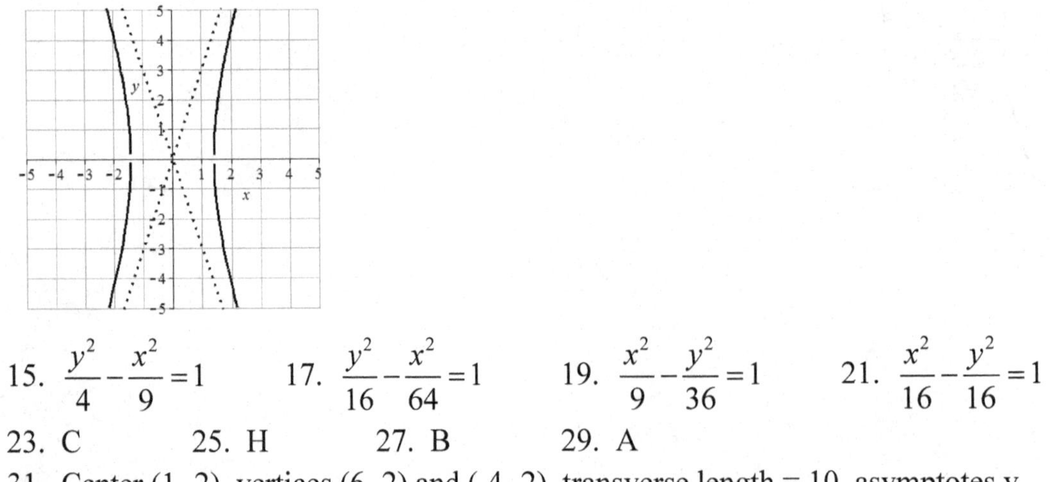

15. $\frac{y^2}{4}-\frac{x^2}{9}=1$ 17. $\frac{y^2}{16}-\frac{x^2}{64}=1$ 19. $\frac{x^2}{9}-\frac{y^2}{36}=1$ 21. $\frac{x^2}{16}-\frac{y^2}{16}=1$

23. C 25. H 27. B 29. A

31. Center (1,-2), vertices (6,-2) and (-4,-2), transverse length = 10, asymptotes y =±2/5(x-1)-2

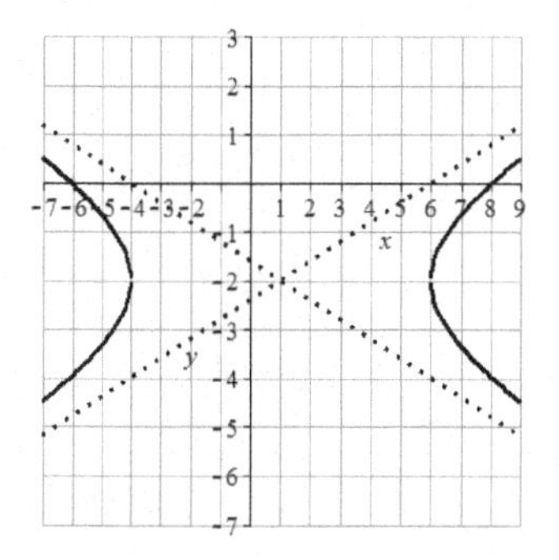

33. Center (-2,1), vertices (-2,4) and (-2,-2), transverse length = 6, asymptotes y =±3(x+2)+1

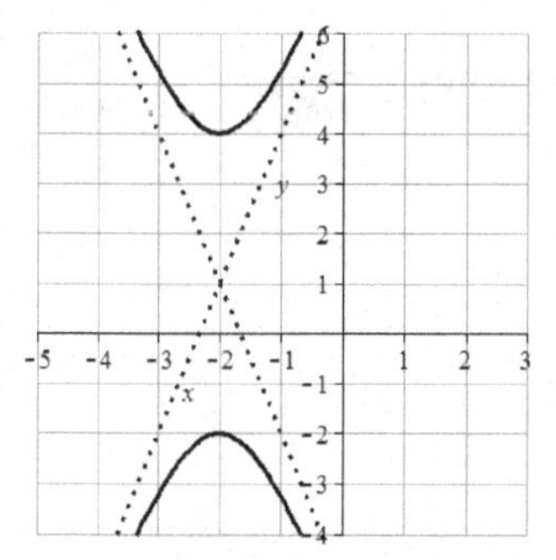

35. Center (1,0), vertices (3,0) and (-1,0), transverse length = 4, asymptotes y =±2(x-1)

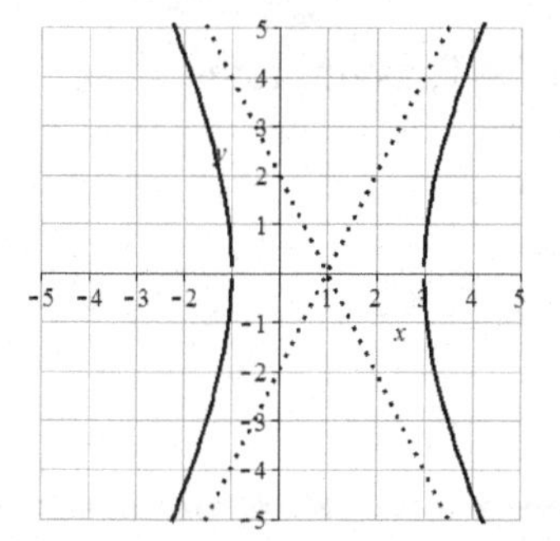

37. Center (-1,2), vertices (-1,4) and (-1,0), transverse length = 4, asymptotes y =±1/2(x+1)+2

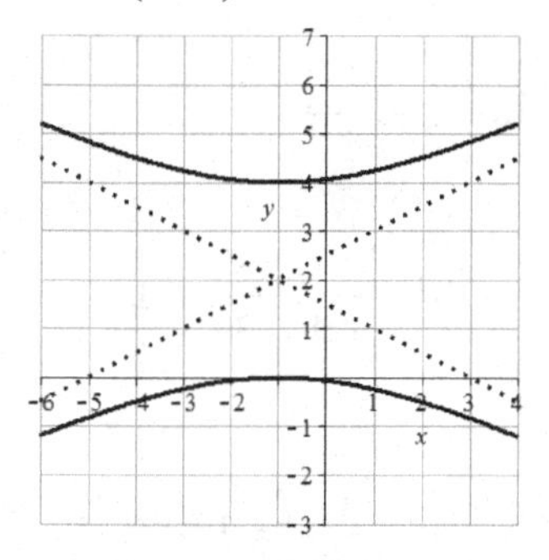

39. Center (-2,1), vertices (0,1) and (-4,1), transverse length = 4, asymptotes y =±3/2(x+2)+1

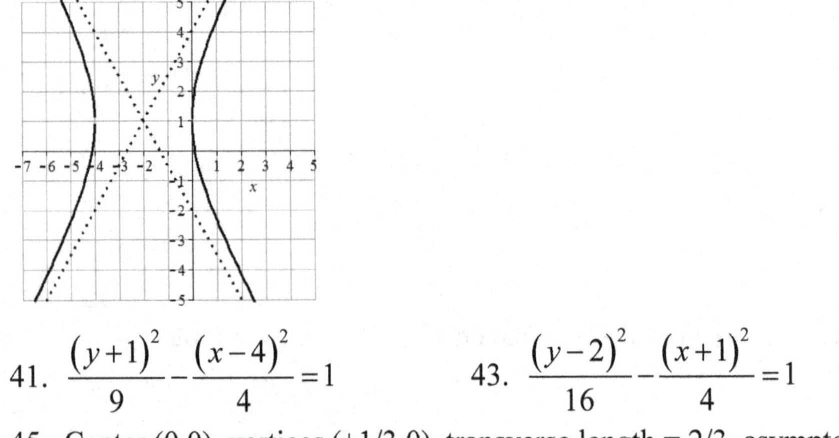

41. $\dfrac{(y+1)^2}{9}-\dfrac{(x-4)^2}{4}=1$ 43. $\dfrac{(y-2)^2}{16}-\dfrac{(x+1)^2}{4}=1$

45. Center (0,0), vertices (±1/3,0), transverse length = 2/3, asymptotes y = ±12x

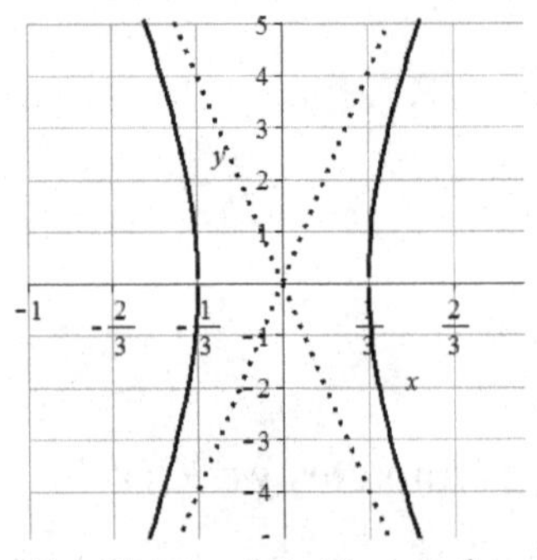

47. Center (-1,1), vertices (-1,3/2) and (-1,1/2), transverse length = 1, asymptotes y = ± 3/2 (x + 1) +1

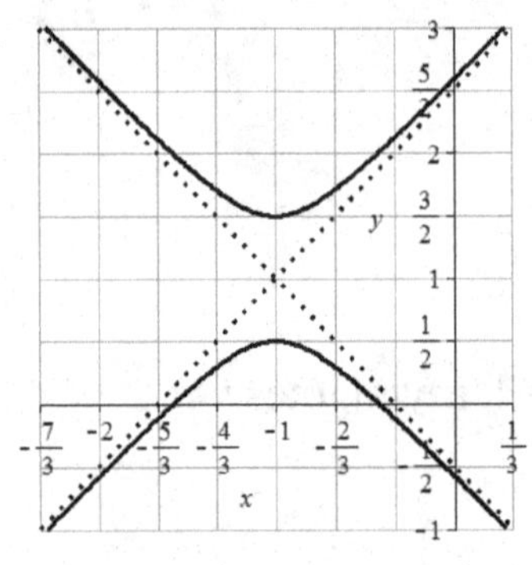

49. Foci (0,±5) 51. Foci (5,6) and (-3,6) 53. Foci (-4,6) and (-4,-4)

55. $\dfrac{x^2}{16}-\dfrac{y^2}{9}=1$ 57. $\dfrac{y^2}{144}-\dfrac{x^2}{25}=1$ 59. $\dfrac{x^2}{225}-\dfrac{y^2}{64}=1$ 61. $\dfrac{x^2}{64}-\dfrac{y^2}{36}=1$

63. $\dfrac{(y-2)^2}{16}-\dfrac{(x-1)^2}{9}=1$ 65. $\dfrac{(x+1)^2}{25}-\dfrac{(y-3)^2}{144}=1$ 67. $\dfrac{x^2}{900}-\dfrac{y^2}{1600}=1$

69. $\dfrac{x^2}{900}-\dfrac{y^2}{14400.3636}=1$

71. $\frac{x^2}{3025} - \frac{y^2}{6975} = 1$

73. $5y^2 - x^2 + 25 = 0$ can be put in the form $\frac{y^2}{5} - \frac{x^2}{25} = -1$. $x^2 - 5y^2 + 25 = 0$ can be put in the form $\frac{y^2}{5} - \frac{x^2}{25} = 1$ showing they are conjugate.

75. $\sqrt{2}$

77. No matter the value of k, the foci are at $\left(\pm\sqrt{6}, 0\right)$

Section 9.3

1. C 3. A

5. Vertex: (0,0). Axis of symmetry: $y = 0$. Directrix: $x = -4$. Focus: (4,0)

7. Vertex: (0,0). Axis of symmetry: $x = 0$. Directrix: $y = -1/8$. Focus: (0,1/8)

9. Vertex: (0,0). Axis of symmetry: $y = 0$. Directrix: $x = 1/16$. Focus: (-1/16,0)

11. Vertex: (2,-1). Axis of symmetry: $x = 2$. Directrix: $y = -3$. Focus: (2,1)

13. Vertex: (-1,4). Axis of symmetry: $x = -1$. Directrix: $y = 3$. Focus: (-1,5)

15. $(y-1)^2 = -(x-3)$

17. $(y-3)^2 = 12(x-2)$

19. $x^2 = 4(y-3)$

21. At the focus, (0,1)

23. 2.25 feet above the vertex.

25. 0.25 ft

27. $\left(\frac{1}{\sqrt{3}}, \frac{2}{\sqrt{3}}\right), \left(\frac{-1}{\sqrt{3}}, \frac{-2}{\sqrt{3}}\right)$

29. $\left(3, \sqrt{2}\right), \left(3, -\sqrt{2}\right), \left(-3, \sqrt{2}\right), \left(-3, -\sqrt{2}\right)$

31. $\left(2\sqrt{2}, 8\right), \left(-2\sqrt{2}, 8\right)$

33. $\left(\sqrt{\frac{5}{3}}, \sqrt{\frac{2}{3}}\right), \left(-\sqrt{\frac{5}{3}}, \sqrt{\frac{2}{3}}\right), \left(\sqrt{\frac{5}{3}}, -\sqrt{\frac{2}{3}}\right), \left(-\sqrt{\frac{5}{3}}, -\sqrt{\frac{2}{3}}\right)$

35. (-64.50476622, 93.37848007) $\approx$ (-64.50, 93.38)

Section 9.4

1. $e = 3$. Directrix: $x = 4$. Hyperbola.

3. $e = 3/4$. Directrix: $y = -2/3$. Ellipse.

5. $e = 1$. Directrix: $x = -1/5$. Parabola.

7. $e = 2/7$. Directrix: $x = 2$. Ellipse.

9. $r = \dfrac{20}{1-5\cos(\theta)}$

11. $r = \dfrac{1}{1+\frac{1}{3}\sin(\theta)}$, or $r = \dfrac{3}{3+\sin(\theta)}$

13. $r = \dfrac{2}{1-\sin(\theta)}$

15. Hyperbola. Vertices at (-9,0) and (-3,0)
Center at (-6,0). $a = 3$. $c = 6$, so $b = \sqrt{27}$
$$\frac{(x+6)^2}{9} - \frac{y^2}{27} = 1$$

17. Ellipse. Vertices at (0,3) and (0,-6)
Center at (0,-1.5). $a = 4.5$, $c = 1.5$, $b = \sqrt{18}$
$$\frac{x^2}{18} + \frac{(y+1.5)^2}{20.25} = 1$$

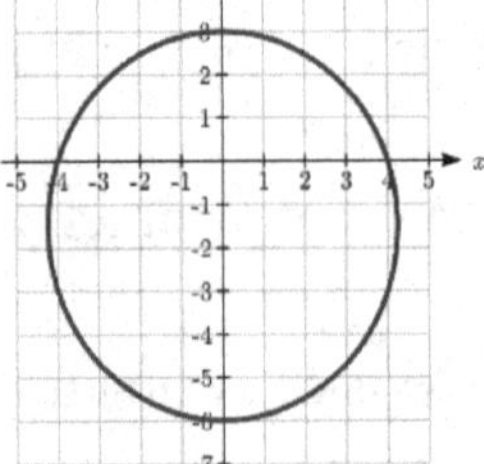

19. Parabola. Vertex at (3,0). $p = 3$.
$$y^2 = -12(x-3)$$

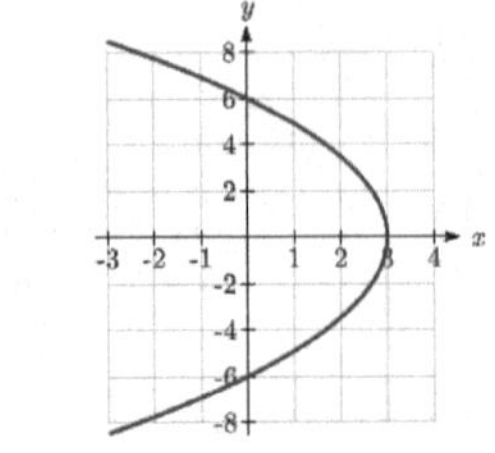

21. a)

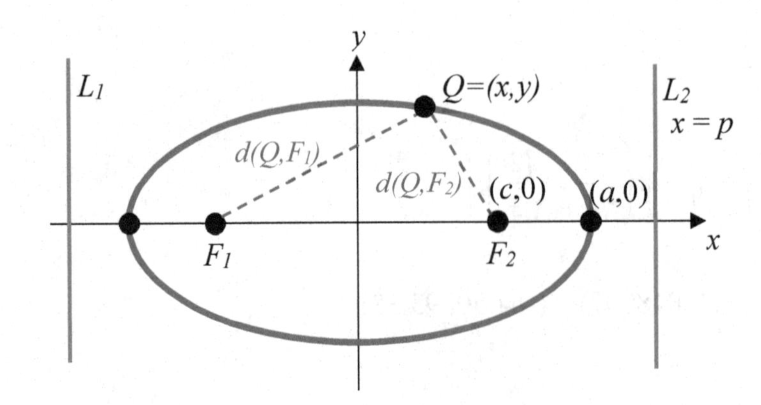

b) $d(Q,L_1) = x-(-p) = x+p$, $d(Q,L_2) = p-x$

c) $d(Q,F_1) = ed(Q,L_1) = e(x+p)$. $d(Q,F_2) = ed(Q,L_2) = e(p-x)$

d) $d(Q,F_1) + d(Q,F_2) = e(x+p) + e(p-x) = 2ep$, a constant.

e) At $Q = (a, 0)$, $d(Q,F_1) = a-(-c) = a+c$, and $d(Q,F_2) = a-c$, so
$d(Q,F_1) + d(Q,F_2) = (a+c)+(a-c) = 2a$

Combining with the result above, $2ep = 2a$, so $p = \frac{a}{e}$.

f) $d(Q,F_2) = a-c$, and $d(Q,L_2) = p-a$

$\frac{d(Q,F_2)}{d(Q,L_2)} = e$, so $\frac{a-c}{p-a} = e$.

$a-c = e(p-a)$. Using the result from (e),

$$a-c = e\left(\frac{a}{e}-a\right)$$

$$a-c = a-ea$$

$$e = \frac{c}{a}$$

Index